Lecture Notes in Computer Science 13332

More information about this series at https://link.springer.com/bookseries/558

Robert A. Sottilare · Jessica Schwarz (Eds.)

Adaptive Instructional Systems

4th International Conference, AIS 2022
Held as Part of the 24th HCI International Conference, HCII 2022
Virtual Event, June 26 – July 1, 2022
Proceedings

 Springer

Editors
Robert A. Sottilare
Soar Technology, Inc.
Orlando, FL, USA

Jessica Schwarz
Fraunhofer FKIE
Wachtberg, Germany

ISSN 0302-9743   ISSN 1611-3349 (electronic)
Lecture Notes in Computer Science
ISBN 978-3-031-05886-8   ISBN 978-3-031-05887-5 (eBook)
https://doi.org/10.1007/978-3-031-05887-5

This Springer imprint is published by the registered company Springer Nature Switzerland AG
The registered company address is: Gewerbestrasse 11, 6330 Cham, Switzerland

Foreword

Human-computer interaction (HCI) is acquiring an ever-increasing scientific and industrial importance, as well as having more impact on people's everyday life, as an ever-growing number of human activities are progressively moving from the physical to the digital world. This process, which has been ongoing for some time now, has been dramatically accelerated by the COVID-19 pandemic. The HCI International (HCII) conference series, held yearly, aims to respond to the compelling need to advance the exchange of knowledge and research and development efforts on the human aspects of design and use of computing systems.

The 24th International Conference on Human-Computer Interaction, HCI International 2022 (HCII 2022), was planned to be held at the Gothia Towers Hotel and Swedish Exhibition & Congress Centre, Göteborg, Sweden, during June 26 to July 1, 2022. Due to the COVID-19 pandemic and with everyone's health and safety in mind, HCII 2022 was organized and run as a virtual conference. It incorporated the 21 thematic areas and affiliated conferences listed on the following page.

A total of 5583 individuals from academia, research institutes, industry, and governmental agencies from 88 countries submitted contributions, and 1276 papers and 275 posters were included in the proceedings to appear just before the start of the conference. The contributions thoroughly cover the entire field of human-computer interaction, addressing major advances in knowledge and effective use of computers in a variety of application areas. These papers provide academics, researchers, engineers, scientists, practitioners, and students with state-of-the-art information on the most recent advances in HCI. The volumes constituting the set of proceedings to appear before the start of the conference are listed in the following pages.

The HCI International (HCII) conference also offers the option of 'Late Breaking Work' which applies both for papers and posters, and the corresponding volume(s) of the proceedings will appear after the conference. Full papers will be included in the 'HCII 2022 - Late Breaking Papers' volumes of the proceedings to be published in the Springer LNCS series, while 'Poster Extended Abstracts' will be included as short research papers in the 'HCII 2022 - Late Breaking Posters' volumes to be published in the Springer CCIS series.

I would like to thank the Program Board Chairs and the members of the Program Boards of all thematic areas and affiliated conferences for their contribution and support towards the highest scientific quality and overall success of the HCI International 2022 conference; they have helped in so many ways, including session organization, paper reviewing (single-blind review process, with a minimum of two reviews per submission) and, more generally, acting as goodwill ambassadors for the HCII conference.

This conference would not have been possible without the continuous and unwavering support and advice of Gavriel Salvendy, founder, General Chair Emeritus, and Scientific Advisor. For his outstanding efforts, I would like to express my appreciation to Abbas Moallem, Communications Chair and Editor of HCI International News.

June 2022   Constantine Stephanidis

HCI International 2022 Thematic Areas and Affiliated Conferences

Thematic Areas

- HCI: Human-Computer Interaction
- HIMI: Human Interface and the Management of Information

Affiliated Conferences

- EPCE: 19th International Conference on Engineering Psychology and Cognitive Ergonomics
- AC: 16th International Conference on Augmented Cognition
- UAHCI: 16th International Conference on Universal Access in Human-Computer Interaction
- CCD: 14th International Conference on Cross-Cultural Design
- SCSM: 14th International Conference on Social Computing and Social Media
- VAMR: 14th International Conference on Virtual, Augmented and Mixed Reality
- DHM: 13th International Conference on Digital Human Modeling and Applications in Health, Safety, Ergonomics and Risk Management
- DUXU: 11th International Conference on Design, User Experience and Usability
- C&C: 10th International Conference on Culture and Computing
- DAPI: 10th International Conference on Distributed, Ambient and Pervasive Interactions
- HCIBGO: 9th International Conference on HCI in Business, Government and Organizations
- LCT: 9th International Conference on Learning and Collaboration Technologies
- ITAP: 8th International Conference on Human Aspects of IT for the Aged Population
- AIS: 4th International Conference on Adaptive Instructional Systems
- HCI-CPT: 4th International Conference on HCI for Cybersecurity, Privacy and Trust
- HCI-Games: 4th International Conference on HCI in Games
- MobiTAS: 4th International Conference on HCI in Mobility, Transport and Automotive Systems
- AI-HCI: 3rd International Conference on Artificial Intelligence in HCI
- MOBILE: 3rd International Conference on Design, Operation and Evaluation of Mobile Communications

List of Conference Proceedings Volumes Appearing Before the Conference

39. CCIS 1582, HCI International 2022 Posters - Part III, edited by Constantine Stephanidis, Margherita Antona and Stavroula Ntoa
40. CCIS 1583, HCI International 2022 Posters - Part IV, edited by Constantine Stephanidis, Margherita Antona and Stavroula Ntoa

http://2022.hci.international/proceedings

Preface

The goal of the Adaptive Instructional Systems (AIS) Conference, affiliated to the HCI International Conference, is to understand the theory and enhance the state-of-practice for a set of technologies (tools and methods) called adaptive instructional systems (AIS). AIS are defined as artificially intelligent, computer-based systems that guide learning experiences by tailoring instruction and recommendations based on the goals, needs, preferences, and interests of each individual learner or team in the context of domain learning objectives. The interaction between individual learners or teams of learners with AIS technologies is a central theme of this conference. AIS observe user behaviors to assess progress toward learning objectives and then act on learners and their learning environments (e.g., problem sets or scenario-based simulations) with the goal of optimizing learning, performance, retention and transfer of learning to work environments. The 4th International Conference on Adaptive Instructional Systems (AIS 2022) encouraged papers from academics, researchers, industry and professionals, on a broad range of theoretical and applied issues related to AIS and their applications. The focus of this conference on instructional tailoring of learning experiences highlights the importance of accurately modeling learners to accelerate their learning, boost the effectiveness of AIS-based experiences, and to precisely reflect their long-term competence in a variety of domains of instruction.

The content for AIS 2022 centered on challenges and approaches in four technical areas of adaptive instruction: 1) modeling and assessing learner performance relative to assigned learning objectives, 2) adapting or tailoring learning content and process for individual learners and teams of learners, 3) designing and developing adaptive instructional content and policy, and 4) evaluating the effectiveness of adaptive instructional systems addressing issues such as sociability, adaptability, recommendations and learning analytics, as well as student engagement. In addition to the papers included in the proceedings, AIS 2022 also explored areas of future research and development through expert panel sessions featuring invited talks.

One volume of the HCII 2022 proceedings is dedicated to this year's edition of the AIS Conference and focuses on topics related to learner modeling and state assessment for adaptive instructional decisions, adaptation design to individual learners and teams, design and development of AIS, as well as to evaluating its effectiveness.

Papers of this volume are included for publication after a minimum of two single–blind reviews from the members of the AIS Program Board or, in some cases, from members of the Program Boards of other affiliated conferences. We would like to thank all of them for their invaluable contribution, support and efforts.

June 2022

Robert A. Sottilare
Jessica Schwarz

4th International Conference on Adaptive Instructional Systems (AIS 2022)

Program Board Chairs: **Robert A. Sottilare,** Soar Technology, Inc., USA and **Jessica Schwarz,** Fraunhofer FKIE, Germany

- Michelle Barrett, Edmentum, Inc., USA
- Benjamin Bell, Eduworks Corporation, USA
- Shelly Blake-Plock, Yet Analytics, Inc., USA
- Bruno Emond, National Research Council Canada, Canada
- Jim Goodell, Quality Information Partners, USA
- Ani Grubišić, University of Split, Croatia
- Xiangen Hu, University of Memphis, USA
- Cheryl I. Johnson, NAWCTSD, USA
- John Edison Muñoz Cardona, University of Waterloo, Canada
- Maria Mercedes Rodrigo, Ateneo de Manila University, Philippines
- Meagan Rothschild, Age of Learning, USA
- Alexander Streicher, Fraunhofer IOSB, Germany
- KP Thai, Age of Learning, USA
- Rachel Van Campenhout, VitalSource, USA
- Joost Van Oijen, Royal Netherlands Aerospace Centre, The Netherlands
- Elizabeth T. Whitaker (retired), Georgia Tech Research Institute, USA
- Thomas Witte, Fraunhofer FKIE, Germany

The full list with the Program Board Chairs and the members of the Program Boards of all thematic areas and affiliated conferences is available online at

http://www.hci.international/board-members-2022.php

HCI International 2023

The 25th International Conference on Human-Computer Interaction, HCI International 2023, will be held jointly with the affiliated conferences at the AC Bella Sky Hotel and Bella Center, Copenhagen, Denmark, 23–28 July 2023. It will cover a broad spectrum of themes related to human-computer interaction, including theoretical issues, methods, tools, processes, and case studies in HCI design, as well as novel interaction techniques, interfaces, and applications. The proceedings will be published by Springer. More information will be available on the conference website: http://2023.hci.international/.

General Chair
Constantine Stephanidis
University of Crete and ICS-FORTH
Heraklion, Crete, Greece
Email: general_chair@hcii2023.org

http://2023.hci.international/

Contents

Learner Modeling and State Assessment for Adaptive Instructional Decisions

Heart Rate Variability for Stress Detection with Autistic Young Adults

Miroslava Migovich[1(✉)], Deeksha Adiani[1], Amy Swanson[2], and Nilanjan Sarkar[1]

[1] Robotics and Autonomous Systems Lab, Vanderbilt University, Nashville, TN 37212, USA
Miroslava.Migovich@vanderbilt.edu
[2] TRIAD, Vanderbilt University Medical Center, Nashville, TN 37203, USA

Abstract. Physiology, such as heart rate viability (HRV), can give meaningful insights about autonomic response to stress. Autism Spectrum Disorder has been linked to atypical physiological responses and poor emotion regulation. Explorations of the differences in physiological response between autistic young adults and neurotypical young adults can provide meaningful information on stress responses in these populations and can be used to create adaptive systems. Stress detection is an important aspect of creating closed-loop systems that can respond and change based on the emotional state of the user. This paper aims to explore HRV as a means of obtaining stress information from physiological data and to explore differences in stress response between autistic young adults and their neurotypical peers using Kubios HRV Premium analysis software during the PASAT-C, a distress tolerance task. Unpaired t tests showed statistically significant ($p < .05$) differences in three stress related indexes: the parasympathetic nervous system index, the sympathetic nervous system index, and the stress index. Preliminary results show validity of HRV for stress insight and provides evidence for physiological differences in stress response between the two groups.

Keywords: Stress detection · Heart rate variability · Adaptive systems · Autism

1 Introduction

Physiological data, specifically heart rate variability (HRV), can give important insight to stress responses from the two branches of the autonomic nervous system (ANS), the parasympathetic nervous system (PNS) and the sympathetic nervous system (SNS) [1–3]. HRV, and other physiological measures, are the body's response to the ANS using sensory and motor neurons to operate between various organs and the central nervous system. During a stressful event, the SNS releases several stress hormones, including cortisol, which cause the organs and motor neurons to respond [4]. These responses happen simultaneously with emotions and stress [5]. Physiological responses to stress are involuntary and are especially useful because they are less susceptible to being masked by an individual [6]. In cases of acute stress, the response is immediate [7] and can be measured. Prior to physiology-based stress detection, questionnaires were used to determine stress levels. However, the questionnaires are subjective and often the

R. A. Sottilare and J. Schwarz (Eds.): HCII 2022, LNCS 13332, pp. 3–13, 2022.
https://doi.org/10.1007/978-3-031-05887-5_1

collection of stress data is delayed [8], furthering the need for physiology-based detection. However, current physiology-based stress detection relies on supervised machine learning approaches that require ground truth labels from a trained behavioral expert or relay on task based stress labels [9–11]. Obtaining ground truth labels is labor intensive and cost prohibitive in many cases and level-based labels may not correctly represent an individual's stress level. Many machine learning approaches also rely on additional inputs beyond heart rate data, such as electrodermal activity and electromyogram data that require additional sensors [12–14]. HRV analysis of stress could be a means of obtaining these stress measures without the need for ground truth labels and reduce the number of necessary sensors.

Autism Spectrum Disorder (ASD) is characterized by differences in social communication and social behaviors, as well as difficulties with emotional regulation, including stress management [15, 16]. The prevalence of autism continues to rise, with the CDC estimating that the prevalence has risen from 1 in 150 in 2000 to 1 in 44 as of 2018 [17]. As the prevalence rises, it becomes even more necessary to understand the differences that characterize autism, including emotional regulation and stress to provide support and accommodations for autistic[1] young adults. Atypical physiological reactivity, including differences in heart rate variability (HRV) have been linked to autism [18]. While research on physiological reactivity in autistic individuals exist, much of the current literature is focused on children and gives little insight to how physiological responses may change in adulthood. In particular, how physiological responses can be utilized in inferring stress in autistic adults during human-computer interaction has not been thoroughly explored.

In recent years, learning technologies have expanded to include adaptive systems that can alter their functionality and interaction based on the needs of an individual or group [19]. Initial studies explored how other emotions could be mapped to physiology to create adaptive games [20, 21] and have expanded to apply to learning technologies and intelligent human-computer interactions [10, 22]. Using these concepts, technologies that harness physiological data for stress detection can be used to create a closed loop[2] system that adapts in response to stress. In turn, the closed loop system allows for "flow" learning to be achieved. Flow theory, as introduced by Csikszentmihalyi [23], refers to the state of mind that can lead to deep learning and high satisfaction. To achieve "flow", the activities must be challenging but should not overmatch existing skills or cause distress [23]. In order to achieve and maintain flow in an adaptive system, stress detection is useful.

In this paper, we explore HRV data for stress detection. While it is known that autistic individuals have heighten emotional responses, to our knowledge, there are no studies that explore differences in HRV measures, such as sympathetic and parasympathetic activity, and stress trends during a distress tolerance task between autistic young adults and NT young adults. To address these questions, we specifically look at trends in

---

[1] Recent surveys with autistic self-advocates suggest a preference for identify first language. In accordance have chosen to adopt identity first (*autistic persons*) language in place of person first (*persons with autism*) language [37].

[2] Closed loop refers to a system in which an operation, process, or mechanism is regulated by feedback [38], in this case the feedback is stress.

HRV based stress between ASD and NT young adults during the Paced Auditory Serial Addition Task (PASAT) [24, 25], a known distress tolerance task using Kubios[3] HRV Premium, a heart rate variability analysis software. In Sect. 2, we detail system design including the methods, and the data analysis using Kubios. Results are presented in Sect. 3 and discussed in Sect. 4.

2   System Design

2.1   Methods

Fig. 1. System design flow

The modified computer version of the PASAT was developed to produce psychological stress in lab settings in a consistent way [25]. Physiological data was collected using the Empatica E4 wristband[4]. The original task code developed by Millisecond Software[5] was adapted in order to add a three-minute baseline and stress self-report between each level. The task presents a single digit that must be added to the digit presented previously. The presentation of the digits can be seen in Fig. 2. The participant is not asked to keep a running total, but instead to only add the last two presented digits. For each incorrect answer, a loud, pitched error sound is played. The task begins with 11 practice trials and proceeds into levels one, two, and three with the time between digits decreasing with each level. Digits are presented every three seconds for three minutes during level one and are presented every two seconds for five minutes during level two. Level three digits are presented in 1.5 s intervals for a duration of 10 min. The Likert self-stress scale is presented between each level.

The Empatica E4 collects blood volume pulse (BVP) data from the photoplethys-mogram (PPG) sensor. The PPG uses combined green and red light to maximize the detection of the pulse wave. The green and red LEDs are oriented toward the skin. This allows the light to be absorbed by the blood and then reflected back. The reflected light is measured by the light receiver. The measurements taken during the green light exposure are used to generate the pulse wave and estimate heartbeats. The measurements during the red light exposure are used to create a reference light level that cancels out motion artefacts [26].

---

[3] www.kubios.com.

[4] www.empatica.com.

[5] www.millisecond.com.

Fig. 2. Presentation of digits for PASAT-C

Prior to the beginning of the task, demographics were collected including gender, height, weight, caffeine intake for the day, and general exercise levels, as these characteristics can affect physiological response and should be taken into account. The E4 was placed on the non-dominant wrist and the participant was asked to reduce motion of that wrist to minimize data loss and noise.

Five autistic young adults (Mean Age = 20.8, SD = 2.5) and five neurotypical young adults (Mean Age = 21, SD = 2.23) completed the task and informed consent was obtained following the approval of the Vanderbilt University Institutional Review Board (IRB). The two groups were age-matched in order to reduce age-related physiological response differences. COVID-19 protocols were followed to reduce exposure including mask usage, social distancing, and disinfecting of all equipment between participants.

2.2 Data Analysis

The BVP obtained from the E4 PPG was analyzed using Kubios HRV Premium. Kubios is a gold-standard heart rate variability software designed to process multiple data formats, including PPG data [27]. Each participant's data were loaded in Kubios along with gender, height, and weight, and segmented into sections based on the timestamps of the baseline and each level, as found from the PASAT-C data, as can be seen in Fig. 3a. While the software does provide automatic noise detection, it was not used as it is generally recommended for long-term ambulatory recordings. As recommended by [2], artefact correction was applied to account for missing, misaligned or extra beat detection and other arrhythmias. A low threshold-based artefact correction algorithm was applied in which each value in a 0.35 s interval is compared to the local average and outliers are replaced with interpolated values using cubic spline interpolation [28]. After artefact correction, several time-domain and frequency domain parameters such as mean time between two successive R-waves (RR interval), root mean square of successive difference between normal heartbeats (RMSSD), low and high frequency peak frequencies, were extracted.

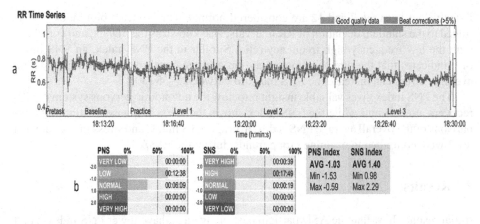

Fig. 3. Kubios output showing a) segmented PPG data based on level and b) PNS and SNS index output for the full duration of the task

The parasympathetic nervous system (PNS) is responsible for "rest and digest" conditions in the body [29] and is known to decrease heart rate by increasing the time interval between successive heart beats, increase changes in RR intervals based on respiration, and decrease the ratio between lower frequency and higher frequency oscillations in the time series [2, 3]. Based on these characteristics, Kubios computes the PNS index based on the mean RR interval, RMSSD, and the Poincaré plot index SD1 in normalized units. A longer mean RR intervals and high RMSSD values both indicate higher PNS activation. The SD1 of the Poincaré plot is linked to RMSSD, the normalized value is used as the third parameter. The parameter values compared to normal population values and then scaled with the standard deviations of normal population before a weighting is applied. A PNS index value of zero means the PNS activity is equal to the normal population average. Positive and negative values show by how many standard deviations are the parameter values above or below the normal population, respectively [30].

The sympathetic nervous system (SNS) drives the "fight or flight" response to stimuli in the body [29], which can be observed by increased heart rate, decreased HRV, and an increased ratio between the lower frequency and the higher frequency oscillations in the HRV data. SNS activation is computed based on mean heartrate interval, Baevsky's stress index [31] and the Poincaré plot index SD2 in normalized units. Kubios calculates a normalized stress index (SI) based on the following equation found in [31]

$$SI = \frac{AMo}{2Mo * MxDMn} \tag{1}$$

where each variable is a mathematical parameter found from a curve of distribution-histogram constructed for variational pulsometry. Mo refers to the mode, or the most frequently observed value in the dynamic line of cardio interval while AMo is the amplitude of mode that measures the number of cardio intervals appropriate to the mode value in percentage. The variation scope ($MxDMn$) reflects the variability degree of cardio interval values in the dynamic line [31]. In order to make the stress index normally distributed, Kubios takes the square root of the SI calculated by Eq. (1). Normalized stress index values between 7 and 12 are considered normal. The Poincaré plot index SD2 is linked to the standard deviation of RR intervals, also known as the SDNN, and correlates with the low frequency/high frequency ratio. Similar to the PNS index, an SNS index of 0 indicates that the three parameters are on average equal to the normal population average. Stress can increase the SNS index to between 5–35 [30].

The PNS index gives valuable insight into how the autonomic nervous system (ANS) recovers after stimuli while the SNS index shows the stress response to stimuli. Comparisons of the overall average PNS index, average SNS index, and stress index per each level were computed to examine group trends during the PASAT-C.

3   Results

Initial results show that the ASD participants, on average, have lower PNS indexes and higher SNS indexes as seen in Fig. 4 when compared to NT participants, showing initial validation of claims that the two groups have varied physiological stress response to the same stimuli.

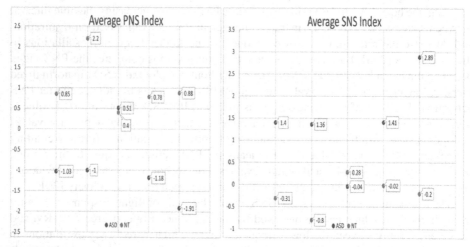

Fig. 4. (a) Average PNS and (b) SNS where each blue dot represents a ASD participant, and each orange dot presents a NT participant

Unpaired two tailed t tests were run for the average PNS, SNS, and stress indexes between the two groups and the difference was found to be statistically significant ($p <$ 0.05) across all three categories. Results of the unpaired t tests can be seen in Table 1.

Average stress index by level was calculated for each group to explore if similar stress trends emerged based on the level of the PASAT-C. Comparison of these averages are shown in Fig. 5. It can be seen that the autistic group had higher stress averages across all levels of the task, with the practice level being the most stressful. The NT group showed the highest level of stress during level 2.

Table 1. Unpaired t test results for average PNS, SNS, and stress index

	Neurotypical		ASD				
	Mean	Standard deviation	Mean	Standard deviation	df	t	p
PNS index	1.02	0.67	−0.92	0.88	8	3.89	.0046**
SNS index	−0.21	0.39	1.40	1.03	8	3.25	.0117*
Stress index	6.37	1.52	11.82	3.70	8	2.83	0.0218*

* $p < .05$, ** $p < .01$

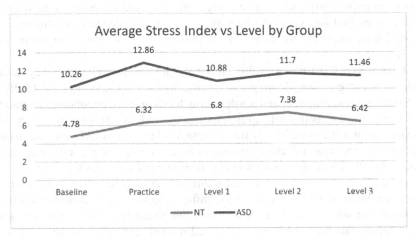

Fig. 5. Average stress index vs level by group

4 Discussion and Conclusion

These results reaffirm that physiological responses to stress varies between the two groups and that autistic young adults are more likely to experience higher levels of stress and lower recovery after a stressful stimulus. This study also proved that HRV alone has the ability to represent these differences. As shown by the unpaired t test, there are statistically significant differences in PNS, SNS, and SI during the duration of the PASAT-C. The low PNS activity and high SNS activity of the autistic group during the task is consistent with findings that this group struggles with emotional regulation. While it is expected that SNS activity would rise during a distress tolerance task, such as the PASAT-C, Fig. 4 shows that, when compared to NT peers, the physiological response of ASD young adults is dominated by their "fight or flight" division of autonomic nervous system while the NT young adults' response is dominated by the "rest and digest" division of the autonomic nervous system. These results also show that the PASAT-C did cause a physiological response in both groups when compared to the general population, as shown by the SNS and PNS activity not being zero for either group. Figure 5 indicates that the ASD group was most stressed during the practice round while the NT group was most stressed during the second level, which shows that not only is the overall stress response is distinct between the groups, but that specific portions of the interaction affected each group differently.

One possible application of HRV stress detection is with a virtual reality interview simulator for autistic young adults. Securing a full-time job is one hallmark of successful transition to adulthood for all young adults [32], however, half of autistic young adults are under-or unemployed [33]. The interview process has been identified by autistic self-advocates and their families as one of the largest barriers to employment [32, 34]. In many circumstances, the autistic candidate possesses the necessary skills and relevant experience, however, is unable to perform well in the interview based on "neurotypical standards" [35]. Overcoming stress during the interview process allows autistic individuals more opportunities to obtain employment. Repeated exposure to the interview process is important for successful interviewing [35]. In order to meet this need, we designed Career Interview Readiness in VR (CIRVR) that is detailed in our previous work [36].

In the context of CIRVR, the results of this paper are significant because they reiterate that when designing a closed-loop system specifically for autistic young adults, different considerations must be made than when designing for the general population. To maintain flow, it is beneficial to both track stress response and to know how to effectively adapt the system so that the challenge is still present while stress is reduced. Maintaining flow allows for the most effective learning to be achieved. For autistic young adults, the concept of flow can also be used to aid them in emotion regulation. For example, due to the differences in stress response, a redirection approach, such as using multiple-choice white board questions or moving to a different topic, may be more helpful for the ASD population while rephrasing the question and giving more time to answer would be more appropriate for the NT group. The redirection approach removes the stressor (the question or topic) thus reducing SNS activity while the rephrasing technique gives more time for the PNS to return to baseline. Further exploration of specific adaptation techniques will be investigated in future work.

While this paper focused specifically on the PNS, SNS, and SI indexes, it also opens the door for deeper exploration of additional HRV features that were not discussed. These initial results validate the use of PPG data alone to extract heart rate variability information that gives insight into stress response for autistic young adults. Kubios provides a number of other time-domain, frequency-domain, nonlinear, and time-varying results that are beyond the scope of this paper.

Future work will explore stress-based adaptation within the CIRVR system in order to ensure that the user is adequately challenged without becoming overly stressed by integrating the PNS, SNS, and SI indexes to inform the real-time system of stress levels. This will allow for exploration of stress trends during the interview process. Further exploration will include how additional HRV features may give insight into stress. Knowledge of interview specific stress trends will allow for accommodations to be made that can reduce the barriers to employment that interviews currently present.

Acknowledgments. This project was funded by a Microsoft AI for Accessibility grant, by the National Science Foundation under awards 1936970 and 2033413 and by the National Science Foundation Research Traineeship DGE 19–22697. The authors would like to thank the participants for their time.

References

1. Hagemann, D., Waldstein, S.R., Thayer, J.F.: Central and autonomic nervous system integration in emotion. Brain Cogn. **52**(1), 79–87 (2003). https://doi.org/10.1016/S0278-262 6(03)00011-3
2. Malik, M., et al.: Heart rate variability. Circulation **93**(5), 1043–1065 (1996). https://doi.org/10.1161/01.CIR.93.5.1043
3. Acharya, U.R., Joseph, K.P., Kannathal, N., et al.: Heart rate variability: a review. Med. Bio. Eng. Comput. **44**, 1031–1051 (2006). https://doi.org/10.1007/S11517-006-0119-0
4. Salai, M., Vassányi, I., Kósa, I.: Stress detection using low cost heart rate sensors. J. Healthc. Eng. 2016 (2016). https://doi.org/10.1155/2016/5136705
5. Friedman, B.H.: Feelings and the body: the Jamesian perspective on autonomic specificity of emotion. Biol. Psychol. **84**(3), 383–393 (2010). https://doi.org/10.1016/j.biopsycho.2009.10.006
6. Shu, L., et al.: A review of emotion recognition using physiological signals. Sensors (Switzerland) **18**(7), 2074 (2018). https://doi.org/10.3390/s18072074
7. Dalmeida, K.M., Masala, G.L.: HRV features as viable physiological markers for stress detection using wearable devices. Sensors **21**(8), 2873 (2021). https://doi.org/10.3390/S21082873
8. Smets, E., de Raedt, W., van Hoof, C.: Into the wild: the challenges of physiological stress detection in laboratory and ambulatory settings. IEEE J. Biomed. Health Inform. **23**(2), 463–473 (2019). https://doi.org/10.1109/JBHI.2018.2883751
9. Migovich, M., Korman, A., Wade, J., Sarkar, N.: Design and validation of a stress detection model for use with a VR based interview simulator for autistic young adults. In: Antona, M., Stephanidis, C. (eds.) HCII 2021. LNCS, vol. 12768, pp. 580–588. Springer, Cham (2021). https://doi.org/10.1007/978-3-030-78092-0_40
10. Bian, D., Wade, J., Swanson, A., Warren, Z., Sarkar, N.: Physiology-based affect recognition during driving in virtual environment for autism intervention. In: PhyCS 2015 - 2nd International Conference on Physiological Computing Systems, Proceedings, pp. 137–145 (2015). https://doi.org/10.5220/0005331301370145
11. Panicker, S.S., Gayathri, P.: A survey of machine learning techniques in physiology based mental stress detection systems. Biocybernetics Biomed. Eng. **39**(2), 444–469 (2019). https://doi.org/10.1016/J.BBE.2019.01.004
12. Zontone, P., Affanni, A., Bernardini, R., Piras, A., Rinaldo, R.: Stress detection through Electrodermal Activity (EDA) and Electrocardiogram (ECG) analysis in car drivers. In: European Signal Processing Conference, vol. 2019-September, September 2019. https://doi.org/10.23919/EUSIPCO.2019.8902631
13. Visnovcova, Z., Calkovska, A., Tonhajzerova, I.: Heart rate variability and electrodermal activity as noninvasive indices of sympathovagal balance in response to stress (2013). https://doi.org/10.2478/acm-2013-0006
14. Pourmohammadi, S., Maleki, A.: Stress detection using ECG and EMG signals: a comprehensive study. Comput. Methods Programs Biomed. **193**, 105482 (2020). https://doi.org/10.1016/J.CMPB.2020.105482
15. American Psychiatric Association. Diagnostic and Statistical Manual of Mental Disorders, May 2013. https://doi.org/10.1176/APPI.BOOKS.9780890425596
16. Mazefsky, C.A.: Emotion regulation and emotional distress in autism spectrum disorder: foundations and considerations for future research. J. Autism Dev. Disord. **45**(11), 3405–3408 (2015). https://doi.org/10.1007/s10803-015-2602-7
17. Data and Statistics on Autism Spectrum Disorder | CDC. https://www.cdc.gov/ncbddd/autism/data.html. Accessed 21 Feb 2022

18. Dindar, K., et al.: Social-pragmatic inferencing, visual social attention and physiological reactivity to complex social scenes in autistic young adults. J. Autism Dev. Disord. 52(1), 73–88 (2022). https://doi.org/10.1007/S10803-021-04915-Y/TABLES/6
19. Benyon, D., Murray, D.: Adaptive systems: from intelligent tutoring to autonomous agents. Knowl.-Based Syst. 6(4), 197–219 (1993). https://doi.org/10.1016/0950-7051(93)90012-I
20. Tijs, T., Brokken, D., IJsselsteijn, W.: Creating an emotionally adaptive game. In: Stevens, S.M., Saldamarco, S.J. (eds.) ICEC 2008. LNCS, vol. 5309, pp. 122–133. Springer, Heidelberg (2008). https://doi.org/10.1007/978-3-540-89222-9_14
21. Frommel, J., Schrader, C., Weber, M.: Towards emotion-based adaptive games: emotion recognition via input and performance features. In: Proceedings of the 2018 Annual Symposium on Computer-Human Interaction in Play (2018). https://doi.org/10.1145/3242671
22. Bian, D., Wade, J., Swanson, A., Weitlauf, A., Warren, Z., Sarkar, N.: Design of a physiology-based adaptive virtual reality driving platform for individuals with ASD. ACM Trans. Accessible Comput. (TACCESS) 12(1), 1–24 (2019). https://doi.org/10.1145/3301498
23. Csikszentmihalyi, M.: Finding flow (1997)
24. Gronwall, D.M.A.: Paced auditory serial addition task: a measure of recovery from concussion. Percept. Mot. Skills 44(2), 367–373 (1977). https://doi.org/10.2466/pms.1977.44.2.367
25. Lejuez, C.W., Kahler, C.W, Brown, R.A.: A modified computer version of the Paced Auditory Serial Addition Task (PASAT) as a laboratory-based stressor (2003). undefined
26. Utilizing the PPG/BVP signal – Empatica Support. https://support.empatica.com/hc/en-us/articles/204954639-Utilizing-the-PPG-BVP-signal. Accessed 21 Feb 2022
27. Tarvainen, M.P., Niskanen, J.P., Lipponen, J.A., Ranta-aho, P.O., Karjalainen, P.A.: Kubios HRV – Heart rate variability analysis software. Comput. Methods Programs Biomed. 113(1), 210–220 (2014). https://doi.org/10.1016/J.CMPB.2013.07.024
28. HRV preprocessing – Kubios. https://www.kubios.com/hrv-preprocessing/. Accessed 06 Feb 2022
29. Tindle, J., Tadi, P.: Neuroanatomy, parasympathetic nervous system. StatPearls, November 2021. https://www.ncbi.nlm.nih.gov/books/NBK553141/. Accessed 06 Feb 2022
30. HRV in evaluating ANS function – Kubios. https://www.kubios.com/hrv-ans-function/. Accessed 01 Nov 2021
31. Baevsky, R.M., Berseneva, A.P.: Use Kardivar system for determination of the stress level and estimation of the body adaptability. Moscow-Prague (2008). https://www.semanticscholar.org/paper/Methodical-recommendations-USE-KARDiVAR-SYSTEM-FOR/74a292bfafca4fdf1149d557348800fcc1b0f33b
32. Anderson, C., Butt, C., Sarsony, C.: Young adults on the autism spectrum and early employment-related experiences: aspirations and obstacles. J. Autism Dev. Disord. 51(1), 88–105 (2021). https://doi.org/10.1007/S10803-020-04513-4/TABLES/2
33. "Autism | U.S. Department of Labor. https://www.dol.gov/agencies/odep/topics/autism. Accessed 28 Oct 2020
34. Burke, S.L., Li, T., Grudzien, A., Garcia, S.: Brief report: improving employment interview self-efficacy among adults with autism and other developmental disabilities using virtual interactive training agents (ViTA). J. Autism Dev. Disord. 51(2), 741–748 (2020). https://doi.org/10.1007/s10803-020-04571-8
35. Mj, S., et al.: Virtual interview training for autistic transition age youth: a randomized controlled feasibility and effectiveness trial. Autism: Int. J. Res. Pract. 25(6), 1536–1552 (2021). https://doi.org/10.1177/1362361321989928
36. Adiani, D., et al.: Career Interview Readiness in Virtual Reality (CIRVR): a platform for simulated interview training for autistic individuals and their employers. ACM Trans. Accessible Comput. 15, 1–28 (2022)

37. Kenny, L., Hattersley, C., Molins, B., Buckley, C., Povey, C., Pellicano, E.: Which terms should be used to describe autism? perspectives from the UK autism community. Autism **20**(4), 442–462 (2016). https://doi.org/10.1177/1362361315588200
38. Closed loop definition & Meaning - Merriam-Webster." https://www.merriam-webster.com/dictionary/closed%20loop. Accessed 15 Feb 2022

Bayesian Cognitive State Modeling for Adaptive Serious Games

Alexander Streicher[1](✉) and Michael Aydinbas[2]

[1] Fraunhofer IOSB, Karlsruhe, Germany
`alexander.streicher@iosb.fraunhofer.de`
[2] TU Darmstadt, Darmstadt, Germany

Abstract. Bayesian modeling of cognitive state is one possible approach to user modeling for use with adaptivity in serious games. Adaptive educational serious games try to keep learners engaged - to keep them in the so called "Flow" channel, i.e., in the right balance between being challenged and entertained. The challenge is to intervene adaptively at the right time. The research question is when to actually adapt, and how to find quantifiable metrics for that. One way to achieve this is to model the users' cognitive state and to adapt to high or low cognitive load, e.g., to apply dynamic difficulty adjustments. Our user modeling approach is based on Hierarchical Bayesian Models (HBM) which are suitable for drawing conclusions about the learner's cognitive state inferred from observable variables. An important aspect is that the approach considers activity stream data such as from the Experience API (xAPI) protocol as input to achieve high interoperability and eased applicability. An evaluation with synthetic data for different user group types shows the feasibility of the approach. The model can explain differences between subjects, between subject groups and between different latent variables such as cognitive load or mental working memory capacity.

Keywords: User modeling · Cognitive modeling · Learner state · Adaptivity · Serious games · Bayesian inference

1 Introduction

A key question for an adaptive education system (AES [15]) is when adaptability should occur, that is, at what point a learner should be best supported. This also concerns serious games, i.e., (digital) games with the characterizing goal to educate and not just to entertain [5,9]. The challenge is to infer the needs of the users by observing how the users interact with the systems. AES could try to determine the cognitive states of the users to trigger the right adaptive responses [14,20]. For example, an adaptive educational system could react when attention decreases, cognitive load increases [14], or when the user seems to be in a repetitive cycle with no real observable progress, or when there are signs of forgetting. Cognitive modeling can provide answers to these questions by

© The Author(s), under exclusive license to Springer Nature Switzerland AG 2022
R. A. Sottilare and J. Schwarz (Eds.): HCII 2022, LNCS 13332, pp. 14–25, 2022.
https://doi.org/10.1007/978-3-031-05887-5_2

focusing on user modeling [14,20]. The principal part of an adaptive system for educational serious games is the user or student model [19] which can include information about the learners' current cognitive states, e.g., cognitive load or stress level, motivation, attention, etc. The model's estimate of the learner's current cognitive state can be used to derive and suggest adaptive measures to help navigating the player through the game on a smooth and engaging path.

The research question of this work is how to create a cognitive user model which allows to estimate latent cognitive variables such as cognitive load, perceived difficulty, or prior knowledge level. Subsequently, AES can use such a model to determine the best time for an adaptive response.

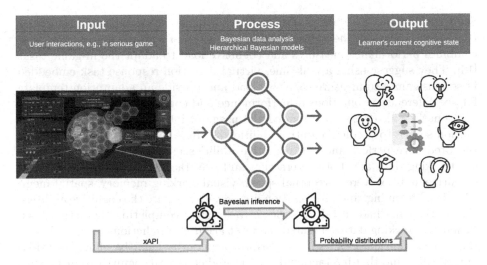

Fig. 1. Input-Processing-Output (IPO) concept for applying generative Bayes models to user models; output are probabilities about user cognitive states.

The contribution of this work is a concept for a cognitive user model which is based on *Hierarchical Bayesian Models* (HBM). The model makes use of the *Cognitive Load Theory* (CLT) [17] to describe the relationship between the characteristics of the learner and the characteristics of the learning material. The model has been validated numerically using synthetic data. A detailed model comparison was conducted to obtain the best model. The final model is able to explain the three observable variables task success, mission score and mission time. In our technical evaluation with synthetic data it can accurately predict those variables with an accuracy of over 90% (mean absolute deviation).

As depicted in Fig. 1, input to the model are activity stream data formatted according to the *Experience API* (xAPI) protocol [1]. This allows for increased interoperability and applicability to other application domains. The xAPI specifies how to capture user interaction events as a stream of activities. Every activity is based on an actor-verb-object triple structure, with additional information such as timestamps, context data or the result outcome of an activity, e.g., an assessment result score. This leads to a machine- and human-readable stream of activities. We make use of xAPI as input for our modeling approach.

The model's output are probability distributions that allow for inferences about the model's parameters and provide the uncertainty associated with those inferences.

To the best of our knowledge, this is the first application of a HBM for realizing cognitive user models for adaptive serious games. Several issues with the final model have been identified and provide directions for further research.

2   Related Work

Seyderhelm et al. (2019) propose a Cognitive Adaptive Serious Game Framework (CASG-F), routed in the cognitive-affective theory of learning with media, that combines performance measures and cognitive load to adapt the in-game tasks [14]. They suggest using a real-time, virtual detection-response task embedded in serious games to measure cognitive load and provide an adaptation template for six different combinations of performance and cognitive load measures [14].

Conati et al. (2020) investigated the usage of interaction data as an information source to predict cognitive abilities [3]. In a user study, they compared the predictive performance for cognitive abilities using only interaction data, eye-tracking data and both interaction and eye-tracking data. The researched cognitive abilities were perceptual speed, visual working memory, spatial memory, visual scanning, and visualization literacy. To measure the cognitive abilities, the participants had to take a series of tests after completing the actual task. While eye-tracking data generated the most accurate predictions, results showed that interaction data can still outperform a majority-class baseline. Additionally, it was found that interaction data can predict several cognitive abilities with better accuracy at the very beginning of the task than eye-tracking data, which is valuable for delivering adaptation early in the task. Left click rate and time to first click were the top two predictors for all cognitive abilities, suggesting the importance of those two features for predicting cognitive abilities. Conati et al. (2020) concluded that adaptation for interactive visualizations tasks could be enabled using solely interaction data [3].

Hallifax et al. (2020) analyzed the effect of combining several learner models to guide the adaptation strategy. They showed that adaptation is more effective when tailored to both player type and motivation, which could improve the intrinsic motivation to engage with the content [8].

Tadlaoui et al. (2018) realized a probabilistic and dynamic learner model in adaptive hypermedia educational systems based on multi-entity Bayesian networks [18]. The model can represent the different actions that the learner can take during their learning path.

One classical example of using dynamic Bayesian networks to model the causes and effects of emotional reactions was given by Conati and Maclaren (2009) [4]. Their diagnostic model targets affect and how emotions are caused by the users' appraisal in a given context, e.g., for goals or preferences.

Gelman et al. (2013) state that checking the model is crucial to statistical analysis [7]. Bayesian prior-to-posterior inferences assume the whole structure of a probability model and can yield misleading inferences when the model is poor [7]. Therefore, good Bayesian analysis should include at least some check of the adequacy of the model's fit to the data and the model's plausibility for the purposes for which the model will be used. The most common way to check the fit of a model is by performing a posterior predictive check—as it was also done here. Posterior predictive distribution denotes a probability distribution over possible values of future data [11].

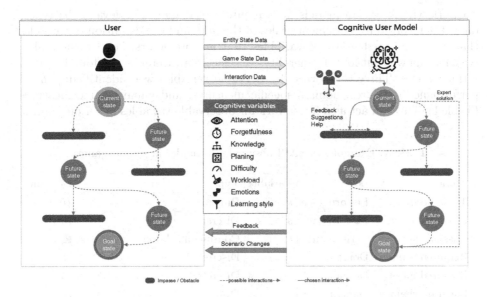

Fig. 2. Application principle for cognitive user models applied to (educational) serious games to estimate various cognitive variables as control input for adaptivity, e.g., adaptive feedback or scenario changes. (Image: Aydinbas 2019 [2])

3   Concept for a Cognitive User Model

A cognitive user or learner model [19] is a model that can make statements about the learner's cognitive state, that is, statements about their mental actions and processes that deal with knowledge acquisition and understanding. In the adaptive cycle by Shute et al. (2012) the learner model builds a connection between the captured user data and the presented learning material that is suitable for the learner [15]. The learner model should allow for a dynamic assessment of the learner's current cognitive state [3]. The model's knowledge about the learner can be leveraged to guide the player through the problem space towards a goal state while avoiding states that are detrimental for the player (e.g., [8,18]).

Figure 2 depicts the idea how to apply cognitive modeling to serious games [2]. On the left is an outline of the user's cognitive states and how it develops while playing a game towards a goal state. As shown on the right, cognitive modeling tries to simulate that with a cognitive user model—the model tries to estimate various cognitive variables as input to adaptive control, e.g., adaptive feedback or scenario changes. This again results in an interaction between the user and the AES (as in a closed-loop feedback adaptive control).

3.1   Observable Variables

To work with Bayesian models, it is required to define the relevant observable data, the involved measurement scales of the data and the definition of variables that are to be predicted and variables that are predictors. For transferability reasons, and without loss of generality, we selected a variable set as found in most serious games. Three categories of observable variables were identified: general performance measures, domain-specific measures, and game-specific measures (Table 1). This can act as a framework for observable variables in serious games.

Table 1. Examples of xAPI observable variables in serious games.

Name	Level	Variable	Type	Unit	Domain
Task success	Performance	k	Binary	–	$\{0, 1\}$
Mission score	Performance	s	Discrete	–	$\mathbb{N}_0$
Mission time	Performance	t	Continuous	Minutes/seconds	$\mathbb{R}_{\geq 0}$
Required rounds	Domain	n_{rnd}	Discrete	–	$\mathbb{N}$
Required hints	Domain	n_{hnt}	Discrete	–	$\mathbb{N}_0$
Location changes	Domain	n_{loc}	Discrete	–	$\mathbb{N}_0$
Dialogues	Domain	n_{dia}	Discrete	–	$\mathbb{N}_0$
Detours	Game	n_{det}	Discrete	–	$\mathbb{N}_0$

General performance measures are mostly domain-independent and can be measured in any application. Games typically report on these measures in an assessment stage, e.g., at the end of a mission or task.

Task success k describes the success of a player for a given task and is either true or false. As an example, in an image interpretation serious game, a task or objective is the correct deployment of an imaging sensor, which can fail if the player has not correctly considered the actual weather conditions or has exceeded the number of rounds available for the mission.

Mission score s is the overall score of the player for the given mission, but is also applicable to any game and any learning task because it is a simple accuracy measure that reflects the number of errors the player made. As an example, this number lies between zero and the maximal number of items the player had to report for a given mission, and typically it is normalized in $[0; 1]$ to be comparable across missions and across different applications.

Mission time t is the amount of time it took the player to finish the mission, but is applicable to any domain or learning task by measuring the time the learner needs to finish a given task.

Domain measures are domain-dependent, but applicable to any similar game of that domain that supports this feature. For example, strategy games use resources of some kind and turn-based games have a measure of rounds. Game measures are the most specific of all variables because they only apply to a particular game and are normally not transferable to other games or applications without reinterpretation. One example of a game measure variable might be detours which captures the number of detours the player has taken during the mission compared to an ideal solution where the player knows exactly what the next steps are and where to go.

3.2 Cognitive User Models for Serious Games

Instead of cognitive modeling frameworks such as Soar 9 [10], which demands a lot of domain and task modeling effort but cannot directly offer inferences about the cognitive state of the learner, we decided to choose a different approach: to build a probabilistic statistical model of the data that only specifies what is really needed for the user modeling task. The Bayesian modeling approach allows for maximal control and flexibility as the model can be as general or as complex as needed. Bayesian models are directly built in a way to infer the state of latent, non-observable variables from observable variables.

After the user has finished a game (or any other learning) session, either successfully or with failure, the collected data is analyzed by the HBM. The model takes predefined prior distributions for each model parameter and the observations as input and calculates (via Bayesian inference) the posterior distributions for all parameters. The latent parameters are variables that can be interpreted as cognitive variables such as motivation or the perceived difficulty.

Because the model produces posterior distributions, which are probability densities, the model gives directly interpretable probabilities as output. It is straight forward to derive point estimates such as mean and mode values as well as highest density intervals from the posterior distributions. At the end of each session, the model can infer the learner's current cognitive state and make statements about their current value with an estimate of uncertainty. Because the posterior is a probability density, it can be reused in another run of Bayesian inference as the prior, replacing the old, non-informative prior with the model's latest belief according to the observed data. This is identical to the prediction-correction step in Bayesian filtering methods, and the principle is depicted in Fig. 3. During each mission the player executes a series of actions, which takes time t. At the end of each mission all user interactions are send to the cognitive user model framework in form of xAPI statements. The framework is given prior distributions for each model parameter and observable data as input and produces, via Bayesian inference, posterior distributions for each model parameter.

This posterior distribution is the base for the computation of point estimates. The posterior of one mission can also serve as the prior for another mission.

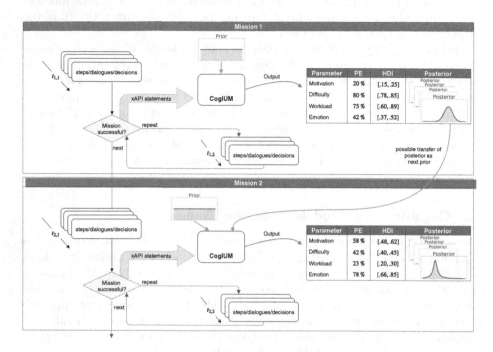

Fig. 3. Realizing a "Cognitive Intelligent User Model" (CogIUM) with HBM. After each round, task or mission, the model's inferences can be used to adapt the game according to the learner's needs, e.g., adjusting the difficulty level.

3.3   A Hierarchical Bayesian User Model

Our final model (Fig. 4) is a fully specified descriptive HBM that is able to model the three observable variables task success, mission score and mission time [2].

The model defines variables for two separate groups (depicted by rectangles): personal variables (indexed with p) that differ between subjects and conceptual variables (indexed with c) that differ between concepts. Concept in this context means a learning concept that the learner should acquire. In addition to group variables, which remain constant within the group, there are also variables that differ between subjects as well as between concepts (indexed with pc). This is the case for the observable variables, as they depend both on the current concept as well as on the current subject.

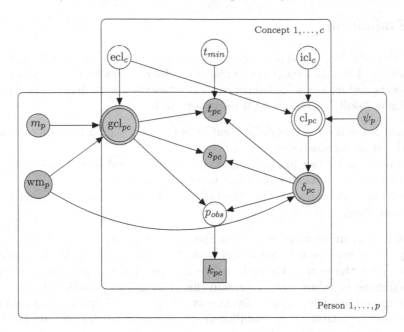

Fig. 4. Simplified graphical model of the best performing HBM (notation according to [6]). The fully specified descriptive model is described in [2] as $\mathcal{M}_4$. (Color figure online)

The HBM models the cause of three observable variables: task success k_{pc}, mission score s_{pc}, and mission time t_{pc}. The latent variables (highlighted in blue) that serve as proxies for cognitive variables - which are the primary focus of the cognitive model - include motivation m_p, prior knowledge ψ_p, germane cognitive load glc_{pc}, and free working memory capacity δ_{pc}.

Besides the observable variables (Fig. 4, gray nodes), all other variables are considered latent and can be the target of inference. Especially, we are interested in variables that can be interpreted as cognitive variables and allow inference about the learner's cognitive state (Fig. 4, blue nodes).

The basic idea of this descriptive model is that cognitive load affects learner performance, which in turn has a direct impact on the observable variables. Higher cognitive load often means an increase in the number of errors, and thus a decrease in performance [12]. Cognitive load is purely defined by the sum of intrinsic and extraneous cognitive load [16]. Germane cognitive load is the proportion of working memory resources that deals with intrinsic cognitive load. Higher extraneous cognitive load means less germane cognitive load, which reduces learning and which can be measured by a decrease in the learner's performance [16]. For example, if the model observes that two different people perform differently when learning the same concept, it can attribute the differences to individual differences in cognitive load and free working memory capacity, which in turn are influenced by the learner's motivation and prior knowledge.

4   Evaluation

To validate the fitness and performance of our models, we generated 14 synthetic data sets [2]. Each data set contains plausible observations for a different number of subjects and missions (called concept). Participants were divided into groups of different "skills". There were three groups in total:

1. "good": a subject successfully completes the task, has a high mission score and a short mission time.
2. "average": a subject has a random chance to succeed in the task, and both has an average mission score and an average mission time.
3. "bad": a subject always fails the task, has a low mission score and a long mission time.

The data sets differ with respect to the number of subjects, the number of concepts, the number of groups, and how the subjects are divided among the groups. With the help of Bayesian inference, the model re-allocates credibility across parameter values to best explain the observations of each data set.

Comparing models is especially important since it is typically the case that more than one reasonable probability model can provide an adequate fit to the data in a scientific problem. A popular metric to quantify predictive performance is the *Mean Squared Error* (MSE) [13].

The best model was able to explain individual differences with more than 90% accuracy for the three observable variables task success, mission score and mission time.

In Table 2 we report the *Mean Absolute Deviation* (MAD), the *Mean of the Squared Deviations* (MSE), the average standard deviation (SD) and the accuracy (ACC). We define accuracy by comparing the MAD estimate with the maximum upper limit, in our case the maximum error of each variable: 1) task success is binary-valued $\{0, 1\}$ hence the maximum error is 1; 2) mission score is in $[0; 1]$ hence the maximum error is 1. Mission time can be any value greater zero, but we limited the session to 5 minutes. Table 2 shows that, on average, task success is reproduced 100% correctly for all subjects, mission score 98.9% and mission time 92.7%.

Table 2. Measures of predictive precision and predictive accuracy for the best model for a single simulated data set. P_i stands for the posterior predictive distribution of the i-th observation.

Observable variable	MAD	SD_{MAD}	MSE	SD_{MSE}	ACC_{MAD}	SD
Task success	0.0	0.0	–	–	1.0	0.46
Mission score	0.011	0.011	0.001	0.001	0.989	0.107
Mission time	0.366	0.203	0.427	0.445	0.927	1.963

$MAD = \frac{1}{n}\sum |y_i - \bar{P}_i|;\ MSE = \frac{1}{n}\sum (y_i - \bar{P}_i)^2;\ ACC = 1 - \frac{MAD}{\text{max. Error}};\ SD = \frac{1}{n}\sum \sigma_{P_i}$

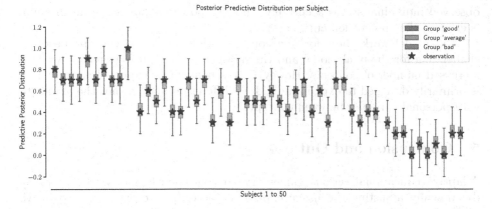

Fig. 5. Comparison between individual observations created by the model and the true observations (marked by a star) for mission score s_{pc}. The simulated data set includes 50 subjects, divided into three groups (good, average, bad). The model's predictions are based on samples drawn from the model's posterior predictive distribution.

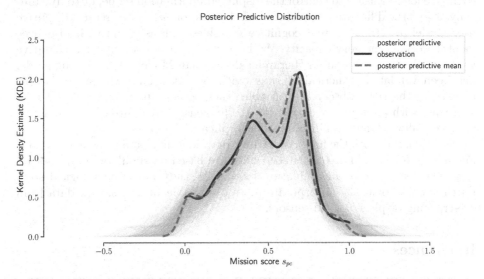

Fig. 6. Comparison between predicted observations (orange) by the model and the true observations (black) for mission score s_{pc}. Distributions were approximated by kernel density estimation for visualization. The predicted observations are based on samples drawn from the model's posterior predictive distribution. The mean posterior predictive distribution is plotted as a dashed blue line. (Color figure online)

As an example, Fig. 5 and Fig. 6 show the results of a posterior predictive graphical check for the observable variable mission score for $n = 50$ subjects divided into the three distinct groups. The model is able to reproduce the observed individual scores accurately, and the posterior predictive mean curve is close to the observation curve.

The model works best for one concept. More than one concept becomes challenging for the final model and the model's performance decreases with an increased number of concepts. However, this behavior is expected as the HBM is primarily designed to explain individual differences and has not the ability to vary personal variables between concepts.

5   Conclusion and Outlook

Adaptive educational serious games try to continuously motivate players by dynamically adjusting the usage experience according to the users' cognitive states, e.g., according to the estimated free working memory or motivation. Cognitive user models can capture cognitive variables such as cognitive load, motivation, attention, etc. Adaptive systems can make use of such models to determine a good point in time when to actually adapt, e.g., a threshold on the cognitive load measure to control the displayed information scope, or to dynamically adjust the difficulty level. The research question is how to create a cognitive user model to estimate latent cognitive variables such as cognitive load or free mental working memory capacity. We have presented a concept for a cognitive user model which is based on Hierarchical Bayesian Models (HBM). The model has been validated numerically using synthetic data. The final model is able to explain the three observable variables task success, mission score and mission time with an accuracy of over 90%. The generic structure of the modeling approach allows transferability to other application fields.

Several issues with the final model have been identified and provide directions for further research. The presented results are based on simulated data, not on empirical data, hence, an evaluation by a user study is the next logical step. Further work could also incorporate more observable variables, e.g., data from eye-tracking or physiological sensors.

References

1. Advanced Distributed Learning (ADL): Experience API (xAPI) Specification, Version 1.0.1. Technical report, Advanced Distributed Learning (ADL) Initiative, U.S. Department of Defense (2013)
2. Aydinbas, M.: Realizing cognitive user models for adaptive serious games. Master thesis, TU Darmstadt, Fraunhofer IOSB (2019)
3. Conati, C., Lallé, S., Rahman, M.A., Toker, D.: Comparing and combining interaction data and eye-tracking data for the real-time prediction of user cognitive abilities in visualization tasks. ACM Trans. Interact. Intell. Syst. 10(2), 1–41 (2020)
4. Conati, C., Maclaren, H.: Empirically building and evaluating a probabilistic model of user affect. User Model. User-Adapted Interact. 19(3), 267–303 (2009)
5. Dörner, R., Göbel, S., Effelsberg, W., Wiemeyer, J.: Serious Games - Foundations, Concepts and Practice. Springer, Cham (2016)

6. Farrell, S., Lewandowsky, S.: Computational Modeling of Cognition and Behavior. Cambridge University Press, Cambridge (2018)
7. Gelman, A., Carlin, J.B., Stern, H.S., Dunson, D.B., Vehtari, A., Rubin, D.B.: Bayesian Data Analysis, 3rd edn. Chapman and Hall/CRC (2014)
8. Hallifax, S., Lavoué, E., Serna, A.: To tailor or not to tailor gamification? An analysis of the impact of tailored game elements on learners' behaviours and motivation. In: Bittencourt, I.I., Cukurova, M., Muldner, K., Luckin, R., Millán, E. (eds.) AIED 2020. LNCS (LNAI), vol. 12163, pp. 216–227. Springer, Cham (2020). https://doi.org/10.1007/978-3-030-52237-7_18
9. Kickmeier-Rust, M., Albert, D.: Educationally adaptive: balancing serious games. Int. J. Comput. Sci. Sport 10(1), 1–10 (2012)
10. Laird, J.E.: The SOAR cognitive architecture. AISB Q. 134(1), 1–4 (2012)
11. Lambert, B.: A Student's Guide to Bayesian Statistics. Sage Publications, New York (2018)
12. van Merriënboer, J.J.G., Sweller, J.: Cognitive load theory and complex learning: Recent developments and future directions. Educ. Psychol. Rev. 17(2), 147–177 (2005)
13. Pelánek, R.: Metrics for evaluation of student models. Int. Educ. Data Min. Soc. 7, 1–19 (2015)
14. Seyderhelm, A.J.A., Blackmore, K.L., Nesbitt, K.: Towards cognitive adaptive serious games: a conceptual framework. In: van der Spek, E., Göbel, S., Do, E.Y.-L., Clua, E., Baalsrud Hauge, J. (eds.) ICEC-JCSG 2019. LNCS, vol. 11863, pp. 331–338. Springer, Cham (2019). https://doi.org/10.1007/978-3-030-34644-7_27
15. Shute, V., Zapata-Rivera, D.: Adaptive educational systems. Adapt. Technol. Training Educ. 7(1), 1–35 (2012)
16. Sweller, J.: Element interactivity and intrinsic, extraneous, and germane cognitive load. Educ. Psychol. Rev. 22(2), 123–138 (2010)
17. Sweller, J.: Cognitive load theory. Psychol. Learn. Motiv. 55, 37–76 (2011)
18. Tadlaoui, M.A., Carvalho, R.N., Khaldi, M.: A learner model based on multi-entity Bayesian networks and artificial intelligence in adaptive hypermedia educational systems. Int. J. Adv. Comput. Res. 8(37), 148–160 (2018)
19. Woolf, B.P.: Building Intelligent Interactive Tutors. Morgan Kaufmann, Burlington (2009)
20. Wray, R.E., Woods, A.: A cognitive systems approach to tailoring learner practice. In: Conference on Advances in Cognitive Systems ACS, vol. 21, p. 18 (2013)

Player State and Diagnostics in Adaptive Serious Games. Suggestions for Modelling the Adversarial Contradiction

Christian Swertz[✉] [ID]

University of Vienna, Universitätsring 1, 1010 Vienna, Austria
christian.swertz@univie.ac.at
https://mediaeducatin.univie.ac.at

Abstract. The premises of theories about Adaptive Serious Games are analyzed. Monistic approaches, which assume a determinacy of humans, and dualistic approaches, in which the ability of humans to self-determine is central, are contrasted. With reference to the concept of Digital Humanism, approaches that focus on the human's ability to self-determine are given preference. With this perspective, existing approaches to Adaptive Serious Games are discussed and suggestions for further development are made. The focus is on the adaptability of adaptive systems. This can be achieved in particular by suitable configuration options and an editor, with which players can determine the adaptivity of an adaptive serious game themselves.

Keywords: Adaptive serious games · Digital humanism · Bildung

1 Introduction

Adaptation is a serious challenge in the design of serious games. Measuring learners state, automatically diagnosing it for adaptation purposes and concluding adaptations is an obvious answer to the challenge (Serbin et al. 2019). An assumption of this answer is that learners are stateful systems. This assumption can be made if data collected about the learner are considered as adequately modeled representations of the learner, since these data can be considered as a stateful system. In this case, the adaption is concluded for the data represented in the stateful system. Fortunately, the correlation between the learner and the data in the learner model is most probably low, since learners are hardly stateful systems. They may act as stateful systems if they wish - but being able to occasionally act like a stateful system and being a stateful system that always acts as a stateful system is not the same.

In order to substantiate this thesis, theories of adaptive serious games will be taken up first. The assumptions underlying these considerations are elaborated. The assumptions are then reflected upon with theories of digital humanism and a positioning of the argumentation presented here in the segment of dualistic theories. Against this background, suggestions for a humanistic design of adaptive digital games are finally developed.

© The Author(s), under exclusive license to Springer Nature Switzerland AG 2022
R. A. Sottilare and J. Schwarz (Eds.): HCII 2022, LNCS 13332, pp. 26–35, 2022.
https://doi.org/10.1007/978-3-031-05887-5_3

2   Premises in Theories of Adaptive Serious Games

The narrative of the history of Adaptive Serious Games is usually started with Abt's (1970) publication on Serious Games. Abt opens his argument with the remark: "As civilizations evolve toward highly technological societies, the ability to use abstractions becomes more and more necessary for people to function effectively" (Abt 1970, 3). There are two surprising twists in this remark: First, the development to highly technological societies is described as an evolutionary process. This implies that the evolutionary process has a goal, namely the development of highly technological societies. Darwin (1859), however, had described evolution as a process that has no goal. According to Darwin's theory, evolution is a non-teleological process.

Contrary to the theory of Darwin, Abt assumes a goal for history. Such a theory is not a description of natural processes, but an ideology. The premise of this ideology is that history (which is the usual word for what Abbot mistakenly calls evolution) has a will that directs the path of humanity. This allows Abbot to seamlessly add the second surprising twist.

Abbot demands that humans should function well. He is able to set this premise because, due to the assumption of humanity's control through history, he believes he knows that it is good for humans to function effectively. The idea that humans may not always want to function well is something Abbot can accommodate in the context of his ideology, as is the fact that it is he who sets the premises - not evolution or history.

Implicitly, Abt does consider people who do not function well - as a disturbance variable. It follows for Abt that it makes sense to regulate people who do not function well in such a way that they function well. This is the motive Abt wants to express in Serious Games: It is about controlling people so that they function well.

This motive also became the basis of the theory of intelligent tutorial systems, which have been extensively researched for decades, mostly without results (Swertz et al. 2017). Both approaches, adaptive serious games and intelligent tutorial systems, are based on the same motive. In both cases, the premises on which this motive is based connect to Wiener's (1965) cybernetic ideology (Barberi 2017), but mostly without explicitly stating this. The connection to cybernetic ideology leads to the fact that the models for Adaptive Serious Games, formulated for instance by Charles and Black (2004) or by Lopes and Bidarra (2011), are very similar to the models for Intelligent Tutorial Systems - it is enough to exchange the word "learner" for "player" to transfer the models into each other.

Players are represented in these models. Thus, learners are viewed as stateful systems. The advantage of viewing learners as stateful systems is that future behavior is assumed to be predictable. It is a mechanistic model. At the same time, it is assumed that the outcome of games, in this case of a serious game, can be calculated in advance.

3 Premise Analysis

Now, here we cannot pursue the understanding of play in pedagogical processes, which has been discussed at least since the influential essays of Schiller (1795). For the purpose of the argument, it is sufficient to note here that with Adaptive Systems, which are based on the motive of controlling people with predictable games in such a way that they function well, these systems are clearly connected to behaviorist learning theories (Skinner 1958). This is particularly evident in Skinner's influential science fiction "Walden II" (Skinner 1948), which describes a society based on controlling people to function well (behavioral engineering).

The premise of humanity's destiny through history and the resulting task of managing people to function well is in direct opposition to humanistic theories.

Humanistic theories have been developed at least since the Renaissance. Accordingly, the theories are diverse in detail. What these theories have in common is that the question of the Conditio Humana, of the essence of man, and thus of man himself, is set as the central question. The premise here is that it is necessary for humans to enlighten themselves about what is special about humans (Schmölz 2020). By enlightening themselves about their humanity, people can reflect on themselves and determine themselves as human beings. The ability to do this is possessed by all human beings (Swertz 2021). In this respect, all human beings are equal. And in this respect humans differ from animals and machines (Schmölz 2020).

With this premise it is possible to criticize social developments and to look at them from one's own point of view. Examples include analyzing digitization as a modern form of colonization (Bon et al. 2022) or reflecting on the notions of agents and intelligence on which artificial intelligence technology is based (Russell 2022). The goal here is to develop policies to ensure that "technologies are designed in accordance with human values and needs" (Werther et al. 2019).

The contrast of digital humanism to cybernetic ideology is obvious: While theorists of cybernetic ideology demand the functioning of humans and the necessary adaptation of humans to technology, digital humanism is about the self-determination of humans and the design of technology according to the needs of humans. The essential difference can be marked with the understanding of history: While for supporters of cybernetic ideology it is history that determines the fate of people and it is therefore necessary to follow the will of history, for supporters of digital humanism it is people who determine history. Thus it becomes necessary to determine history responsibly.

4 Performative Retorsion

The logical basis for this difference, which has been discussed for millennia, is the understanding of performative retorsion. A performative retorsion is a figure of argumentation for which it must be presupposed that, first, logic is possible, second,

$$\neg(A \wedge \neg A)$$

(the theorem of contradiction) holds, third, judgments have a propositional content, and fourth, judgments are expressions of performative acts.

An illustrative argument based on retorsion is that "one cannot not communicate, because all communication (not just with words) is behavior, and just as one cannot not behave, one cannot not communicate" (Watzlawick et al. 1967, 53). The argument can be justified by saying that the assertion I am behaving now is not a behavior. By making this assertion, the assertor is contradicting himself or herself because he or she is behaving in the act of articulating. In contrast, the assertion that one can communicate cannot be refuted with a performative retorsion.

This does not prove that it is possible to communicate. It is proven that it cannot be meaningfully denied that one communicates. This marks a limit of what can be proven: It can be meaningfully disputed that it is precisely with this proposition that one has to start - other approaches are always possible with self-limitation of one's own approach. But it cannot be denied that it is possible to start in this way. Likewise it is true for the proposition: There are no true propositions, that with the articulation of the proposition the recognition of the proposition as truth is demanded, but at the same time it is demanded not to recognize exactly that. Therefore the proposition is false. This does not prove that there are true propositions, but it does prove that the existence of true propositions cannot be denied: It is not possible "to deny validity at all [transl. C. S.]" (Hönigswald 1927, 148).

What is required then is to make a beginning: "No theory, not even a philosophical theory, can begin without a basic assumption, without an axiom, because a beginning from nothing is not possible [transl. C. S.]" (Meder 2016, 179). Theories, if one accepts this premise, are always incomplete (Gödel 1931, 174). Because this assumes that there is always something outside the theories, such theories are called dualistic theories. The alternative is to assume that it is possible to fully comprehend the world with a theory. Such theories are therefore called monistic theories (Schaffer 2018). People who adhere to a monistic theory assume that the world can be fully known and is therefore determined. What happens is therefore not dependent on the will of humans, but on the will of nature, history or, in the case of the monotheistic variant, a God. Since this will itself cannot be analyzed and shaped, but can only be taken note of through observation, humans have no choice but to follow this will. In contrast, people who adhere to a dualistic theory have the possibility to choose and thus to shape their own future. The future is thus not thought of as predetermined, determined and calculable, but as open.

People with an open future cannot now be meaningfully modeled as stateful systems. Interestingly, people cannot be modeled as stateless systems either. The reason for this is simple: humans ultimately cannot be modeled at all. It is only possible to design models that people can use for their purposes. For this reason, probabilistic models are too simple, in addition to mechanistic models such as the behavioristic model. With the cybernetic concept of information (Shannon and Weaver 1964), which is based on the concept of entropy, it is assumed that the prediction of the probability of the occurrence of an information is possible. This works well if averages are measured, which is very useful in dice or card games, for example. But even the prediction of a single case is not possible: Even if six sixes have already been rolled in a row, the probability of a six on the next roll is 1/6. In the case of adaptation in learning processes, case-by-case decisions for individual learners are necessary. These decisions cannot be predicted in individual cases.

In the case of players, there is also the fact that - unlike a die - they can change their behavior in an incalculable way. Therefore, behavior can only be predicted if players choose to behave predictably. People can do that - but it is not legitimate to force them to do so and thus treat them as a mechanical or probabilistic system.

The problem of uncertainty is well known and is described by the uncertainty principle. The uncertainty principle shows that the state of a system can not be predicted precisely from initial conditions (that is: the learner model). However, the uncertainty principle assumes that in the case of electrons, for example, there is a list of properties such as location, trajectory, velocity and energy (Heisenberg 1927), which must be taken into account in the model. Humans, however, are capable of changing this list of properties. Speaking in the vocabulary of game theory, people can make the rules of the game the subject of negotiation. This can be seen even in young children, who often creatively invent the rules of the game as they play. Formulated from the point of view of modeling theory, learners thereby change the model. From a dualistic point of view, this cannot be modeled.

Since this difference has been discussed for many millennia in many cultures, it makes empirical sense to work with the thesis that there is exactly the monistic and the dualistic alternative and no others. There are only two basic axioms: The monistic axiom and the dualistic axiom. Because theorists who set the monistic axiom must understand man as determined and thus externally determined (and therefore reject, for example, human rights), and theorists who set the dualistic axiom must understand man as free and thus self-determined (and therefore can justify, for example, human rights), the dualistic axiom is set here.

5   Adaptivity

This has consequences for the understanding of learning: While for theorists who adhere to a monistic theory, pedagogy must be about the insertion of human beings into existing, known and modelable conditions, for theorists who adhere to a dualistic theory, it is about human beings being able to determine and model the conditions in which they live and, in doing so, also determine themselves. In the German-speaking world, the latter is referred to as bildung rather than learning, while the former is referred to as education. The word bildung is not translatable and is therefore used in the following.

It must be the goal of the design of Adaptive Serious Games to create an occasion for human bildung. The challenge here is that there is no underlying notion of what proper bildung is. The challenge can be seen in behaviorist theories, in which it is claimed that the learner is central and that learner-centered instruction is used. By this is meant that learning behavior is analyzed (this is what learner-centeredness means) and then learners are taught in such a way that goals are achieved efficiently and effectively. This does not mean that learners can determine their own goals and choose and design their own paths to those goals. To determine goals, contents, methods and media by oneself characterizes a person who has bildung.

That is precisely what it must be about: People must have the opportunity to determine their own goals and choose their own paths in order to create - by all means themselves - an invitation to educate themselves. So adaptive systems must not be about people being adapted by adaptive systems. It must be about people being able to determine themselves and adapt algorithmic systems for this purpose.

One consequence of this is that it is meaningless to measure whether people have achieved pre-determined goals - that is not a relevant criterion. What is relevant, on the other hand, is to measure whether people have experienced the learning experience as an occasion to educate themselves. In terms of adaptivity, one goal may thus be that people can adapt systems like serious games for their purposes. For this purpose, assistance systems that make it casually easy for people to adapt systems to their purposes, and that they can then use as long as they do not wish to change their behavior, make perfect sense. It is obvious that modeling such systems is a complex challenge because the model parameters cannot be fully known.

6   Adaptive Serious Games

Consequences from the previous considerations will be discussed below on the basis of the excellent overview by Streicher and Smeddinck (2016). First, attention must be drawn to a difference: While Streicher and Smeddick state as one goal to replicate teaching by a personal trainer with an adaptive system, this is not possible from the perspective developed here. This is because replicating the behavior of humans with machines is only possible if one adheres to a monistic worldview. Elsewhere in the same paper, however, Streicher and Smeddick suggest that adaptive systems should be understood as assistance systems - a reasonable and responsible perspective.

An option to deal with this perspective can be derived from the fact that in contrast to physical elements, learners are able to actively configure the model. This can be made possible through HCI. The principle is to make adaptations only with learner consent and to allow learners to change the configuration. This takes into account another difference between physical elements and learners: Learners occasionally change their minds. These changes are not predictable, but can be actively expressed by learners if learners are allowed to configure the model.

Even with this, the limits of Turing-powerful machines cannot be exceeded. However, it is possible for learners to analyze the limits of the media they are using at any given time and then choose which medium to use now. Thus, when learners choose Adaptive Serious Games, they must also choose to act within the limits of Turing-powerful machines.

This makes it clear at first that Adaptive Serious Games cannot be the universal solution for media in learning processes. It is an option. And this option is already valuable because the change between different media is an important occasion for processes of Bildung. Therefore, increasing the media ensemble is good because it allows more switching between media.

For Adaptive Serious Games, the focus then needs to be on adaptation by learners. For this, it is helpful to understand the use of an Adaptive Serious Game as communication. If adaptive serious games are understood as communication models that describes the communication between developer and player, a limitation becomes visible. The limitation is the fact that it is a unidirectional asynchronous communication. To model this communication situation appropriately, learners can be supported in modeling adaptations themselves. In this case, learners communicate with themselves. This turns the unidirectional asynchronous communication into a bidirectional synchronous communication. This can be achieved by offering an adaption editor.

A first proposal for an adaptation editor can be developed with the considerations presented here, based on the development model presented by Streicher and Smeddick. This model contains a step in which the scope of adjustment automation, adjustment frequency, adjustment extent, adjustment visibility, and adjustment control is to be defined. In each case, a scale from "high" to "low" is provided on which developers must decide how to design the system. It is quite conceivable to design this not as a decision for the developers, but to offer it to the players as a setting. For this, a set of a few sliders would suffice, allowing players to parameterize the system. The same goes for accessibility, for example, which is best parameterized by the players themselves. It is obvious to also offer the players to automatically adjust these parameters in the course of the game.

This suggestion can be supplemented with a second suggestion. To also offer players the possibility to modify the parameter set, an adaptation editor is needed. The usability of such an editor could be improved by a visual adaptation language (Gómez et al. 2012), which could be used to link the available data and the adaptable elements with different algorithms. Incidentally, this does not only apply to players - it would also be helpful for developers and researchers to have such a system, which would facilitate rapid prototyping, at their disposal.

A third suggestion can be developed from the cold start problem. To deal with the cold start problem, there are several possibilities: First, it is possible to ask the player for an initial parameterization (Streicher and Smeddinck 2016). To do this, the player must be given the opportunity to enter parameters. A possibility not discussed so far is to suggest that the player generates the initial paramters by using the system. This allows the learner to decide what behavior the game should be adapted to, then demonstrate it, and thus parameterize the system. Third, it is possible to offer the learner to try out the system first and then enter the parameters or generate them by demonstration when he has decided how to train the adaptive system. Fourth, it is possible to offer to work with default parameters that are generated from existing data of other players (Streicher and Smeddinck 2016) or can be set by the developers. These four options should not be fixed, but should be available at any time, because otherwise players will be restricted in their decision to change their behavior. Thus, if learners choose to turn off the adaptation system and leave it turned off, this should be possible as well as keeping default parameters, turning off the adaptation system later, or resetting parameters generated by demonstration or later recordings followed by a new demonstration, manual configuration, or training phase.

This makes it possible not only to give players a sense of autonomy while they are determined by being controlled by giving them a sense of autonomy, but to actually support them in acting autonomously. And it also becomes possible to view learner motivation not as a means of controlling players, but as an expression of the purposes set by the players themselves. People who have Bildung can determine their own learning. Automatic changes to the parameters should never be made - the player's consent must always be sought (Streicher and Smeddinck 2016). This is necessary because collected data often cannot be interpreted reliably. A simple example is a learner's response delayed by 180 s compared to the average. The cause may be a comprehension problem - or the trip to the coffee machine. If players are made aware that there were delays and then asked if they should be taken into account, they can determine this by a simple input.

A fourth suggestion relates to the goals pursued by the game. Streicher and Smeddick write, "The ultimate goal of an adaptive educational game is to support players in achieving progress toward individual learning goals" (Streicher and Smeddinck 2016). One consequence of this sensible perspective is that players must be able to set the goals themselves. To this end, at the beginning of a game, the player must be offered a choice list of goals that can be achieved with the game, so that the player can first decide whether the game is appropriate for his or her goals and, if so, specify which goals he or she wants to achieve. According to Streicher and Smeddick, the explicit specification of goals falls within the domain of adaptability. It is important to note that players can decide not to play at any time. After all, if a game has to be played within an imposed framework, it is not a game but work (Huizinga 2019). The contradiction between work and play is implicitly taken into account by Streicher and Smeddick by emphasizing the relevance of the player's life. For a serious game is formally work when it is serious, and becomes a game only when it is experienced as a game.

In Streicher and Smeddick's terminology, the proposals boil down to focusing on the adaptability of adaptive systems. To endow adaptive systems with adaptability is to allow players to adapt the adaptive system to their needs. It is clear that when players take advantage of the adaptability of an Adaptive System, they themselves become experts in the field of Adaptive Serious Games. And this is exactly the goal: the transfer of knowledge. In this case, this passing on of knowledge about Adaptive Serious Games is done through the learning method: the method is the message. And that is certainly not the worst way.

References

Barberi, A.: Medienpädagogik als Sozialtechnologie im digital-kybernetischen Kapitalismus? Kybernetik, Systemtheorie und Gesellschaftskritik in Dieter Baackes Kommunikation und Kompetenz. MedienPädagogik: Zeitschrift für Theorie und Praxis der Medienbildung **27**, 173–209 (2017). https://doi.org/10.21240/mpaed/27/2017.04.07.X

Bon, A., et al.: Decolonizing technology and society: a perspective from the global south. In: Werthner, H., Prem, E., Lee, E.A., Ghezzi, C. (eds.) Perspectives on Digital Humanism. TU Wien (2022). https://dighum.ec.tuwien.ac.at/perspectives-on-digital-humanism/decolonizing-technology-and-society-a-perspective-from-the-global-south/

Charles, D., Black, M.: Dynamic player modelling: a framework for player-centred digital games. In: Proceedings of the International Conference on Computer Games: Artificial Intelligence, Design and Education, pp. 1–7 (2004)

Darwin, C.: On the Origin of Species by Means of Natural Selection, or the Preservation of Favoured Races in the Struggle for Life. Johan Murray (1859)

Gödel, K.: über formal unentscheidbare Satze der Principia Mathematica und verwandter Systeme I. Monatshefte Für Mathematik Und Physik **38**, 173–198 (1931)

Gómez, M., Mansanet, I., Fons, J., Pelechano, V.: Moskitt4SPL: tool support for developing self-adaptive systems. In: Ruíz, A., Iribarne, L. (eds.) JISBD 2012, pp. 453–456 (2012)

Heisenberg, W.: Über den anschaulichen Inhalte der quantentheoretischen Kinematik und Mechanik. Z. Phys. **17**, 1–26 (1927)

Hönigswald, R.: Über die Grundlagen der Pädagogik. Ein Beitrag zur Frage des pädagogischen Universitäts-Unterrichts (2. umgearbeitete Auflage). Ernst Reinhardt (1927)

Huizinga, J.: Homo ludens: Vom Ursprung der Kultur im Spiel (H. Nachod, Trans.; 26. Auflage). Rowohlt Taschenbuch Verlag (2019)

Lopes, R., Bidarra, R.: Adaptivity challenges in games and simulations: a survey. IEEE Trans. Comput. Intell. AI Games **3**(2), 85–99 (2011). https://doi.org/10.1109/TCIAIG.2011.2152841

Meder, N.: Philosophische Grundlegung von Bildung als einem komplexen Relationengefüge. In: Verständig, D., Holze, J., Biermann, R. (eds.) Von der Bildung zur Medienbildung. MG, vol. 31, pp. 179–210. Springer, Wiesbaden (2016). https://doi.org/10.1007/978-3-658-10007-0_10

Russell, S.: Artificial intelligence and the problem of control. In: Werthner, H., Prem, E., Lee, E.A., Ghezzi, C. (eds.) Perspectives on Digital Humanism. TU Wien (2022). https://dighum.ec.tuwien.ac.at/perspectives-on-digital-humanism/artificial-intelligence-and-the-problem-of-control/

Schaffer, J.: Monism. In: Zalta, E.N. (ed.) The Stanford Encyclopedia of Philosophy (Winter 2018). Metaphysics Research Lab, Stanford University (2018). https://plato.stanford.edu/archives/win2018/entries/monism/

von Schiller, F.: Ueber die ästhetische Erziehung des Menschen in einer Reyhe von Briefen (1. Auflage). J. G. Cotta (1795). http://www.deutschestextarchiv.de/book/show/schiller_erziehung01_1795

Schmölz, A.: Die Conditio Humana im digitalen Zeitalter: Zur Grundlegung des Digitalen Humanismus und des Wiener Manifests. MedienPädagogik: Zeitschrift für Theorie und Praxis der Medienbildung, pp. 208–234 (2020). https://doi.org/10.21240/mpaed/00/2020.11.13.X

Serbin, V.V., et al.: Multicriteria model of students' knowledge diagnostics based on the doubt measuring level method for E and M learning. Mob. Inf. Syst. **2019**, 1–11 (2019). https://doi.org/10.1155/2019/4260196

Shannon, C., Weaver, W.: The Mathematical Theory of Communication. University of Illinois Press (1964)

Skinner, B.F.: Walden Two. Hackett Publishing (1948)

Skinner, B.F.: Teaching machines. Science **128**(3330), 969–977 (1958)

Streicher, A., Smeddinck, J.D.: Personalized and adaptive serious games. In: Dörner, R., Göbel, S., Kickmeier-Rust, M., Masuch, M., Zweig, K. (eds.) Entertainment Computing and Serious Games. LNCS, vol. 9970, pp. 332–377. Springer, Cham (2016). https://doi.org/10.1007/978-3-319-46152-6_14

Swertz, C.: Bildung, Verantwortung und digitale Daten. Medienimpulse **59**(3), 1–40 (2021). https://doi.org/10.21243/MI-03-21-12

Swertz, C., Schmölz, A., Barberi, A., Forstner, A.: The history of adaptive assistant systems for teaching and learning. In: Fuchs, K., Henning, P.A. (eds.) Computer-Driven Instructional Design with INTUITEL: An Intelligent Tutoring Interface for Technology-Enhanced Learning, p. 1824. River Publishers (2017)

Watzlawick, P., Beavin, J.H., Jackson, D.J.: Pragmatics of Human Communication: A Study of Interactional Patterns, Pathologies, and Paradoxes. W. W. Norton and Co., Inc. (1967)

Werther, H., et al.: Vienna Manifesto on Digital Humanism. DIGHUM - Perspectives on Digital Humanism (2019). https://dighum.ec.tuwien.ac.at/perspectives-on-digital-humanism/vienna-manifesto-on-digital-humanism/

Wiener, N.: Cybernetics: Or Control and Communication in the Animal and the Machine, 2nd edn. MIT Press, Cambridge (1965)

Adaptive Learner Assessment to Train Social Media Analysts

Ethan Trewhitt[1]([⊠]) [iD], Elizabeth T. Whitaker[1] [iD], Thomas Holland[1],
and Lauren Glenister[2] [iD]

[1] Georgia Tech Research Institute, Atlanta, GA 30318, USA
ethan.trewhitt@gtri.gatech.edu
[2] Soar Technology, Inc., Ann Arbor, MI 48105, USA

Abstract. The Twiner project provides a practice environment to support a learner in analysis and interpretation of an ongoing situation through a social media data framework. Within this framework exists a host practice environment complemented by a set of modular components for facilitating training, known as the GTRI Learner Assessment Engine, or *GLAsE*. GLAsE maps the evaluated learner activity to individual components of analysis, i.e., the learning objectives. It estimates learner proficiency for individual learning objectives based on observations of the learner actions while performing the learning tasks. GLAsE represents the state of the world using a Learner Model that provides the current state of the learner's domain competency, and a Curriculum Model that represents the concepts, activities, and skills of an expert analyst in this domain. GLAsE may produce advice and hints for the learner at multiple points in the assessment process, which are then passed to the learner via the practice environment as a form of direct feedback for the learner's performance. Through these capabilities, GLAsE reduces the time necessary for domain experts to dedicate to novice training and increases the agility and responsiveness of the training system.

Keywords: Learner modeling · Student modeling · Intelligent tutoring systems · Social media Analysis

1 Introduction

The *Twiner* project provides a practice environment to support a learner in analysis and interpretation of an ongoing situation through a social media data framework. The Twiner *framework* consists of a host practice environment complemented by a set of modular components for facilitating training, known as the GTRI Learner Assessment Engine, or GLAsE. Internal components of GLAsE consist of the five components shown in the blue center of Fig. 1, namely:

- the **Plan Recognizer**, which matches a novice learner's steps against expert plans to identify and assess those steps,
- the **Critic**, which provides quantitative scores for the learner's plans and steps,

- the **Curriculum Element Scorer**, which associates quantitative scores for the relevant elements within the Curriculum Model,
- the **Learner Modeler**, which is responsible for updating the Learner Model as the novice learner's performance is evaluated by other components, and
- the **Advisor/Hint Generator**, which provides feedback to the learner based on their performance.

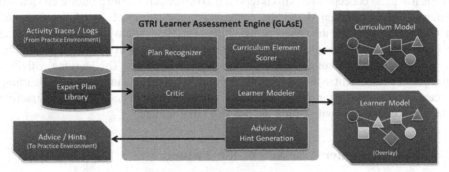

Fig. 1. GLAsE design showing GTRI components within the Twiner system architecture. Blue indicates GLAsE elements. Purple indicates models. Red indicates external data.

GLAsE is designed to integrate with the overall Twiner framework by mapping the evaluated learner activity to individual components of analysis i.e., the learning objectives. GLAsE estimates learner proficiency for individual learning objectives based on observations of the learner actions while performing the learning tasks.

2 Background

2.1 Student Modeling in an Intelligent Tutoring System

As part of an intelligent tutoring system, a student model uses student performance and behaviors observed by the learning system to provide evidence of mastery levels associated with elements in a domain curriculum. The mastery levels are increased or decreased depending on the student's behaviors. This approach uses both the identification of the student's missing information and the identification of the student's incorrect information (bugs) to assess the learner's state of mastery–see [1]. The intelligent tutoring system can then use these assessments to recognize the relative progress of the student within a domain, then schedule learning activities to maximize student progress.

This work uses a common approach to student modeling in which the student model is built as an overlay representation superimposed over a curriculum model, thus using the same representation as the curriculum model [2]. The curriculum model, which represents the domain knowledge and learning material, enumerates and indexes the concepts being taught by the intelligent tutoring system.

2.2 Social Media Analysis

A curriculum model consists of an explicit representation of the concepts and skills in the domain. In this work, the domain practice environment exists to train social media analysts, and the concepts and skills to be mastered are indexed within the curriculum model. These concepts and skills relate to recognizing patterns or anomalies in data on various social media platforms, performing network graph analysis, monitoring real-time information operations, drawing conclusions and providing recommendations, and so on [3]. Each of these concepts and skills is represented in the curriculum model as curriculum elements. The Curriculum Model also contains representations of the relationships and interconnections among the concepts and skills. The Curriculum Elements include the concepts that are to be taught or experienced through the student's interaction with the practice environment through experiencing learning activities. The learning activities will be stored as content in the catalog of Learning Activities and indexed by the learning concepts, represented as curriculum elements, which are experienced through interaction with each Learning Activity within the practice environment.

3 GLAsE: Learner Assessment for Twiner

GLAsE represents the state of the world using a Learner Model that provides the current state of the learner's domain competency, and a Curriculum Model that represents the concepts, activities, and skills of an expert analyst in this domain.

The *Learner Model* maintains an estimate of learner proficiency based on the demonstrated skills of learners, and it records the assessments that are mapped to identified curriculum elements and compute and update proficiency estimates for the indicated curriculum elements. It supports learning by being available to other system components to guide the choices of learning activities, and by being available to the learner and other users who need to be aware of the proficiency states of the learner. The Learner Model can be queried to retrieve the proficiency estimates of a given learner for each leaf node/element in the Curriculum Model, along with aggregated scores for each higher-level node in the Curriculum Model.

The *Curriculum Model* represents a hierarchical collection of curriculum elements which describe the relationships and interconnections among domain-specific concepts, skills, elements of workflows, and problem-solving approaches. It contains an explicit representation of the concepts and skills related to recognizing patterns or anomalies in the data, analyzing the situation and drawing conclusions. The Curriculum Model includes the concepts that are to be taught or experienced through the student's interaction with the practice environment through experiencing learning activities. Numerical weights indicate the relative contributions of sub-skills to skills as defined by experts.

4 GLAsE Model and Component Overview

The GLAsE set of modular components complement the practice environment of Twiner to facilitate training. The following describes how the GTRI components interact with this practice environment:

- The **Curriculum Model** represents the concepts, activities, and skills of an expert analyst in this domain in a hierarchical structure, with relative weights for element siblings and support for prerequisites when appropriate.
- The **Learner Model** provides the current state of the learner's domain competency using reasoning techniques that are guided by learning theories and teaching theories known to be appropriate for intelligent tutoring.
- The **Learner Modeler** updates the Learner Model each time the learner's performance is evaluated by other components and provides a friendly API for reading/writing the Learner Model data.
- The **Plan Recognizer** matches a learner's steps against expert plans to identify and assess those steps. It uses a natural language semantic similarity metric to match human-readable steps performed by the learner with known expert plan steps.
- The **Critic** that adjudicates the quality of learner content vs. expert plans and recognizes known bugs/mistakes.
- The **Curriculum Element Scorer** provides quantitative scores for the relevant elements within the Curriculum Model corresponding to each unit of work performed by the novice.
- The **Advisor/Hint Generator** provides feedback to the learner based on their performance, pointing out missing steps, common mistakes, or other areas of potential improvement.

Important distinction: We use the term *Learner Model* for the database that provides a snapshot of the learner's competency and *Learner Modeler* for the software module that reasons about the data and updates the Learner Model. Each model and component is described in greater detail below.

4.1 Curriculum Model

The Curriculum Model is an explicit representation of the concepts and skills related to recognizing patterns or anomalies in the data, analyzing the situation and drawing conclusions. It contains representations of the hierarchical relationships among the concepts and skills. This includes the concepts that are to be learned or experienced through the learner's interaction with the practice environment through a set of learning activities, tasks, or games.

Example Concepts for Twiner Curriculum. Some examples of skills to be learned or demonstrated through interaction with the practice environment (based on Twiner team discussions) are:

- Recognizing hidden relationships among actors
- Identifying which actors are pushing the deceptive content or influence campaigns
- Understanding how deceptive content is being received
- Understanding how deceptive content is being countered
- Pinpointing where the actors are
- Identifying who is supporting or working against particular actors
- Recognizing what problems this activity poses for their own troops and mission

The Curriculum Model is built from curriculum elements, where each element is either a topic, concept, or skill. Every curriculum element is indexed with the learning objectives that it supports, and the curriculum elements are hierarchically arranged as a tree structure. The highest level of the hierarchy specifies the type of learning that is taking place when that learning objective is being satisfied:

- Analysis Approach
- Tool Training
- Understand Important Concepts (Concepts)
- Compound Tasks (Skills)

This hierarchical representation is anchored by an abstract root node, with a set of children that represent the high-level topics covered by the entire system. All nodes may have children of their own, which represent the conceptual structure of those other topics/concepts/skills. Each topic/concept/skill contains the following properties:

- Name
- Curriculum element group
- Prerequisites (list of other curriculum elements)
- LAs that teach this concept
- LAs that evaluate this concept
- Learning objectives that are supported by the concept or skill

We have organized the learning objectives for the Twiner project into a curriculum hierarchy of topics, concepts, and skills. Figure 2 depicts an abstraction of this hierarchy for a generic domain.

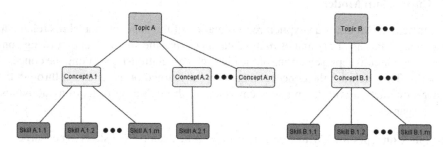

Fig. 2. Abstraction of the curriculum model hierarchy

Each of these types of learning has several subcomponents and each of those subcomponents has a set of sub-objectives until after several levels, every branch of the hierarchy has reached a leaf node. Since we are using an overlay representation of the learner model, the learner model uses this same representation and each node in the learner model contains a value which represents the proficiency of the student in the curriculum element or learner objective represented by that node.

4.2 Learner Model

A learner model (also called a student model), is a component of an intelligent tutoring system with the purpose of providing a representation of the state of the learner's competency for information and use by the learner and the instructor for guiding the learning activities and experience of the learner in the practice environment [4]. The Learner Model in GLAsE is represented as a curriculum overlay model with additional annotations and background information. Items in the Learner Model are represented in the Curriculum Model as learning objectives. The contents of the Learner Model include:

- Mastery scores for each of the learning objectives that represent the model's best estimate of the state of the learner competency for that objective.
- Background information on the learner which is valuable for understanding what learning experiences the learner might need, both by a human and by computer-based tutor.

Learner Model as an Overlay Representation. An overlay representation is an approach to learner modeling in which the student model is built on top of a curriculum model, i.e., using the same representation as the curriculum model [2]. The curriculum is a form of the domain model and is a representation, enumeration, or indexing of the concepts being taught by the learning system. The learner model uses student performance and behaviors in the learning system to provide evidence of mastery levels associated with elements in the curriculum. The mastery levels are increased or decreased, depending on the student's behaviors. This approach uses both the identification of the learner's *missing* information and the identification of the learner's *incorrect* information (bugs) to assess the learner's state of mastery; see [1]. In GLAsE, this enables straightforward comparison between changes made to the curriculum by an expert and the specific structure of scores attributed to the learner.

Proficiency Scoring Theory. The purpose of the Learner Model in GLAsE is to provide a state of the learner's competency or proficiency for information and use by the learner and instructor to guide the learning activities and experience of the learner in the practice environment. The Learner Model is represented as a curriculum overlay model with additional annotations and background information. The contents of the Learner Model consist of items represented in the Curriculum Model as learning objectives: Concepts, Skills, Problem-solving approaches, and in the case of the Twiner system, it includes some tool training. This Learner Model is called an overlay model because the Learner Model has the same representation as the expert domain knowledge, i.e., the Curriculum Model—it is overlaid on that representation.

The Learner Model is a representation of:

- Mastery or proficiency scores for each of the learning objectives that represent the model's best estimate of the state of the learner proficiency for that objective.
- Background information on the learner
- Demographic information
- Educational level and specialties

- Professional experience

In developing the Twiner proficiency calculation, we provide the following required/desired characteristics:

- For each curriculum element, track and estimate the proficiency level
- Track proficiency based on the activities and experiences that the learner has had with each curriculum element
- Estimate the space of activities related to the curriculum element

The curriculum characterizes the space of activities that it takes to learn and show proficiency. Each activity that may be evaluated includes a numeric "coverage" value that represents the fraction of the curriculum element coverage that is provided for that activity. The proficiency represented is across all of the learner's interactions with the system, that is, it includes interactions with multiple scenarios, across multiple days, and multiple interaction systems.

In this discussion the term "learning activity" is meant to refer to a simple aggregation of a collection of the most basic key or mouse clicks that represents a meaningful intent or plan of the learner. These are the inputs to Learner Assessment for the Twiner system:

- Catalog of concepts in which students should have and demonstrate competence/proficiency
- Catalog of learning activities that evaluate and/or provide experience which can lead to competence/proficiency
- Mappings of which learning activities evaluate which concepts
- Fraction of the curriculum element which is experienced in each activity. This is meant to provide the capability for the system to require multiple experiences with a particular curriculum element in order to develop total proficiency.
- Mappings of which learning activities provide experience with which curriculum elements
- Definitions of the constraints regarding the selection and ordering of concepts to provide to the learner to interact with and experience. This recognizes that the Twiner system may not have any capability to reorder experiences, but this may allow the assessment to recognize when the learner has not yet experienced some activities which may be needed to understand later activities
- Definitions of the constraints regarding the selection and ordering of learning activities. Again, we may not have control, but this may help the system to understand when the learner is missing some information necessary to move ahead in the analysis.

4.3 Learner Modeler

The Learner Modeler is the component responsible for updating the Learner Model in response to new information about the learner's proficiency in one or more curriculum elements. It infers and models the learner's strengths and weaknesses by analysis of logged activities, then updates the *Learner Model* to represent the current state of the learner's knowledge and skills.

Learner Assessment Process (This process is described, recognizing that Twiner exists in a practice environment):

1. Wherever possible, evaluate student competence/proficiency in concepts to populate Learner Model
2. Identify deficient concepts
3. Prioritize deficient concepts
4. Select concepts to experience that adhere to concept constraints (Twiner may not be able to direct the learner to particular concepts or activities, but may be able to suggest certain activities)
5. When possible, direct the attention of the learner within the practice environment toward a learning activity that provides experience in one or more deficient concepts and adheres to learning activity constraints.
6. Provides access to selected learning activity within the practice environment
7. Use metrics provided by completed learning activity to update Learner Model.

Cumulative Weighted Proficiency Scoring Implementation. The proficiency algorithm, known as cumulative weighted proficiency scoring (CWPS) and shown in Fig. 3, is used to calculate the proficiency score for each leaf node in the curriculum element hierarchy. The CWPS is calculated using:

- The seed value (the estimated level of proficiency before interaction with the system)
- The recency of the performance—items that the learner has performed more recently should be weighted more highly than performance on activities that were performed longer ago (with the assumption that the learner has learned and that more recent activities more accurately reflect what the learner knows (the learner may have learned or forgotten)
- The percentage of the space that has been experienced by the learner
- The learner's score on each activity in which the learner was evaluated for that curriculum element

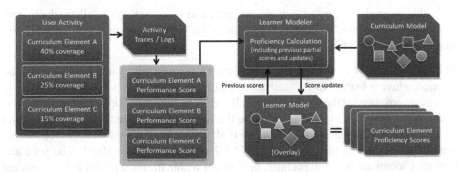

Fig. 3. Cumulative weighted proficiency scoring (CWPS)

Other details of the CWPS calculation:

- Each activity-to-curriculum element mapping has a % of coverage from 0–100.
- Seed scores have a coverage amount as well, e.g. 25%.
- The most recent 100 points' worth of CE coverage are used for evaluation.
- For each activity, the learner's trace score for that activity is scaled by the % coverage of that activity.
- If the oldest included activity exceeds the 100-point recency cutoff, the coverage of that activity *prior to* the cutoff is used to scale the learner's score for that activity.
- If a learner has not yet experienced activities totaling 100% coverage, their greatest possible score is the total % coverage so far.

Proficiency Example 1: Importance of Coverage. In this example we show how a learner with a seed score can complete 3 activities but the total percentage coverage of the seed score and all activities still does not reach 100%. In this case, the learner cannot reach a perfect score even with perfect trace scores. As shown in the table below, the learner has reached a score of 70.25% out of 80% possible (sum of all seed and activity coverage so far) (Table 1).

Table 1. First example proficiency score calculation results using CWPS with insufficient activities logged to reach full proficiency

Activity	Activity coverage	Learner's trace score	Net effect on CE score (coverage * trace)
Seed	25%	89%	22.25%
Activity 1	10%	60%	6.0%
Activity 2	15%	90%	13.5%
Activity 3	30%	95%	28.5%
Learner's current CE proficiency score: (sum of column)			**70.25% (80.0% possible)**

In this example, because coverage amounts add up to only 80%, all available proficiency scores are used to their full potential. The overall proficiency is the weighted sum of the individual trace scores.

Proficiency Example 2: Decay of Learner Information. In this example we examine the situation where a learner has a seed score and has completed 5 activities but the total percentage coverage of the seed score and all activities exceeds 100. In this case, the most recent 100% coverage is used to compute the learner's proficiency score. The learner could hypothetically reach a perfect score if all recent trace scores were perfect. In the example below, the learner has reached 94.5% out of a possible 100%. The seed and Activity 1 scores are not used because they are not within the most recent 100% coverage and are thus grayed-out in the table. Activity 2 crosses the 100% coverage boundary with 5% "inside" the most recent 100%. Therefore only 5% of the learner's Activity 2 trace score is used in the learner's final proficiency score (Table 2).

Table 2. Second example proficiency score calculation results using CWPS with sufficient activities logged to exceed full proficiency

Activity	Activity Coverage	Learner's Trace Score	Net Effect on CE Score (coverage * trace)
Seed	25%	55%	0%
Activity 1	10%	78%	0%
Activity 2	15%	80%	4.0% (5% out of 15% coverage used)
Activity 3	30%	95%	28.5%
Activity 4	40%	100%	40.0%
Activity 5	25%	88%	22.0%
Learner's current CE proficiency score: (sum of column)			**94.5% (100.0% possible)**

This proficiency calculation accounts for the decay of learner information if it is not used over time and the possible need for review or practice. If the learner has performed well on a given curriculum element on some activities and then shows lower performance, the most recent scores take precedent once enough activities have been evaluated.

GLAsE collects and represent mis-learnings or bugs that can be used to identify the need for a correction message or example [1]. GLAsE will flag a bug that is represented in a curriculum element. It will then lower the proficiency level and add the flag. The learner's proficiency score represents both the extent to which they've shown success in demonstrating the material (proficiency level), but additionally the learner may mis-learn some concepts and thus be penalized (bug).

Hierarchical Proficiency Scoring. The hierarchical representation provided for the curriculum allows a read-out of the leaf node concepts as well as a proficiency calculation for the parent nodes, including intermediate nodes. Leaf node proficiency scores are calculated using the algorithm above. Intermediate node scores are the weighted average of direct descendants and so on up the tree. Figure 4 shows an abstraction of weights for child elements in the hierarchy as a relative proportion of their parent element. In this figure, each child's relative size within its parent is proportional to its weight, which will determine the extent to which that child's score affects the weighted average when the parent's score is calculated.

4.4 Plan Recognizer

This component is responsible for identifying a set of learner-generated step phrases, written in a natural style as the learner notes their work process in their virtual notebook. The primary data source for this component is the Plan Library, which contains a collection of expert plans represented as a set of tasks/steps. The Plan Recognizer uses a semantic similarity model to compare each of the learner's steps (entered via the Practice Environment) against the optimal set of steps defined in each expert plan to determine

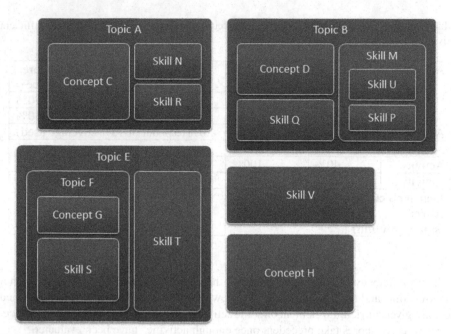

Fig. 4. Visualization of the relative weight of each child element as a proportion of its parent element, which forms the basis of the hierarchical scoring approach

which plan best matches the learner's work. In addition to detailing the steps to achieve a plan goal, the expert plan includes a set of Curriculum Element identifiers that indicate which topics, skills, or concepts are in use by this plan.

Once the plan is matched to a sufficient threshold against an expert plan, the Critic determines whether the novice plan is sufficiently compatible with the expert plan. If required steps are missing, or undesired superfluous steps are recognized, the learner's proficiency score will be diminished. A score is produced for the quality of the learner's work and the score, along with the Curriculum Elements associated with the expert plan, are passed to the Curriculum Element Scorer and eventually the Learner Model for the learner is updated to reflect the latest understanding of the learner's proficiency.

Mapping Learner Actions to Expert Plans. Low-level learner events and actions must be mapped to Curriculum Element scores to be used by the Learner Modeler. A sequence of these events or actions can be thought of as a plan to accomplish some higher-level goal or intent which is represented in the Curriculum Model. In order to recognize the higher-level goal or curriculum element it is proposed that there be a library of plans (expert plans, satisficing plans, and buggy plans) which can be compared with the sequence of steps or actions of the learner. A similarity metric is used to determine which plan or plans in the library most closely match the learner's step sequence. The similarity metric should be able to compare any two plans to see which steps match, to hypothesize that there might be missing step(s), to hypothesize that there might be extra step(s), or to hypothesize that there might be incorrect step(s).

There are several confounding situations that could occur that would make the matching of the learner plan with the plan in the plan library. These situations are:

- **Partially ordered plan steps:** The learner may not always execute plan steps in the same sequence, especially if some of the steps are not interdependent. Contrast this with a totally ordered plan where there is only one sequence of plan steps.
- **Multiple concurrent goals:** A learner may have several goals that are being pursued in parallel. Simply having observations that support one goal hypothesis doesn't mean there might not be another goal for which the same actions provide support. This could result in steps for two goals that intermingled.
- **Actions can have multiple effects:** One action could be used to support two concurrent goal (for example consider a search for social network connections that could be used to identify super spreaders and in addition to identify a person who initiated a particular piece of information).
- **Need to carry along multiple hypotheses:** Because of the above considerations, we need to be able to carry several interpretations of the learner's actions and a way to make a final determination for updating the student mode.

Some simplifying assumptions have been made for the first version of this mapping mechanism. The current prototype does not include detail for multiple concurrent goals, and there are no actions that are part of two different plans.

Elements of Expert Plan Model Schema:

- Objective of the plan, e.g. "Identify Superspreaders and their Spheres of Influence"
- Steps that must be taken to achieve the desired objective (hierarchy, tree structure), e.g.:

1. Load messages from file
2. Perform centrality analysis
3. Note most central users
4. Follow most central users' messages
5. Perform popularity analysis of top-shared central user messages

- Tool or tools that can achieve each step, e.g. ORA Pro [5]. If multiple tools can achieve the step, identify all of them.
- (Instrumentable) Action taken by the learner within the tool (e.g. open "Reports" menu, select "Centrality report") to achieve each plan step. If multiple actions can achieve a step, identify all of them.
- Represent the optionality of each step.
- Explicitly represent steps that indicate known bugs/mistakes learner may make when performing a plan and identify a rule to recognize each bug/mistake. E.g. "the system recognizes that the learner is trying to perform Plan Y, but also executes Step M, and the expert model for Plan Y says that Step M is a mistake, therefore register Step M as a bug."

Plans can reference other plans. Steps are just single-item plans (degenerate plans). Requirements of the mapping mechanism consist of:

- Per-tool mapping (and mechanisms) of learner activity to one or more medium-level tasks
- Mapping of medium-level tasks to Curriculum Elements (with coverage %)
- Mechanisms for judging the sufficiency of learner activity toward the tasks they are attempting to perform
- Task hierarchy
- Design for how to handle errors of omission: if we have a system to evaluate a task after it's complete, we need a way to handle tasks that weren't attempted. Score deduction for missing tasks.
- Score deduction for extra/missing steps
- Score deduction for incorrect steps ("bugs")

Plan Recognition Implementation. To compare learner plans and expert plans, the Plan Recognizer uses a semantic similarity metric generated by comparing two large numeric vectors that represent each sentence pair: one step from the learner and one step from the selected expert plan. This similarity metric is the dot product of the two vectors. For each step's text phrase, the vector is produced using version 4 of Google's Universal Sentence Encoder [6] running in TensorFlow, which uses a pretrained model to create a 512-float vector for any arbitrary English-language sentence. An example of the confusion matrix of two sets of plan steps is show in Fig. 5.

Once the similarities are evaluated, the linear sum assignment problem is solved using the modified Jonker-Volgenant algorithm [7] provided by the popular SciPy library. This finds the combination of step pairs (one each from the learner and from the expert) in which the overall sum is (in this case) maximized. Any leftover steps from the learner are either extra (ignored) or erroneously superfluous (a bug), whereas any leftover steps from the expert are either optional (ignored) or missing (a bug). By combining the step pair scores and adjusting for any erroneous or missing steps, and overall score is generated for the learner's plan against the expert plan.

4.5 Critic

The Critic evaluates learner task steps against a library of expert task steps to identify the task(s) being performed by the learner, judging the performance of those tasks as compared to past experts. The Critic also determines when a task transition occurs, and performs evaluation of the work products for the ending task. Outputs of the Critic consist of:

- Recognized task(s)
- Identified extra or missing step(s)
- Identified bug(s)

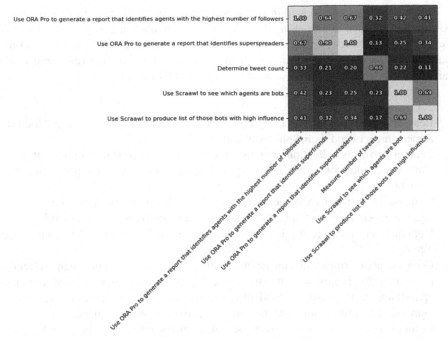

Fig. 5. Confusion matrix for five learner plan steps (left) and six expert plan steps (bottom)

When plans are presented by the learner as a collection of task steps, the Critic will evaluate the overall quality of the plan when compared to the corresponding expert plan. This means that the learner will receive credit for any steps that match the expert plan, but will lose credit for missing required steps and for superfluous steps that are indicated as mistakes in the expert plan.

Additionally, individual steps may be scored by the Critic, whether part of a plan or individually. This aspect of the Critic's design is hypothetical and has not yet been implemented.

4.6 Curriculum Element Scorer

The Curriculum Element Scorer is the primary enabler of the Learner Modeler (the component used to build the Learner Model). As the learner performs tasks and the Critic evaluates the accuracy of those tasks against expert performance, the Curriculum Element Scorer takes an explicit mapping of tasks to curriculum elements and provides new LearningTrace values, or update the current trace values. Each learning objective that is addressed in the learner's latest activities are scored on the performance level of the results. Concept proficiency scores are logged as part of the Learner Model for reading by other components or by humans. The Curriculum Element Scorer outputs:

- LearningTraces to be saved in Learner Model
- Updates to the latest LearningTrace in the Learner Model

The scoring mechanism within the scorer makes use of expert problem-solving knowledge to be used in evaluating the student's performance and identifying the level of mastery exhibited for skills and concepts used in the activities, and makes use of the catalog of evaluation metrics, which is organized by concept. This information is made available to the Learner Modeler and used to update the Learner Model.

Example mapping concept of operations:

6. Learner decides to perform a set of tasks in service of a named goal, e.g. "Identify Superspreaders and their Spheres of Influence"
7. Learner performs collection of actions representing an effort to achieve that goal
8. Component monitors learner activity and creates data objects representing that activity
9. Component compares that activity against a collection of expert plan models, e.g. the expert plan for "Identify Superspreaders and their Spheres of Influence". Quantitatively measure the learner's activity similarity against all known expert plans.
10. Component determines that the learner is attempting to perform that plan. Select the expert plan that is most similar to the learner's activity, then threshold that similarity against a minimum required to identify the learner's activity as the selected expert plan, otherwise determine that the learner's activity is unrecognized.
11. Component evaluates the "correctness" of learner's actions for the detected plan, potentially the same quantitative process as step 4, or possibly with leeway added in areas where the learner might have the flexibility to deviate from the expert plan without that representing an "incorrect" approach.
12. Component produces one or more updates to the Learner Model that indicate how well the learner is following the expert plan. Since each expert plan is mapped to one or more Curriculum Elements, it is now possible to assign scores to those elements based on the evaluation(s) created by step 6.

4.7 Advisor/Hint Generator

Advice and hints may be produced for the learner at multiple points in the assessment process. For example, the Plan Recognizer supports explicit hints to present to the learner when they omit important steps in a recognized expert plan, such as "Don't forget to check for superspreaders using the report in ORA Pro or Scraawl" if the learner skips the step in which that task is performed. These hints are returned when the plan is passed to GLAsE for analysis. In future versions, this will be pushed to the practice environment via the shared message-passing queue.

5 GLAsE Implementation Details

The diagram in Fig. 6 shows an exploded view of the GLAsE components and models, the data shared among them and with other Twiner components.

GLAsE is implemented in Python using the Flask web service framework to provide a REST API for manipulating the Learner Model and evaluating plans as shown in Fig. 7. The details of this REST API are included in the ICD in the appendices.

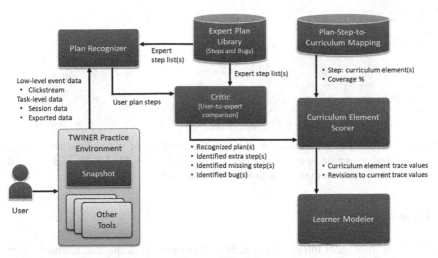

Fig. 6. Data flow into and among GLAsE components and models

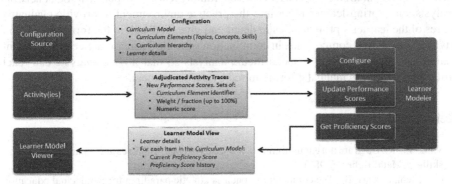

Fig. 7. GLAsE Learner Model software interfaces at a high level

In addition to manual updates via REST, GLAsE monitors an Apache Kafka message-passing service for certain types of computer-parsable statements in the practice environment that may indicate improvements in a learner's proficiency. When these statements are added to the queue via the practice environment, GLAsE is notified and will update the proficiency model accordingly.

Additionally, GLAsE provides a simple interface to each Learner Model, viewable in a web browser for diagnostic purposes. Figure 8 shows a screenshot of one example learner's scores over time.

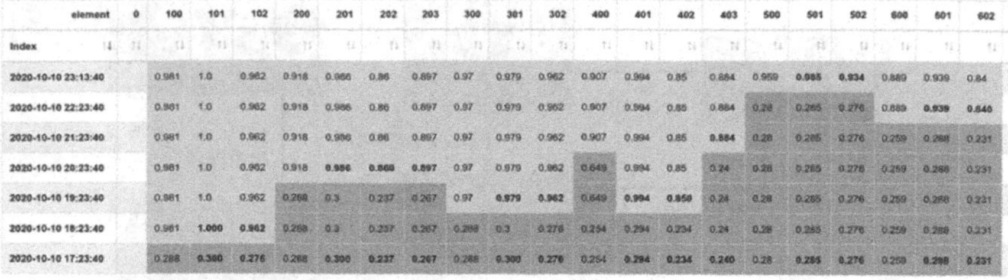

element	0	100	101	102	200	201	202	203	300	301	302	400	401	402	403	500	501	502	600	601	602
Index																					
2020-10-10 23:13:40		0.981	1.0	0.962	0.918	0.986	0.86	0.897	0.97	0.979	0.962	0.907	0.994	0.85	0.884	0.959	0.985	0.934	0.889	0.939	0.84
2020-10-10 22:23:40		0.981	1.0	0.962	0.918	0.986	0.86	0.897	0.97	0.979	0.962	0.907	0.994	0.85	0.884	0.28	0.285	0.276	0.889	0.939	0.840
2020-10-10 21:23:40		0.981	1.0	0.962	0.918	0.986	0.86	0.897	0.97	0.979	0.962	0.907	0.994	0.85	0.884	0.28	0.285	0.276	0.259	0.288	0.231
2020-10-10 20:23:40		0.981	1.0	0.962	0.918	0.986	0.860	0.897	0.97	0.979	0.962	0.649	0.994	0.85	0.24	0.28	0.285	0.276	0.259	0.288	0.231
2020-10-10 19:23:40		0.981	1.0	0.962	0.288	0.3	0.237	0.267	0.97	0.979	0.962	0.649	0.994	0.850	0.24	0.28	0.285	0.276	0.259	0.288	0.221
2020-10-10 18:23:40		0.981	1.000	0.962	0.288	0.3	0.237	0.267	0.288	0.3	0.276	0.254	0.294	0.234	0.24	0.28	0.285	0.276	0.259	0.288	0.231
2020-10-10 17:23:40		0.288	0.300	0.276	0.288	0.300	0.237	0.267	0.288	0.300	0.276	0.254	0.294	0.234	0.240	0.28	0.285	0.276	0.259	0.288	0.231

Fig. 8. Screenshot of example learner scores over time. Most recent scores are in the top row, and each column indicates a single curriculum element's numeric code. The UI also includes a listing of the curriculum elements to match with their codes, not shown).

6 Conclusions

GLAsE is an intelligent tutoring approach and software package that enables adaptive training to help novice professionals work toward an expert level of performance. The Twiner implementation of GLAsE focuses on training learners to perform social network analysis, monitoring learner input into the practice environment to provide evaluative scores of the learner's proficiencies in detail and give feedback in the form of hints and advice. The GLAsE components, in combination with the Twiner practice environment, enable automated training of social media analysts so that novice analysts can reach proficiency with less manual effort required by expert trainers.

References

1. VanLehn, K.: Bugs are not enough: empirical studies of bugs, impasses and repairs in procedural skills. J. Math. Behav. (1982)
2. Goldstein, I., Carr, B.: The computer as coach as athletic paradigm for intellectual education. In: Proceedings of the 1977 Annual Conference. ACM (1977)
3. Boulianne, S.: Social media use and participation: a meta-analysis of current research. Inf. Commun. Soc. **18**, 524–538 (2015)
4. VanLehn, K.: Student modeling. Found. Intell. tutoring Syst. **55**, 78 (1988)
5. Carley, K.M.: ORA: A Toolkit for Dynamic Network Analysis and Visualization (2014)
6. Cer, D., et al.: Universal sentence encoder for English. In: Proceedings of the Conference on Empirical Methods in Natural Language Processing: System Demonstrations, EMNLP 2018 (2018)
7. Crouse, D.F.: On implementing 2D rectangular assignment algorithms. IEEE Trans. Aerosp. Electron. Syst. **52**(4), 1679–1696 (2016). https://doi.org/10.1109/TAES.2016.140952

A Review of Automatic Detection of Learner States in Four Typical Learning Scenarios

Guanfeng Wang, Chen Gong, and Shuxia Wang[✉]

Northwestern Polytechnical University, Xi'an, China
shuxiaw@nwpu.edu.cn

Abstract. Artificial intelligence technology has already been applied in the education scene, and the automatic detecting technology of learning state has attracted the attention of many researchers. This paper summarizes the main types of learning state that researchers pay attention to at present, including affect, engagement, attention, and cognitive load. Based on four typical learning scenarios: computer-based learning, mobile learning, traditional classroom-based learning, and individual computer-free learning, this paper discusses the shortcomings and development trends of detecting hardware and methods used in this field, and the social problems in obtaining a large amount of personal privacy data.

Keywords: Learning state · Education scene · Automatic detection · Artificial intelligence

1 Introduction

With the continuous progress of modern educational technology, educators will not only impart knowledge, but also pay attention to the changes of learners' learning states, including affect, cognition, motivation, and behavior [1]. They make corresponding adjustments and feedbacks according to learners' learning states, to alleviate the bad learning states, realize personalized knowledge transfer, and improve teaching quality. They will also analyze the overall performance of learners in the whole learning activities, to adjust their future teaching strategies and teaching content. With the gradual application of information technology in the field of education, new learning scenarios and learning methods are also emerging. Among them, online learning can provide high-quality courses on a large scale and open to learners around the world with extremely low thresholds. With its excellent flexibility and convenience, online learning provides people with a large number of learning and teaching options. The extensive application of various MOOCs (Massive Open Online Courses) learning platforms represented by Chinese universities MOOC, Coursera, and edX makes the previous Quality education resources limited by time and space can be shared by learners around the world. Other learning methods include autonomous learning [2], mobile learning [3], digital game-based learning [4, 5], immersive learning [6], and so on.

In the current scenario of online learning or intelligent tutoring system without teachers, teachers cannot directly obtain the learning states of learners, so they cannot

© The Author(s), under exclusive license to Springer Nature Switzerland AG 2022
R. A. Sottilare and J. Schwarz (Eds.): HCII 2022, LNCS 13332, pp. 53–72, 2022.
https://doi.org/10.1007/978-3-031-05887-5_5

make corresponding adjustments according to students' feedback. Therefore, in the era of artificial intelligence, researchers try to use effective computing technology to detect learners' learning states [7]. State detecting technology can be divided into invasive and noninvasive in terms of the use of sensors. Invasive detecting requires the monitored object to passively wear various sensors, which affects its normal learning and work. The contact of the detected object makes the user feel more comfortable and causes less interference to the user's learning process. In terms of detecting methods, we focus on such features as noninvasiveness, universality, low cost, portability, and inconspicuous. Combining affective computing can expand the use of educational technologies and provide additional opportunities to improve overall distance learning outcomes, as well as provide new opportunities for personalized and low-cost teaching [8, 9].

The automatic detection of learning states based on affective computing has gradually become an active research topic. In the past two decades, researchers have gradually applied technologies such as effective computing to the field of education, but many attempts have remained in the laboratory environment. More in-depth research is necessary to make these results applicable to real-world learning environments. Because research studies on automatic detection of learning states are so far more extensive and immature than other fields. Up to date, few reviews have been conducted to synthesize the results of previous studies on automatic detection technology of learning states in the field of education. The literature review conducted by Felicia et al. [10] only focused on the method of using portable technology to monitor students' states in the traditional classroom-based learning environment, and did not summarize the automatic detecting method of learning states in other typical learning environments. Elaheh et al. [11] focused on the application of effective computing in the field of education and did not summarize specific detecting methods, nor did they consider learning states other than effect.

Our current research examines and analyzes the research findings in the field of learning states theory and automatic detection of learning states in the past two decades, to classify these studies according to learning states and detecting methods, and discusses the possibility of various technologies applied to the field of automatic detection of learning states in different learning scenarios. Based on the existing findings in the field of learning states detection, this review summarizes the development in the field from the perspective of learning states theory and detecting methods, and forecasts the future technology development trend. It is hoped that this review would help researchers identify areas that have been investigated or require further investigation and introduce practitioners to new developments in the field of automatic detection of learning states.

2   Review Method

This study refers to the systematic literature review guidelines of Kitchenham et al. [12] and we formulated the following research steps: formulate the research questions; identify inclusion and exclusion criteria; search the literature; gather information from the studies; conclude and report the research results. We developed the following research questions to help us manage the literature review and achieve research objectives:

(Q1) What are the noninvasive learning state detecting methods used in the selected papers, and what are their advantages and disadvantages?

(Q2) What learning states theory is used in the selected paper, and what learning states are considered?

(Q3) In which typical learning scenarios is the detecting method used for the selected paper suitable?

In this study, we reviewed relevant literature in conferences, journals, and books in the past three decades on the theory of learning states, because many of the early theories of learning states are very classical and have been used by researchers to this day. As for the automatic detecting method of learning states, we mainly choose the articles in conferences and journals in the last ten years, because since 2010, the research results such as affective computing have been increasingly applied in the field of education. Table 1 summarizes the inclusion and exclusion criteria, and the final included literature should satisfy all criteria.

Table 1. Inclusion and exclusion criteria.

Inclusion criteria	Exclusion criteria
Written in English	Non-English
Full-text papers	Uncomplicated studies
Published in the selected databases	Non-target domain
	Duplicated studies

The literature search focused on sources including library databases, such as Ei, Compendex, Web of Science, IEEE Xplore, Springer Link, and the web-based search engines of Google Scholar. These different sources enable a comprehensive search of any potentially relevant references for inclusion in this review. The main phrases and keywords used in literature retrieval include: "learner affects recognition", "learner states detection", "automated detection of engagement", "attention detection", "affective computing in learning environment". By searching the references that have been included in the literature, we also found some documents that meet the requirements.

A search was performed on selected databases using defined phrases and keywords, and 247 studies were searched. The titles, keywords, and abstracts of these articles were checked for inclusion and exclusion criteria. After examining the literature unrelated to the objectives and research questions of this study, 162 studies were excluded and 85 studies remained. The review methodology flowchart of this study is illustrated in Fig. 1.

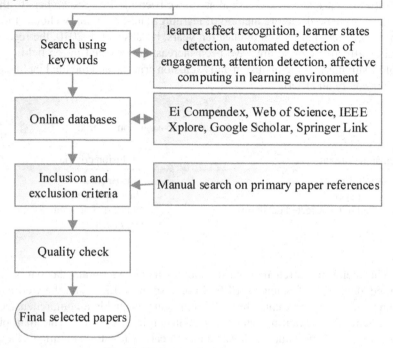

Fig. 1. Review methodology

3 Learning States

3.1 Affect

In the field of effective computing, researchers focus on several basic human emotions, such as six basic emotions [13], including anger, disgust, fear, happiness, sadness, and surprise. However, some researchers have found that during the learning process, some nonbasic emotions such as engagement, boredom, confusion, and frustration play a more important role [14–16], and their frequency of occurrence is higher than that of basic emotions. However, some scholars agree that these nonbasic emotions are not included in the category of emotions [17], and some people refer to these effective states accompanied by epistemic processes as epistemic affective states [18].

Affect is fundamentally important for human learning and development [19, 20], and it has a substantial influence on human cognitive processes, including perception, attention, learning, memory, reasoning, and problem-solving [21]. Efforts in deeper learning activities involve complex coordination of cognitive processes and effective states. The cognitive processes that underlie generative generation, causal reasoning, problem diagnosis, conceptual comparison, and coherent interpretation generation are accompanied by effective states, such as anger, Frustration, anger, and anger sometimes occur when learners make mistakes, struggle with troubled deadlocks, and experience failures. On the other hand, when tasks are completed, challenges are conquered, insights are revealed, and major discoveries are experienced, positive emotional states such as flow, joy, and excitement [22]. Experimental emotion studies have found that emotions affect a variety of cognitive processes that help learning [23]. Positive emotions promote learners' learning effects [24]. Confused emotions have certain bene-fits for learning [25], but students who experience anxiety or anger tend to have lower learning efficiency [24]. Therefore, it is necessary to obtain and ensure the effective health of learners to obtain targeted and timely feedback to adjust the effective states that hinder learning.

3.2 Engagement

The term of engagement is interpreted differently by different research communities [26], but most definitions consider engagement to include attentional and emotional involvement in tasks. Many scholars agree that there are four types of engagement in student learning activities, which are behavioral engagement, emotional engagement, cognitive engagement, and agentic engagement [27]. These forms of engagement are manifested through students' behavior, emotions, and motivation. In the study of the impact of effect on the learning process, researchers have considered engagement as an emotional state [14, 22], but more scholars agree that engagement is a behavioral state and can be considered the most fundamental state. Any other affective state, such as boredom, confusion, sleepiness, etc. gets reflected in the engagement levels of a learner [28].

In an educational environment, engagement and learning are often interrelated, and maintaining participation will improve learners' performance [27, 29]. In real learning scenarios, teachers improve their learning ability by keeping learners in engagement state [30]. Providing a high engagement flow experience has been shown to be positively cor-related with learning outcomes [31]. Therefore, when learners' engagement decreases, timely feedback and adjustment should be provided to maintain high engagement, which can improve the learning effect of learners.

3.3 Attention/Mind Wandering

Attention and mind wandering can be considered as opposite states, and they are also two states that researchers in the field of learning states detection often pay attention to. The essence of attention is the accumulation and concentration of consciousness. We can delineate attention into overt attention and convert attention defined by the selection mechanism used by the learner himself. Overt attention involves an attention mechanism that physically moves the sensory organs to select a specific type of neural processing [32]. Overt attention is often associated with external stimuli and is by far the most commonly studied form of attention because sensory organ movements provide measurable value for analysis [33]. Human attention is closely related to one's efficiency in work and study. Among various mental states, attention is one of the most important and indicative states that affects user efficiency, productivity, and even creativity [34]. Therefore, the detection of attention is one of the significant ways to understand the learning state of learners. Many studies on the detection of attention divide attention into multiple levels, such as high attention, medium attention, and low attention, etc. [7, 33].

Mind wandering is the involuntary shift of attention from task-related thoughts to task-unrelated thoughts. Studies have shown that mind wandering is a common phenomenon, and it occurs 20–50% of the time in the whole learning process, depending on different environments and tasks [35, 36]. Mind wandering has been shown to be negatively correlated with the performance of various tasks requiring conscious control in many cases [37]. For example, mind-wandering leads to an increase in error rate during signal detection tasks [38], a decrease in recall rate during memory tasks [39], and worse comprehension during reading tasks [40]. The detection of mind wandering is most closely related to the field of attention detection. By detecting learners' state of attention or mind wandering, an appropriate intervention can be conducted to reduce the negative effects of mind wandering, which is conducive to improving the learning effect.

3.4 Cognitive Load

Cognitive load theory was first proposed by John Sweller, who defined cognitive load as the total amount of information that can be processed during human information processing, mainly including the total amount of information stored and processed by working memory [41]. According to Sweller's theory, the sources of cognitive load in the learning process include intrinsic cognitive load, extraneous cognitive load, and germane cognitive load. The intrinsic cognitive load is determined by the knowledge content itself, while the extraneous cognitive load is determined by the teaching design and teaching methods. The germane cognitive load is related to the deeper information processing of learners. However, the three types of cognitive load are interrelated. According to the cognitive load theory, effective learning can only happen if the cognitive load is controlled within the range of working memory, otherwise it will lead to overload of cognitive load and learning failure [42].

Cognitive load theory is considered to have had a significant impact on the field of learning and teaching [43]. Compared with traditional teaching tasks, teaching tasks based on cognitive load theory have been proved to be more efficient because they require less training time and less psychological burden to achieve the same or better learning and transfer performance [44]. Cognitive load is not only seen as a by-product of the learning process, but is a major factor determining the success of teaching interventions. Therefore, it is necessary to study the cognitive mechanism related to specific learning behaviors. Many studies have been conducted to measure cognitive load during learning [45], proving the importance of real-time cognitive load detection during learning.

Table 2. Summary of noninvasive learning state monitoring methods in four learning scenarios.

States	Related factors	Influence
Affect	Basic emotions Non-basic emotions	Cognitive processes Learning effects Learning efficiency
Engagement	Behavior Emotions Motivation	Learning performance Learning ability
Attention/mind wandering	Consciousness Selection mechanism	Efficiency Productivity Creativity
Cognitive load	Working memory Information processing	Learning efficiency

Learning state can be analyzed from various perspectives of learners. This study only summarizes the four states of affect, engagement, attention, and cognitive load that researchers pay most attention to. Table 2 shows the related factors of the four learning states and their influence on learners.

4   Typical Learning Scenarios

(See Fig. 2).

Fig. 2. Four typical learning scenarios. (a) Computer-based Learning, (b) Mobile Learning, (c) Traditional Classroom-based Learning, (d) Individual Computer-free Learning

4.1 Computer-Based Learning

Desktop computer and laptop-based learning environments are the most common online learning environments. Compared with other learning scenarios, this learning environment has more signal channels to obtain learners' states. Since the general computer will be equipped with a camera, keyboard, mouse, and other devices, and may be equipped with additional microphones and other sensors.

Detecting learning states through the camera is one of the most versatile and portable noninvasive detection methods, and there have been a lot of related studies. Facial expression recognition is one of the common methods of effective computing, and algorithms are emerging to optimize its recognition effect [46–48]. Many studies have applied facial expression recognition in the educational environment of online learning, for example, identifying the effective states of students during learning through facial expressions [49–54], using facial expressions to identify students' understanding of the entire learning process [55], detecting attention or engagement from learner's facial and head pose characteristics [7, 28, 56, 57]. Body gesture and behavior are also important channels for learners' states that can be obtained through cameras. Studies have been conducted to identify learners' affect based on facial expressions and gestures in videos [52, 53], and to identify attention by acquiring head pose or body gesture with ordinary cameras [58, 59]. Ordinary cameras can be used to obtain the sight information of the eyes to identify the learner's attention [58, 59] and to obtain eye movements and pupil information to identify the user's cognitive load [60]. Ordinary infrared web cameras are also a measure of cognitive load [61]. However, due to the performance limitations of ordinary cameras, there are still few studies in this field.

The mouse and keyboard are devices that ordinary computers are equipped with and are also ideal noninvasive tools for obtaining learner information and identifying their learning states. There are various ways to obtain learner states through the keyboard and mouse, and the types of states detected vary. Some researchers use the characteristics of the experimenter tapping the keyboard or moving the mouse to monitor the effective state, engagement and stress state of the experimenter [7, 62–65]. Others use special physiological mouse to obtain the user's photoplethysmography (PPG) signal to detect the learner's cognitive and effective states [66]. Although the method of acquiring learner states through the keyboard and mouse is only effective when they are used frequently, and their detection results are not as good as other sensors', these studies will provide research for multimodal learning status detection. A new dimension.

In addition to common devices on ordinary computers, to detect learners' states more accurately and effectively, additional sensors can be used for states detection. Somatosensory cameras are often applied to states detection based on body gesture and movement, among which Microsoft Kinect camera is the most widely used sensor [67–69]. Its research in the field of education includes detecting learners' effective states, engagement, and attention [33, 54, 70]. Pressure sensors have also been used in posture-based learner's interest level detection [53, 71]. Eye tracker can capture accurate and comprehensive eye-related signals of learners, including fixation frequency, blink frequency, line of sight change and pupil dilation, etc. These signals reflect learners' real-time stress state [63], attention state [72–74] and partial effective state [75]. Headphones are often used in online learning. Special headphones can capture the user's electroencephalogram (EEG), which can be used to identify various learning states [76]. The microphone enables learners to interact with teachers or intelligent tutoring systems to realize voice interaction in online learning, and the voice and language information captured by the microphone can reflect the effective state of learners [77–81]. The additional sensors provide a new dimension for the study of multimodal learning states detection, but also reduce the universality of the detecting method and increase its use cost.

4.2 Mobile Learning

Mobile learning enables students and teachers to make use of the smart wireless devices (such as mobile phones, tablet computers, etc.) that are widely used today to realize interactive teaching activities, as well as information exchange in education and technology more convenient and flexible. These smart wireless devices have sensors such as cameras, microphones, touch screens, and accelerometers, providing multiple channels for capturing learner states signals.

Touch screen is the main sensor for learners to interact with the smart wireless device. The learner's finger sliding and typing behavior when using mobile devices reflects part of his learning state at that time. By analyzing the reader's finger activities to detect their cognitive states, and then automatically adjusting the text playback mode, reading efficiency can be improved [82]. Analyzing the user's key stroke dynamics on mobile devices [83] can also identify the user's current effective states. It is also possible to analyze the finger sliding and typing behavior at the same time to identify learner states [84].

The front camera of smart mobile devices provides an opportunity to identify effective states through learners' facial expressions in mobile learning. There are researches on facial recognition for mobile devices [85, 86], and many learning applications have added facial expression-based effective recognition into the system [87, 88].

In addition to the above main channels, more researches tend to use multiple sensors for multimodal state detection in mobile learning. Using a rear camera to detect heart rate and a front camera to detect facial expressions can identify learners' affect and cognitive states [86, 89, 90]. Speech recognition combined with facial expression recognition can improve the accuracy rate of affect detection on mobile devices [85, 91]. Some studies also use accelerometers and other sensors to acquire learners' movements and other signals for multimodal effect detection [88, 92]. In addition to real-time identification of learners' various learning states, long-term objective sensor data recorded by mobile devices can also be used to assess learners' long-term stress, emotional, and mental health status [93].

4.3 Traditional Classroom-Based Learning

Traditional classroom-based multi-person learning is still the most important way of learning. However, in a multi-person classroom, teachers cannot observe each student's learning states comprehensively and objectively. With the development of artificial intelligence technology, the concept of intelligent classroom has also been proposed. By identifying the learning state of learners, the intelligent classroom system can assist learners' learning and the interaction between learners and teachers.

The cameras are the most ideal learning state monitoring devices in the classroom. Cameras are low-cost, noninvasive devices that can obtain rich information. At present, cameras have been installed in most classrooms. However, it is more difficult to detect the learner states in the classroom than the above two scenarios, because multiple people are detected at the same time in the classroom, which increases the difficulty in algorithm and calculation [94]. Learner's facial expression information obtained by the camera in the classroom reflects the learner's current state of affect, attention, and engagement [95–98], and the student's attendance can also be detected by matching the student's face information. Cameras in the classroom can also detect students' learning states from multi-channel, such as facial expression and body gesture. Although multi-channel detection is more accurate and reliable, it is more difficult to realize multi-channel detecting of multiple people [52, 99, 100].

Ordinary cameras are the main channel for state detecting, and some other sensors can also be used in traditional classroom scenarios. Kinect is another tool to monitor learner behavior, and learners' behaviors in the classroom reflect changes of their engagement [101]. Effective recognition by speech using microphones in the classroom has also been attempted [102, 103]. In addition, pressure sensors can also be used for sitting posture detecting in classroom scenes [71]. Although these sensors are not practical for multi-person state detecting, they are expensive to use and difficult to deploy, but they are very meaningful ideas and attempts. In classes that students are encouraged to move freely, the sociometric badges worn by students can measure their relationships and interactions, maximizing the effect of team learning [10, 104].

4.4 Individual Computer-Free Learning

In the traditional paper-based learning scenario, learners study alone without the assistance of mobile phones or computers. Although the learner in this scenario does not have the distance teaching of a teacher or the guidance of an intelligent system, by recording their learning status, the education researcher or learner can analyze the recorded data to further the learning and cognitive process in this scenario to research or adjust.

In this scenario, the camera is still the preferred noninvasive detecting device. As the learner is sitting in a fixed position to learn, the methods of detecting learning states through the camera in computer-based learning scenarios can be used, for example, by identifying expressions to detect the learners' effective states [50, 55], cognitive process and engagement [98], by obtaining the body and head posture change to detect the learners' effective states and attention [52, 58], and by obtaining eye movement data to detect the learners' cognitive load, and so on.

In other devices, Kinect can also be used for expression recognition and gesture recognition, and the gesture recognition rate is more accurate than ordinary cameras [67], and pressure sensors can obtain learner sitting information [53]. Although there are some devices that can also be used for learning state recognition, their use in this scenario is not good (Fig. 3 and Table 3).

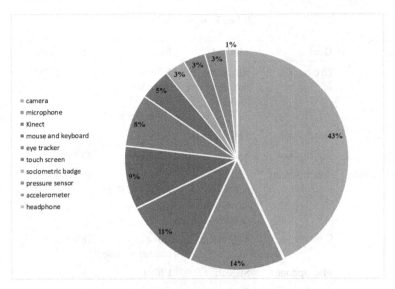

Fig. 3. Proportion of devices used in the article

Table 3. Summary of noninvasive learning state monitoring methods in four learning scenarios.

Scenario	Device	Method	State	References
Computer-based Learning	Camera	Expression, body gesture, head pose, behavior, reactivity and movement of the eye	Effect, attention, engagement, cognitive load	[7, 28, 46–61]
	Mouse, keyboard	Keystroke dynamics, mouse activities, PPG	Effect, engagement, stress	[7, 62–66]
	Kinect	Body gesture, head pose, behavior	Effect, attention, engagement	[33, 54, 67–70]
	Pressure sensor	Body gesture	Interest level	[53, 71]

(continued)

Table 3. (*continued*)

Scenario	Device	Method	State	References
	Eye tracker	Reactivity and movement of the eye	Effect, attention, stress, cognitive load	[63, 72–75]
	Headphone	PPG	Effect	[76]
	Microphone	Speech	Effect	[77–81]
Mobile Learning	Touch screen	Finger activities, keystroke dynamics	Effect, cognitive state	[82–84]
	Camera	Expression	Effect, cognitive state	[85–90]
	Microphone	Speech	Effect	[85, 91]
	Accelerometer	Body gesture, behavior	Effect	[88, 92]
Traditional Classroom-based Learning	Camera	Body gesture, head pose, behavior	Effect, attention, engagement	[52, 94–100]
	Kinect	Body gesture, head pose, behavior	Engagement	[101]
	Microphone	Speech	Effect	[102, 103]
	Sociometric badge	Behavior	Relationship and interaction	[10, 104]
Individual Computer-free Learning	Camera	Expression, body gesture, head pose, behavior, reactivity and movement of the eye	Effect, attention, engagement, cognitive load	[50, 52, 55, 58, 60, 98]
	Kinect	Body gesture, head pose, behavior	Effect, attention	[67]
	Pressure sensor	Body gesture	Effect, attention	[53]

5 Discussion

In the use of sensors to detect learning states, the characteristics of sensors such as noninvasiveness, versatility, low cost, portability, and unobtrusiveness of the sensors are the key considerations of researchers. Compared to the other devices mentioned, the camera is the most widely used device in the natural environment. There are still some shortcomings in image-based learning states detection, and multimodal detection through expressions, body poses, and eye movement information improves the reliability of detection. While ideally, the lack of unbiased observational tools available and increased adoption of intelligent detecting devices in a variety of learning scenarios could contribute to future advances in technology for teaching and learning environments.

In terms of the design of recognition algorithms, current recognition algorithms are usually divided into performance-based type and real-time type. Performance-based algorithms focus on recognition quality and high recognition accuracy, but run slowly, and are often used for subsequent analysis of the entire learning process. Real-time algorithms pay attention to recognition speed and running speed, but their recognition accuracy is low, and they are always used in scenarios that require real-time feedback to learners. At present, most of the recognition algorithms are only tested in the laboratory environment, and the reliability of the recognition algorithms is still questioned because there are more disturbances and emergencies in the natural learning environment.

The detection of learner states through multiple devices is based on the premise of obtaining a large amount of personal privacy data. On the one hand, the abuse of artificial intelligence technology may lead to large-scale privacy invasion. On the other hand, most of the current applications of artificial intelligence use cloud computing. If the security of personal privacy data stored in the cloud is not guaranteed, it will lead to great privacy disclosure risks.

Acknowledgement. This work is partly supported by National Key R&D Program of China (Grant No. 2019YFB1703800), Natural Science Basic Research Plan in Shanxi Province of China (Grant No. 2016JM6054), the Programme of Introducing Talents of Discipline to Universities(111 Project), China(Grant No. B13044).

Data Availability. The data used to support the findings of this study are available from the corresponding author upon request.

Conflicts of Interest. The authors declare that there are no conflicts of interest regarding the publication of this paper.

References

1. D'Mello, S., Taylor, R., Graesser, A.: Monitoring affective trajectories during complex learning. In: Proceedings of the 29th Annual Meeting of the Cognitive Science Society, pp. 203–208, 01 January 2007
2. Boekaerts, M.: Cognitive load and self-regulation: attempts to build a bridge. Learn. Instr. **51**, 90–97 (2017)

3. Imtinan, U., Chang, V., Issa, T.: Common mobile learning characteristics-an analysis of mobile learning models and frameworks. In: IADIS International Conference on Mobile Learning 2013, ML 2013, Lisbon, Portugal, 14–16 March 2013, pp. 3–11. IADIS (2013)
4. Boyle, E.A., Hainey, T., Connolly, T.M., et al.: An update to the systematic literature review of empirical evidence of the impacts and outcomes of computer games and serious games. Comput. Educ. **94**, 178–192 (2016)
5. Easterday, M.W., Aleven, V., Scheines, R., et al.: Using tutors to improve educational games: a cognitive game for policy argument. J. Learn. Sci. **26**(2), 226–276 (2017)
6. Dede, C.: Immersive interfaces for engagement and learning. Science **323**(5910), 66–69 (2009). (in English)
7. Sun, H.J., Huang, M.X., Ngai, G., et al.: Nonintrusive multimodal attention detection. In: 7th International Conference on Advances in Computer-Human Interactions, ACHI 2014, Barcelona, Spain, 23–27 March 2014, pp. 192–199. International Academy, Research and Industry Association, IARIA (2014)
8. Cabada, R.Z., Estrada, M.L.B., Hernández, F.G., et al.: An affective and Web 3.0-based learning environment for a programming language. Telemat. Inform. **35**(3), 611–628 (2018)
9. Caballé, S.: Towards a multi-modal emotion-awareness e-learning system. In: 2015 International Conference on Intelligent Networking and Collaborative Systems, pp. 280–287. IEEE (2015)
10. Goh, F., Carroll, A., Gillies, R.M.: A review of the use of portable technologies as observational aids in the classroom. Inf. Learn. Sci. **120**(3/4), 228–241 (2019)
11. Yadegaridehkordi, E., Noor, N.F.B.M., Ayub, M.N.B., et al.: Affective computing in education: a systematic review and future research. Comput. Educ. **142**, 103649 (2019)
12. Keele, S.: Guidelines for performing systematic literature reviews in software engineering. Technical report, ver. 2.3 EBSE Technical Report. EBSE (2007)
13. Elfenbein, H.A., Ambady, N.: On the universality and cultural specificity of emotion recognition: a meta-analysis. Psychol. Bull. **128**(2), 203–235 (2002). https://doi.org/10.1037/0033-2909.128.2.203. (in English)
14. Dmello, S., Calvo, R.A.: Beyond the basic emotions: what should affective computing compute? In: 31st Annual CHI Conference on Human Factors in Computing Systems: CHI EA 2013, Paris, France, 27 April–2 May 2013, vol. 2013, pp. 2287–2294. Association for Computing Machinery (2013)
15. Russell, J.A.: Core affect and the psychological construction of emotion. Psychol. Rev. **110**(1), 145–172 (2003). (in English)
16. Graesser, A., McDaniel, B., Chipman, P., et al.: Detection of emotions during learning with AutoTutor. In: Proceedings of the 28th Annual Meetings of the Cognitive Science Society, pp. 285–290. Citeseer (2006)
17. Silvia, P.J.: Looking past pleasure: anger, confusion, disgust, pride, surprise, and other unusual aesthetic emotions. Psychol. Aesthet. Creat. Arts **3**(1), 48 (2009)
18. Pekrun, R.: Academic emotions. Handb. Motiv. Sch. **2**, 120–144 (2016)
19. Calvo, R.A., D'Mello, S., Gratch, J.M., et al.: The Oxford Handbook of Affective Computing. Oxford University Press, USA (2015)
20. Woolf, B., Burleson, W., Arroyo, I., et al.: Affect-aware tutors: recognising and responding to student affect. Int. J. Learn. Technol. **4**(3–4), 129–164 (2009)
21. Tyng, C.M., Amin, H.U., Saad, M.N.M., Malik, A.S.: The influences of emotion on learning and memory. Front. Psychol. **8**, 1454 (2017). https://doi.org/10.3389/fpsyg.2017.01454. (in English)
22. D'Mello, S., Graesser, A.: Dynamics of affective states during complex learning. Learn. Instr. **22**(2), 145–157 (2012). https://doi.org/10.1016/j.learninstruc.2011.10.001
23. Clore, G.L., Huntsinger, J.R.: How emotions inform judgment and regulate thought. Trends Cogn. Sci. **11**(9), 393–399 (2007). https://doi.org/10.1016/j.tics.2007.08.005

24. Pekrun, R.: Emotions as drivers of learning and cognitive development. In: Calvo, R.A., D'Mello, S.K. (eds.) New Perspectives on Affect and Learning Technologies, pp. 23–39. Springer, New York (2011). https://doi.org/10.1007/978-1-4419-9625-1_3

25. D'Mello, S., Lehman, B., Pekrun, R., Graesser, A.: Confusion can be beneficial for learning. Learn. Instr. **29**, 153–170 (2014). https://doi.org/10.1016/j.learninstruc.2012.05.003

26. Peters, C., Castellano, G., de Freitas, S.: An exploration of user engagement in HCI. In: Proceedings of the International Workshop on Affective-Aware Virtual Agents and Social Robots, p. 9. ACM (2009)

27. Christenson, S.L., Reschly, A.L., Wylie, C. (eds.): Handbook of Research on Student Engagement. Springer, Boston (2012). https://doi.org/10.1007/978-1-4614-2018-7

28. Kamath, A., Biswas, A., Balasubramanian, V.: A crowdsourced approach to student engagement recognition in e-learning environments. In: IEEE Winter Conference on Applications of Computer Vision, WACV 2016, Lake Placid, NY, USA, 7–10 March 2016. Institute of Electrical and Electronics Engineers Inc. (2016)

29. Kahn, W.A.: Psychological conditions of personal engagement and disengagement at work. Acad. Manag. J. **33**(4), 692–724 (1990)

30. Papadopoulos, F., Corrigan, L.J., Jones, A., et al.: Learner modelling and automatic engagement recognition with robotic tutors. In: 2013 5th Humaine Association Conference on Affective Computing and Intelligent Interaction, ACII 2013, Geneva, Switzerland, 2–5 September 2013, pp. 740–744. IEEE Computer Society (2013)

31. Czikszentmihalyi, M.: Flow: The Psychology of Optimal Experience. Harper & Row, New York (1990)

32. Geisler, W.S., Cormack, L.K.: Models of overt attention. In: Oxford Handbook of Eye Movements, pp. 439–454 (2011)

33. Stanley, D.: Measuring attention using Microsoft Kinect (2013)

34. Howard-Jones, P., Murray, S.: Ideational productivity, focus of attention, and context. Creat. Res. J. **15**(2–3), 153–166 (2003)

35. Killingsworth, M.A., Gilbert, D.T.: A wandering mind is an unhappy mind. Science **330**(6006), 932 (2010)

36. Smilek, D., Carriere, J.S.A., Allan Cheyne, J.: Out of mind, out of sight: eye blinking as indicator and embodiment of mind wandering. Psychol. Sci. **21**(6), 786–789 (2010). https://doi.org/10.1177/0956797610368063

37. Randall, J.G., Oswald, F.L., Beier, M.E.: Mind-wandering, cognition, and performance: a theory-driven meta-analysis of attention regulation. Psychol Bull **140**(6), 1411 (2014)

38. Smallwood, J., Davies, J.B., Heim, D., et al.: Subjective experience and the attentional lapse: task engagement and disengagement during sustained attention. Conscious. Cogn. **13**(4), 657–690 (2004)

39. Smallwood, J., Schooler, J.W.: The restless mind (2013)

40. Feng, S., D'Mello, S., Graesser, A.C.: Mind wandering while reading easy and difficult texts. Psychon. Bull. Rev. **20**(3), 586–592 (2012). https://doi.org/10.3758/s13423-012-0367-y

41. Sweller, J.: Human cognitive architecture. In: Handbook of Research on Educational Communications and Technology, pp. 369–381 (2008)

42. Paas, F., Renkl, A., Sweller, J.: Cognitive load theory: instructional implications of the interaction between information structures and cognitive architecture. Instr. Sci. **32**(1/2), 1–8 (2004). https://doi.org/10.1023/B:TRUC.0000021806.17516.d0

43. Ozcinar, Z.: The topic of instructional design in research journals: a citation analysis for the years 1980–2008. Australas. J. Educ. Technol. **25**(4) (2009)

44. Paas, F., Tuovinen, J.E., Tabbers, H., Van Gerven, P.W.M.: Cognitive load measurement as a means to advance cognitive load theory. Educ. Psychol. **38**(1), 63–71 (2010). https://doi.org/10.1207/S15326985EP3801_8

45. Skulmowski, A., Rey, G.D.: Measuring cognitive load in embodied learning settings. Front. Psychol. **8**, 1191 (2017)
46. Jeon, J., Park, J.-C., Jo, Y., et al.: A real-time facial expression recognizer using deep neural network. In: 10th International Conference on Ubiquitous Information Management and Communication, IMCOM 2016, Danang, Vietnam, 4–6 January 2016. Association for Computing Machinery, Inc., ACM SIGAPP (2016)
47. Verma, B., Choudhary, A.: A framework for driver emotion recognition using deep learning and Grassmann manifolds. In: 21st IEEE International Conference on Intelligent Transportation Systems, ITSC 2018, Maui, HI, USA, 4–7 November 2018, vol. 2018, pp. 1421–1426. Institute of Electrical and Electronics Engineers Inc. (2018)
48. Hasani, B., Mahoor, M.H.: Facial expression recognition using enhanced deep 3D convolutional neural networks. In: 30th IEEE Conference on Computer Vision and Pattern Recognition Workshops, CVPRW 2017, Honolulu, HI, USA, 21–26 July 2017, vol. 2017, pp. 2278–2288. IEEE Computer Society (2017)
49. Sawyer, R., Smith, A., Rowe, J., et al.: Enhancing student models in game-based learning with facial expression recognition. In: 25th ACM International Conference on User Modeling, Adaptation, and Personalization, UMAP 2017, Bratislava, Slovakia, 9–12 July 2017, pp. 192–201. Association for Computing Machinery, Inc. (2017)
50. McDaniel, B.T., D'Mello, S.K., King, B.G., et al.: Facial features for affective state detection in learning environments. In: Proceedings of the 29th Annual Cognitive Science Society (2007)
51. Liu, S., Wang, W.: The application study of learner's face detection and location in the teaching network system based on emotion recognition. In: 2nd International Conference on Networks Security, Wireless Communications and Trusted Computing, NSWCTC 2010, Wuhan, Hubei, China, 24–25 April 2010, vol. 1, pp. 394–397. IEEE Computer Society (2010)
52. Bosch, N., et al.: Using video to automatically detect learner affect in computer-enabled classrooms. ACM Trans. Interact. Intell. Syst. **6**(2), 1–26 (2016). https://doi.org/10.1145/2946837
53. Kapoor, A., Picard, R.W.: Multimodal affect recognition in learning environments. In: 13th ACM International Conference on Multimedia, MM 2005, Singapore, Singapore, 6–11 November 2005, pp. 677–682. Association for Computing Machinery (2005)
54. Psaltis, A., Apostolakis, K.C., Dimitropoulos, K., et al.: Multimodal student engagement recognition in prosocial games. IEEE Trans. Games **10**(3), 292–303 (2018)
55. Yang, D., Abeer Alsadoon, P.W.C., Prasad, A.K., Singh, A.E.: An emotion recognition model based on facial recognition in virtual learning environment. Procedia Comput. Sci. **125**, 2–10 (2018). https://doi.org/10.1016/j.procs.2017.12.003
56. Kaur, A., Mustafa, A., Mehta, L., et al.: Prediction and localization of student engagement in the wild. In: 2018 International Conference on Digital Image Computing: Techniques and Applications, DICTA 2018, Canberra, ACT, Australia, 10–13 December 2018, pp. APRS; Australian Government, Department of Defence, Defence Science and Technology Group; Canon Information Systems Research Australia Pty Ltd (CiSRA); IAPRO; IEEE; UNSW Canberra. Institute of Electrical and Electronics Engineers Inc. (2018)
57. Monkaresi, H., Bosch, N., Calvo, R.A., et al.: Automated detection of engagement using video-based estimation of facial expressions and heart rate. IEEE Trans. Affect. Comput. **8**(1), 15–28 (2017)
58. Asteriadis, S., Karpouzis, K., Kollias, S.: Visual focus of attention in non-calibrated environments using gaze estimation. Int. J. Comput. Vision **107**(3), 293–316 (2014)

59. Yang, F., Jiang, Z., Wang, C., et al.: Student eye gaze tracking during MOOC teaching. In: 2018 Joint 10th International Conference on Soft Computing and Intelligent Systems (SCIS) and 19th International Symposium on Advanced Intelligent Systems (ISIS), pp. 875–880 (2018)

60. Fridman, L., Reimer, B., Mehler, B., et al.: Cognitive load estimation in the wild. In: 2018 CHI Conference on Human Factors in Computing Systems, CHI 2018, Montreal, QC, Canada, 21–26 April 2018, vol. 2018. Association for Computing Machinery, ACM SIGCHI (2018)

61. Chen, S., Epps, J., Chen, F.: An investigation of pupil-based cognitive load measurement with low cost infrared webcam under light reflex interference. In: 2013 35th Annual International Conference of the IEEE Engineering in Medicine and Biology Society, EMBC 2013, Osaka, Japan, 3–7 July 2013, pp. 3202–3205. Institute of Electrical and Electronics Engineers Inc. (2013)

62. Bixler, R., D'Mello, S.: Detecting boredom and engagement during writing with keystroke analysis, task appraisals, and stable traits. In: 18th International Conference on Intelligent User Interfaces, IUI 2013, Santa Monica, CA, USA, 19–22 March 2013, pp. 225–233. Association for Computing Machinery (2013)

63. Wang, J., Huang, M.X., Ngai, G., et al.: Are you stressed? Your eyes and the mouse can tell. In: 7th International Conference on Affective Computing and Intelligent Interaction, ACII 2017, San Antonio, TX, USA, 23–26 October 2017, vol. 2018, pp. 222–228. Institute of Electrical and Electronics Engineers Inc. (2017)

64. Lali, P., Naghizadeh, M., Nasrollahi, H., et al.: Your mouse can tell about your emotions. In: 4th International Conference on Computer and Knowledge Engineering, ICCKE 2014, Azadi Square, Mashhad, Iran, 29–30 October 2014, pp. 47–51. Institute of Electrical and Electronics Engineers Inc. (2014)

65. Shikder, R., Rahaman, S., Afroze, F., et al.: Keystroke/mouse usage based emotion detection and user identification. In: 2017 International Conference on Networking, Systems and Security, NSysS 2017, Dhaka, Bangladesh, 5–8 January 2017, pp. 96–104. Institute of Electrical and Electronics Engineers Inc. (2017)

66. Fu, Y., Leong, H.V., Ngai, G., Huang, M.X., Chan, S.C.F.: Physiological mouse: toward an emotion-aware mouse. Univ. Access Inf. Soc. **16**(2), 365–379 (2016). https://doi.org/10.1007/s10209-016-0469-9

67. Saha, S., Datta, S., Konar, A., et al.: A study on emotion recognition from body gestures using Kinect sensor. In: 3rd International Conference on Communication and Signal Processing, ICCSP 2014, Melmaruvathur, Tamil Nadu, India, 3–5 April 2014, pp. 56–60. Institute of Electrical and Electronics Engineers Inc. (2014)

68. Piana, S., Staglianò, A., Odone, F., Camurri, A.: Adaptive body gesture representation for automatic emotion recognition. ACM Trans. Interact. Intell. Syst. **6**(1), 1–31 (2016). https://doi.org/10.1145/2818740

69. Kaza, K., et al.: Body motion analysis for emotion recognition in serious games. In: Antona, M., Stephanidis, C. (eds.) UAHCI 2016. LNCS, vol. 9738, pp. 33–42. Springer, Cham (2016). https://doi.org/10.1007/978-3-319-40244-4_4

70. Wei, H., Scanlon, P., Li, Y., et al.: Real-time head nod and shake detection for continuous human affect recognition. In: 2013 14th International Workshop on Image Analysis for Multimedia Interactive Services, WIAMIS 2013, Paris, France, 3–5 July 2013. IEEE Computer Society (2013)

71. Mota, S., Picard, R.W.: Automated posture analysis for detecting learner's interest level. In: Conference on Computer Vision and Pattern Recognition Workshop, CVPRW 2003, Madison, WI, USA, 16–22 June 2003, vol. 5. IEEE Computer Society (2003)

72. Bixler, R., Blanchard, N., Garrison, L., et al.: Automatic detection of mind wandering during reading using gaze and physiology. In: ACM International Conference on Multimodal Interaction, ICMI 2015, Seattle, WA, USA, 9–13 November 2015, pp. 299–306. Association for Computing Machinery, Inc. (2015)

73. Bixler, R., D'Mello, S.: Automatic gaze-based user-independent detection of mind wandering during computerized reading. User Model. User-Adap. Inter. **26**(1), 33–68 (2015). https://doi.org/10.1007/s11257-015-9167-1

74. Yonetani, R., Kawashima, H., Matsuyama, T.: Multi-mode saliency dynamics model for analyzing gaze and attention. In: 7th Eye Tracking Research and Applications Symposium, ETRA 2012, Santa Barbara, CA, USA, 28–30 March 2012, pp. 115–122. Association for Computing Machinery (2012)

75. D'Mello, S., Olney, A., Williams, C., et al.: Gaze tutor: a gaze-reactive intelligent tutoring system. Int. J. Hum. Comput. Stud. **70**(5), 377–398 (2012)

76. Girardi, D., Lanubile, F., Novielli, N.: Emotion detection using noninvasive low cost sensors. In: 7th International Conference on Affective Computing and Intelligent Interaction, ACII 2017, San Antonio, TX, USA, 23–26 October 2017, vol. 2018, pp. 125–130. Institute of Electrical and Electronics Engineers Inc. (2017)

77. Yang, L., Jiang, D., Han, W., et al.: DCNN and DNN based multi-modal depression recognition. In: 2017 Seventh International Conference on Affective Computing and Intelligent Interaction (ACII), pp. 484-489 (2017)

78. Kim, J., Truong, K.P., Englebienne, G., et al.: Learning spectro-temporal features with 3D CNNs for speech emotion recognition. In: 7th International Conference on Affective Computing and Intelligent Interaction, ACII 2017, San Antonio, TX, USA, 23–26 October 2017, vol. 2018, pp. 383–388. Institute of Electrical and Electronics Engineers Inc. (2017)

79. Chen, K., Yue, G., Yu, F., Shen, Y., Zhu, A.: Research on speech emotion recognition system in e-learning. In: Shi, Y., van Albada, G.D., Dongarra, J., Sloot, P.M.A. (eds.) ICCS 2007. LNCS, vol. 4489, pp. 555–558. Springer, Heidelberg (2007). https://doi.org/10.1007/978-3-540-72588-6_91

80. Tzinis, E., Potamianos, A.: Segment-based speech emotion recognition using recurrent neural networks. In: 7th International Conference on Affective Computing and Intelligent Interaction, ACII 2017, San Antonio, TX, USA, 23–26 October 2017, vol. 2018, pp. 190–195. Institute of Electrical and Electronics Engineers Inc. (2017)

81. Slater, S., Ocumpaugh, J., Baker, R., et al.: Using natural language processing tools to develop complex models of student engagement. In: 7th International Conference on Affective Computing and Intelligent Interaction, ACII 2017, San Antonio, TX, USA, 23–26 October 2017, vol. 2018, pp. 542–547. Institute of Electrical and Electronics Engineers Inc. (2017)

82. Demmans Epp, C., Munteanu, C., Axtell, B., et al.: Finger tracking: facilitating noncommercial content production for mobile E-reading applications. In: 19th International Conference on Human-Computer Interaction with Mobile Devices and Services, MobileHCI 2017, Vienna, Austria, 4–7 September 2017. Association for Computing Machinery, Inc., ACM Special Interest Group on Computer-Human Interaction (SIGCHI) (2017)

83. Ghosh, S., Ganguly, N., Mitra, B., et al.: Evaluating effectiveness of smartphone typing as an indicator of user emotion. In: 7th International Conference on Affective Computing and Intelligent Interaction, ACII 2017, San Antonio, TX, USA, 23–26 October 2017, vol. 2018, pp. 146–151. Institute of Electrical and Electronics Engineers Inc. (2017)

84. Ciman, M., Wac, K., Gaggi, O.: iSensestress: assessing stress through human-smartphone interaction analysis. In: 2015 9th International Conference on Pervasive Computing Technologies for Healthcare (PervasiveHealth), pp. 84–91 (2015)

85. Wu, Y.-H., Lin, S.-J., Yang, D.-L.: A mobile emotion recognition system based on speech signals and facial images. In: 2013 17th International Computer Science and Engineering Conference, ICSEC 2013, Bangkok, Thailand, 4–6 September 2013, pp. 212–217. IEEE Computer Society (2013)

86. Pham, P., Wang, J.: Predicting learners' emotions in mobile MOOC learning via a multimodal intelligent tutor. In: Nkambou, R., Azevedo, R., Vassileva, J. (eds.) Intelligent Tutoring Systems, pp. 150–159. Springer, Cham (2018). https://doi.org/10.1007/978-3-319-91464-0_15

87. Zatarain-Cabada, R., Barrón-Estrada, M.L., Alor-Hernández, G., Reyes-García, C.A.: Emotion recognition in intelligent tutoring systems for android-based mobile devices. In: Gelbukh, A., Espinoza, F.C., Galicia-Haro, S.N. (eds.) MICAI 2014. LNCS (LNAI), vol. 8856, pp. 494–504. Springer, Cham (2014). https://doi.org/10.1007/978-3-319-13647-9_44

88. Benta, K.-I., Cremene, M., Vaida, M.-F.: A multimodal affective monitoring tool for mobile learning. In: 14th RoEduNet International Conference - Networking in Education and Research, RoEduNet NER 2015, Craiova, Romania, 24–26 September 2015, pp. 34–38. Institute of Electrical and Electronics Engineers Inc. (2015)

89. Pham, P., Wang, J.: AttentiveLearner: improving mobile MOOC learning via implicit heart rate tracking. In: Conati, C., Heffernan, N., Mitrovic, A., Verdejo, M.F. (eds.) AIED 2015. LNCS (LNAI), vol. 9112, pp. 367–376. Springer, Cham (2015). https://doi.org/10.1007/978-3-319-19773-9_37

90. Pham, P., Wang, J.: AttentiveLearner2: a multimodal approach for improving MOOC learning on mobile devices. In: André, E., et al. (eds.) Artificial Intelligence in Education, pp. 561–564. Springer, Cham (2017). https://doi.org/10.1007/978-3-319-61425-0_64

91. Rachuri, K.K., Musolesi, M., Mascolo, C., et al.: EmotionSense: a mobile phones based adaptive platform for experimental social psychology research. In: Proceedings of the 2010 ACM Conference on Ubiquitous Computing, UbiComp 2010, pp. 281–290. Association for Computing Machinery (2010)

92. Barron-Estrada, M.L., Zatarain-Cabada, R., Aispuro-Gallegos, C.G.: Multimodal recognition of emotions with application to mobile learning. In: 18th IEEE International Conference on Advanced Learning Technologies, ICALT 2018, Bombay, India, 9–13 July 2018, pp. 416–418. Institute of Electrical and Electronics Engineers Inc. (2018)

93. Wang, R., Chen, F., Chen, Z., et al.: Studentlife: assessing mental health, academic performance and behavioral trends of college students using smartphones. In: 2014 ACM International Joint Conference on Pervasive and Ubiquitous Computing, UbiComp 2014, Seattle, USA, 13–17 September 2014, pp. 3–14. Association for Computing Machinery, Inc. (2014)

94. Noroozi, F., Kaminska, D., Corneanu, C., et al.: Survey on emotional body gesture recognition. IEEE Trans. Affect. Comput. 12, 505–523 (2018)

95. Tseng, C.-H., Chen, Y.-H.: A camera-based attention level assessment tool designed for classroom usage. J. Supercomput. 74(11), 5889–5902 (2017). https://doi.org/10.1007/s11227-017-2122-7

96. Ayvaz, U., Guruler, H.: Real-time detection of students' emotional states in the classroom

97. Orencilerin Sinif ici Duygusal Durumlarinin Gercek Zamanli Tespit Edilmesi. In: 25th Signal Processing and Communications Applications Conference, SIU 2017, Antalya, Turkey, 15–18 May 2017. Institute of Electrical and Electronics Engineers Inc. (2017)

98. Gupta, S.K., Ashwin, T.S., Guddeti, R.M.R.: Students affective content analysis in smart classroom environment using deep learning techniques. Multimed. Tools Appl. 78(18), 25321–25348 (2019)

99. Whitehill, J., Serpell, Z., Lin, Y., et al.: The faces of engagement: automatic recognition of student engagementfrom facial expressions. IEEE Trans. Affect. Comput. 5(1), 86–98 (2014)

100. Lim, J.H., Teh, E.Y., Geh, M.H., et al.: Automated classroom monitoring with connected visioning system. In: 9th Asia-Pacific Signal and Information Processing Association Annual Summit and Conference, APSIPA ASC 2017, Kuala Lumpur, Malaysia, 12–15 December 2017, vol. 2018, pp. 386–393. Institute of Electrical and Electronics Engineers Inc. (2017)
101. Bosch, N., D'Mello, S.K., Baker, R.S., et al.: Detecting student emotions in computer-enabled classrooms. In: 25th International Joint Conference on Artificial Intelligence, IJCAI 2016, New York, NY, USA, 9–15 July 2016, vol. 2016, pp. 4125–4129. International Joint Conferences on Artificial Intelligence (2016)
102. Burnik, U., Zaletelj, J., Koir, A.: Kinect based system for student engagement monitoring. In: 1st IEEE Ukraine Conference on Electrical and Computer Engineering, UKRCON 2017, Kyiv, Ukraine, 29 May–2 June 2017, pp. 1229–1232. Institute of Electrical and Electronics Engineers Inc. (2017)
103. Kim, Y., Soyata, T., Behnagh, R.F.: Towards emotionally aware AI smart classroom: current issues and directions for engineering and education. IEEE Access 6, 5308–5331 (2018)
104. Nikopoulou, R., Spyrou, E., Vernikos, I., et al.: Emotion recognition from speech: a classroom experiment. In: 11th ACM International Conference on PErvasive Technologies Related to Assistive Environments, PETRA 2018, Corfu, Greece, 26–29 June 2018, pp. 104–105. Association for Computing Machinery (2018)
105. Waber, B.N., Olguin Olguin, D., Kim, T., et al.: Understanding organizational behavior with wearable sensing technology (2008). SSRN 1263992

Learner Modeling in Conversation-Based Assessment

Diego Zapata-Rivera[✉] and Carol M. Forsyth

Educational Testing Service, Princeton, NJ 08541, USA
{dzapata,cforsyth}@ets.org

Abstract. Conversation-based assessments (CBAs) make use of dialogue systems to create situations that can be used to gather evidence of the decisions learners make on performance-based assessments such as computer-based scenarios, simulations, or other forms of technology rich tasks. CBAs are adaptive in nature providing subsequent questions, feedback and eliciting additional information from learners about the given topic. However, the adaptivity can always be improved with additional evidence of the learner's knowledge, skills, and other attributes (KSAs) which can be encompassed in a learner model. Learner models based on CBA's data can then be used to support and improve adaptive sequencing of questions and conversations as well as adaptive feedback. This information can also aid in selecting appropriate conversations and additional tasks. These improvements may result in more engaging and relevant conversations and tasks. In this paper, we discuss some of the challenges and opportunities for learner modeling in dialogue systems.

Keywords: Learner modeling · Dialogue systems

1 Introduction

Conversation-based assessments (CBAs) make use of advances in dialogue systems with natural language tutorial conversations [e.g., 2, 26, 39] to create situations for learners to demonstrate their knowledge, skills, and other attributes (KSAs). CBAs have been used in combination with other types of tasks in computer-based scenarios, simulations, or other forms of technology rich environments [67]. CBAs can adapt the nature of the conversation between the human student and one or more artificial agents based on the responses provided by the learner and evidence gathered from other activities (e.g., multiple-choice questions). CBAs are designed to measure learners' knowledge, skills, and other attributes (KSAs), which can aid the system in adapting to the student's particular needs [62].

Adaptive Instructional Systems (AISs) that make use of dialogue systems can benefit from a learner model [32]. Although there is not a formal definition of a learner model, the main idea is that a learner model is an algorithmic representation of the current state of the student's KSAs, thus informing adaptivity and potential learning interventions. Specifically, learner models can be used to maintain an up-to-date representation of the

R. A. Sottilare and J. Schwarz (Eds.): HCII 2022, LNCS 13332, pp. 73–83, 2022.
https://doi.org/10.1007/978-3-031-05887-5_6

learner's KSAs [27, 44, 60]. Within CBAs, evidence can be aggregated from multiple sources and used to update the learner's profile maintained in the system (i.e., the learner's KSAs). Integrating different sources of evidence can result in better understanding of learner interactions on a fine-grained level in dialogue-based learning and assessment systems [21, 66].

Adaptive features on AISs can be described as a two-loop model that includes macro adaptation where the best possible task to match the student's need is selected and micro adaptation where different aspects of the task and associated feedback are adjusted [56]. Learner models have been successfully implemented and used to support learning [e.g., 3, 7, 10, 12, 42, 44, 49]. However, dialogue systems including natural language conversations for tutoring have not been as well investigated, though preliminary evidence from Katz et al. [32] suggests that learner models in these contexts can increase learning compared to comparable controls.

In this paper, we explore opportunities and challenges for the use of learner models in dialogue systems. We elaborate on how these models can be used to integrate evidence from various sources and guide the selection of natural language conversations for learning and assessment.

2   Evidence from Dialogue Systems

Dialogue systems that allow for natural language conversations between humans and artificial agents have made significant headway in increasing learning gains for students in a variety of domains [23]. Although many dialogic systems exist, we will focus on Autotutor, an Intelligent Tutoring System (ITS) where humans learn via conversations with an artificial agent tutor, as this is the impetus for the dialogic components in our current CBAs. Autotutor has aided students in improving learning within multiple domains by over a letter grade compared to control groups (for a review see [23]). Autotutor creates conversations based on the analysis of expert human tutor conversations with human students [25]. The resulting framework is called the Expectation-Misconception Dialogue Framework as the students are asked questions by an artificial agent with pre-defined expected correct responses as well as answers with pre-defined misconceptions. Specifically, students are guided to answer an overall question which is broken down into multiple expectations with scaffolding moves to aid the student in better understanding the given topic during the natural language tutoring session with the artificial agent. The scaffolding moves include pumps (e.g., "anything else?"), hints (broad clues), prompts (questions or statements requiring a single word or phrase as an answer), assertions (stating the correct answer) and misconception correction (where the correct answer is given). Furthermore, positive, negative and neutral feedback are provided as well as other common discourse moves to maintain the illusion of artificial intelligence [23].

The Autotutor framework has been expanded to allow for conversations between two or more artificial agents and a human, referred to as trialogues. These trialogues were first introduced in the gamified learning environment, OperationARA/OperationARIES! [39]. With these trialogues, new layers of adaptive instruction were made possible. For example, students who did not have sufficient prior-knowledge to ask a deep question would then interact with the agents in a vicarious learning mode, where they simply watched the two artificial agents interact with one yes or no question to keep the student engaged. This idea was based on the work of Driscoll, Craig, Gholson, Ventura, and Hu [17]. Additional pedagogical moves and modalities were also employed and are described in detail in [24]. However, to decipher the best pedagogical conversational mode, multiple choice questions, as additional sources of evidence, were implemented to gauge the learners' current level of knowledge on the given topic. Post-hoc analysis showed the importance of using additional sources of evidence including but not limited to responses to multiple-choice questions and evidence from conversations in predicting student learning [19]. However, no complete learner model was ever created and instantiated to track students' current KSA's via multiple sources of evidence in real time.

Therefore, an over-arching learner model that is continuously updated based on various sources of evidence could aid in creating a more comprehensive picture of the students understanding and better address the next conversation or learning module the student should progress on. Learner models can include a variety of aspects about cognitive, non-cognitive aspects of the learner such as socioemotional factors as well as information about the context of learning [1, 50, 65].

Some attempts were made early on for affect detection with Autotutor, but an integrated learning model was never achieved, and the system required many multi-modal sources [15, 16], thus making the transition of this system to real classrooms difficult. However, this work offered a great opportunity for expanding our understanding of learners. For example, researchers explored and created detectors for emotions in various contexts [6] and other non-cognitive factors that could be used to improve learner models and inform the development of adaptive features.

CBAs also make use of multiple sources of evidence including conversations with artificial agents and other types of tasks to gather evidence of learners' KSAs [62]. The next section expands on issues regarding evidence identification and aggregation in relation to the use of learner model information by dialogue systems and other AIS components to create an adaptive and personalized experience for the learner.

3 Evidence in CBAs

Evidence-Centered Design [ECD; 40, 41] is a principled methodology for assessment design which applies evidentiary arguments to represent the link between tasks and constructs. By following ECD principles, it is possible to create assessment design structures that are usually found in learner models [64].

CBAs are designed following ECD principles [62, 65]. Therefore, every discourse move provided by an agent to a student is classified into general categories and aligned to over-arching KSAs. This level of pre-planning allows for evidence to support learner model inferences which may not be possible with the larger amounts of scaffolding in the learning systems of Autotutor. Specifically, with theoretical grounding and the aligning of evidence to support learner model inferences, CBAs can provide relevant information for teachers and students (e.g., via OLMs).

Learner models can be used to support the learning process in digitalized learning and assessment environments. For example, learner models in AISs can be used to trigger specific dialogic conversations (See Fig. 1). Evidence collected through conversations and other types of tasks can be aggregated and made available as part of a *learner model*. Consider a student interacting with a CBA designed to measure science inquiry skills [59, 63]. Students can be asked to show data collection skills by placing seismometers on a virtual volcano, collect data about seismic events (presented as a data table), select data, annotate data and use these data to support hypothesis.

Evidence about students' KSAs (in this case data collection) can be gathered by *detectors from multiple sources* of evidence based on data from the students' interaction with a digitalized learning or assessment environment such as a CBA (in this case, the seismometers, the data table, and open response answers). This information informs the *evidence model* which can then provide evidence to support assessment claims about students' KSAs in learner model [40]. Also, a conversation (or a set of conversations) can be triggered specific to the student's needs as reflected in the learner model. That is, d*ialogue engines* can be programmed to trigger these conversations based on *learner model* information. In addition, this information can be shared with students and teachers as part of an *Open Learner Model* [OLM; 8, 31], which would likely be presented in the form of an interactive report or dashboard to the student or teacher. It's important to note dialogues and other types of tasks can produce evidence for more than one KSA at a time. As students interact with different learning modules different sets of detectors would be used to collect the evidence necessary to update the learner model. Learning modules could be part of a single AIS or several AISs (see Fig. 1).

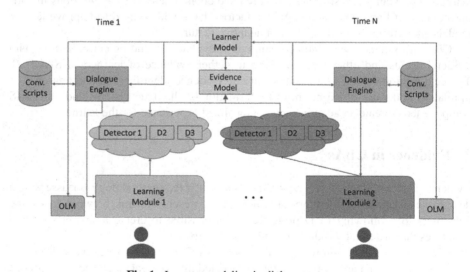

Fig. 1. Learner modeling in dialogue systems

3.1 Integrating Process and Response Data

The process and response data resulting from CBA's can come from multiple sources of evidence requiring sophisticated analysis. As previously mentioned, it's imperative that the evidence gleaned aligns to a model of learning that makes theoretical sense and has empirical backing to ensure that we accurately, reliably and validly interpret evidence provided by students. To accomplish this goal, we follow ECD principles and apply Educational Data Mining (EDM) and psychometric processes to create a theoretically-grounded data mining [19, 66].

These complex analyses are often required as the data is not merely one piece of evidence to measure KSAs for a student in these conversational simulated environments. Simulated learning and assessment environments that include conversations, whether between two or more humans or artificial agents and humans, by nature include process and response data [66]. Specifically, the process of a conversation includes not only what is said but also how and why the person states it [29]. For example, in the context of collaborative problem solving (CPS), one could make a statement that is important to keep the social "glue" or to contribute knowledge about the problem at hand [5]. Therefore, in this instance each discourse move can be tightly connected to over-arching skills aligning to social and cognitive dimensions that encompass the construct of CPS in a systematic manner. Analyses must account for this careful linking to create meaningful output from simulated environments with multiple sources of evidence in both human-to-human and human-to-agent conversations. This allows for ease of interpretation of the results while simultaneously avoiding s spurious correlations. To remain aligned with the principles of ECD, interpretation of evidence from process and response data should be theoretically grounded. This evidence in turn can be used to support inferences maintained in the learner model.

Human-to-human conversations can be analyzed in the context of CPS. Specifically, conversations resulting from multiple humans interacting with a simulated environment have been analyzed in conjunction with additional event and performance data to cluster students using a theoretically-grounded data mining approach [5, 18]. In this context, all sources of evidence were first aligned to the overarching constructs related to social and cognitive processes (aligned with ECD). With this approach, we can ensure the features extracted are evidence of the intended construct. Theoretical grounding was necessary to make adequate interpretations of the analyzed evidence. Therefore, after clustering students with a hierarchical clustering algorithm (EDM approach), we were able to find profiles of students, but only profiles that aligned to theory and external measures were accepted as valid. For example, one profile that commonly emerged was entitled *social loafers*, as these participants performed a minimum number of actions and did not contribute to solving the problem, but rather allowed their teammates to carry the burden of most of the work. This profile was accepted as there is a long history of this phenomenon in the social sciences literature [e.g., 34]. This is important as clustering algorithms do not by nature determine meaning and allow for a lot of human intervention (assigning number of clusters, tagging, etc.). Therefore, by incorporating theory, we ensure that results are interpretable. Furthermore, by correlating with external measures, we can create an argument for validity, which contributes to improving the learner model.

Similarly, in human to agent conversations, we explored the various processes of conversations with scaffolding along with other features such as answers to multiple-choice questions, timing, and pressing the hint button in *OperationARIES!*, a conversational system for teaching research methodology. Principles of cognitive psychology were employed in the design of the system. For example, the concept of discrimination or detecting the signal vs. the noise was incorporated because it has a history of aiding student learning [e.g. 4, 56] and was implemented using strategic conversations in an applied module, multiple-choice questions, and determination of flawed vs. not-flawed research cases via a final judgment made in the form of answering a binary question. Features representing this construct (as well as others) were determined based on their alignment to the construct as well as their ability to meet the base principles of analysis including magnitude, sensitivity and reliability. Features derived from this approach were then able to predict learning on a fine-grained level in an interpretable manner aligned to existing cognitive theory [18, 20].

Finally, Zapata-Rivera et al. [66] describe the process of applying psychometric and EDM processes to analyze process and response data from a CBA designed to measure science inquiry skills (the volcano example mentioned above). Pieces of evidence included responses to multiple choice and constructed response questions, conversations with artificial agents, notes on data collected provided by students, and the number and placement of seismometers used to measure volcanic activity. Process data features such as number of annotations and number of seismometers used were selected based on their relevance to the construct (science inquiry skills) following ECD principles.

The blended-approaches mentioned above take into account the combination of theory and careful linking of evidence to construct along with psychometric and data mining techniques to make inferences that could be incorporated into a sophisticated learner model representing the student's KSAs.

3.2 Interpretability of Learner Models

Supporting appropriate use of learner model information by users (e.g., student and teachers) and internal adaptive mechanisms is an important property of AISs. Understanding how learner information is used to update the learner model and trigger adaptive mechanisms (e.g., additional tasks, feedback and types of conversations) can contribute to maintaining and improving trust in these systems [45, 46, 58].

Approaches to improving learner model interpretability include the use of learner modeling representations and inference mechanisms that can be easily interpreted [11, 31, 64], approaches to support the interpretability of machine learning approaches [11, 28], and approaches for reducing bias in AI [37, 38, 57].

OLMs have shown promise in supporting students learning [35, 43] and improving motivation [e.g., 9, 30, 55], knowledge awareness [48, 51] and self-reflection [54, 60]. Dialogue systems have been used as a mechanism to interact with learner models [13, 14, 33] and support teacher understanding of score report information [22].

OLMs can support revision and negotiation of learner model information. These features become useful when keeping teachers and students in the loop as a mechanism for keeping the learner model up to date [8]. OLMs can maintain and integrate different views of the learner model and use these views to support student learning [60].

Finally, Kay, Zapata-Rivera & Conati [31] elaborate on the effort learners make to understand the system (i.e., scrutability). AISs can support users though the creation and evaluation of guidance mechanisms, interfaces, reports or dashboards that facilitate user scrutiny, control, and trust in the system. Some of the questions these tools can answer include: What does the system "believe" (e.g., a set of inferences) about me? What data and processes support these beliefs? How is the data used to make those inferences? Who has access to these data and inferences?

4 Discussion

Conversations designed following ECD principles, provide evidence of relevant KSAs. This information can be used to create a learner model that can be made available to other components of the system (e.g., through a server-based platform). Some challenges and opportunities for learner models in dialogue systems include:

- Conversations and other types of tasks in CBAs. Dialogue systems can benefit from evidence gathered via other types of tasks. This evidence can be gathered and maintained in a learner model. Work designing and evaluating CBAs has shown the potential for integrating various sources of evidence [66]. There has been work connecting conversations to other types of tasks (e.g., conversations triggered from student selection of a choice in a multiple-choice item; [36]). This approach uses the learner's selection as a starting point for a conversation that may involve the reasons why the choice is correct or wrong. In addition, when the choice is a metacognitive statement (e.g., responding "I do know know"), the conversation can be used to provide instruction to learners and prevent learner disengagement.
- Triggering conversations based on learning model information. Information in a learner model can be used to help select which conversations (or parts of a conversation) are appropriate for a learner in a particular context. This type of adaptive mechanism becomes convenient when confirming learner model information (i.e., by engaging a learner when additional evidence about learner's KSAs is required) before providing feedback or recommending additional tasks.
- Leveraging conversations in an ecosystem of assessment and learning tools. By deploying a learner model, it is possible to provide dialogue services across assessment and learning tools. This requires the creation of learning platforms that makes use of standards for the integration and sharing of learning model information [52, 53] and mechanisms to facilitate the use of learners' data from multiple sources [47].

  Approaches to OLM using conversations. Conversations with artificial agents can be used to guide exploration to learner model information. These conversations can be an effective way to support understanding and appropriate use of learner model by teachers and learners [8, 14, 22, 61].

5 Future Work

Future work in this area includes the development of mechanisms that facilitate the integration and use of conversations in AISs. These mechanisms should support the use of learner model information and reuse of conversation components across AISs. As more naturalistic conversations are developed and integrated into AISs, the need for approaches that support scalability and adoption of these systems becomes a priority.

Access to learner model information including cognitive, socioemotional aspects and information about the context of learning as well as learner model mechanisms designed to support trust and data security in AISs will contribute to achieving these goals. Results from research on OLM can inform the development of communication approaches aimed at providing learners and teachers with information about how learners' data are used by AISs to adapt their interactions (e.g., conversations with agents). Access to learner model information by learners and teachers can be a critical aspect in the adoption of these systems.

References

1. Abyaa, A., Idrissi, M.K., Bennani, S.: Learner modelling: systematic review of the literature from the last 5 years. Educ. Technol. Res. Dev. **67**, 1–39 (2019)
2. Adamson, D., Dyke, G., Jang, H.J., Rosé, C.P.: Towards an agile approach to adapting dynamic collaboration support to student needs. Int. J. AI Educ. **24**(1), 91–121 (2014)
3. Aleven, V., McLaughlin, E.A., Glenn, R.A., Koedinger, K.R.: Instruction based on adaptive learning technologies. In: Mayer, R.E., Alexander, P.A. (eds.) Handbook of Research on Learning and Instruction, 2nd edn., pp. 522–560. Routledge, New York (2016)
4. Anderson, J.R., Corbett, A.T., Koedinger, K.R., Pelletier, R.: Cognitive tutors: lessons learned. J. Learn. Sci. **4**, 167–207 (1995)
5. Andrews-Todd, J., Forsyth, C., Steinberg, J., Rupp, A.A.: Identifying profiles of collaborative problem solvers in an online electronics environment. In: Boyer, K.E., Yudelson, M. (eds.) Proceedings of the 11th International Conference on Educational Data Mining, pp. 239–245. International Educational Data Mining Society, Buffalo (2018)
6. Baker, R.S., D'Mello, S.K., Rodrigo, M.M.T., Graesser, A.C.: Better to be frustrated than bored: the incidence, persistence, and impact of learners' cognitive–affective states during interactions with three different computer-based learning environments. Int. J. Hum. Comput. Stud. **68**(4), 223–241 (2010)
7. Blessing, S.B., Gilbert, S.B., Ourada, S., Ritter, S.: Authoring model-tracing cognitive tutors. Int. J. Artif. Intell. Educ. **19**(2), 189–210 (2009)
8. Bull, S.: There are open learner models about! IEEE Trans. Learn. Technol. **13**(2), 425–448 (2020)
9. Bull, S., Brna, P., Pain, H.: Extending the scope of the student model. User Model User-Adap. Inter. **5**, 45–65 (1995)
10. Conati, C., Kardan, S.: Student modeling: supporting personalized instruction, from problem solving to exploratory open-ended activities. AI Mag. **34**(3), 13–26 (2013)
11. Conati, C., Porayska-Pomsta, K., Mavrikis, M.: AI in education needs interpretable machine learning: lessons from Open Learner Modelling. arXiv (2018). http://arxiv.org/abs/1807.00154
12. Desmarais, M.C., d Baker, R.S.J.: A review of recent advances in learner and skill modeling in intelligent learning environments. User Model. User-Adap. Inter. **22**(1–2), 9–38 (2012). https://doi.org/10.1007/s11257-011-9106-8
13. Dimitrova, V.: STyLE-OLM: interactive open learner modelling. Int. J. Artif. Intell. Educ. **13**(1), 35–78 (2003)
14. Dimitrova, V., Brna, P.: From interactive open learner modelling to intelligent mentoring: STyLE-OLM and beyond. Int. J. Artif. Intell. Educ. **26**(1), 332–349 (2015). https://doi.org/10.1007/s40593-015-0087-3

15. D'Mello, S.K., Graesser, A.: Multimodal semi-automated affect detection from conversational cues, gross body language, and facial features. User Model. User-Adap. Inter. **20**(2), 147–187 (2010). https://doi.org/10.1007/s11257-010-9074-4

16. D'Mello, S., Graesser, A.: Dynamics of affective states during complex learning. Learn. Instr. **22**(2), 145–157 (2012)

17. Driscoll, D.M., Craig, S.D., Gholson, B., Ventura, M., Hu, X.: Vicarious learning. Effects of overhearing dialog and monologue-like virtual tutoring sessions. J. Exp. Psychol.: Hum. Learn. Mem. **6**, 588–598 (2003)

18. Forsyth, C.M., Andrews-Todd, J., Steinberg, J.: Are you really a team player?: profiles of collaborative problem solvers in an online environment. In: Rafferty, A.N., Whitehill, J., Cavalli-Sforza, V., Romero, C. (eds.) Proceedings of the 13th International Conference on Educational Data Mining (EDM 2020), pp. 403–408 (2020)

19. Forsyth, C.M., Graesser, A.C., Millis, K.: Predicting learning in a multi-component serious game. Technol. Knowl. Learn. **25**, 251–277 (2020)

20. Forsyth, C.M., Graesser, A.C., Pavlik, P., Millis, K., Samei, B.: Discovering theoretically grounded predictors of shallow vs. deep- level learning. In: Stamper, J., Pardos, Z., Mavrikis, M., McLaren, B.M. (eds.) Proceedings of the 7th International Conference on Educational Data Mining (EDM 2014), pp. 229–232 (2014)

21. Forsyth, C.M., Peters, S., Moon, J., Napolitano, D.: Assessing scientific inquiry based on multiple sources of evidence. Presented at the Annual Meeting of the American Educational Research Association, Toronto, Canada, April 2019

22. Forsyth, C.M., Peters, S., Zapata-Rivera, D., Lentini, J., Graesser, A., Cai, Z.: Interactive score reporting: an AutoTutor-based system for teachers. In: André, E., Baker, R., Hu, X., Rodrigo, M.M.T., du Boulay, B. (eds.) AIED 2017. LNCS (LNAI), vol. 10331, pp. 506–509. Springer, Cham (2017). https://doi.org/10.1007/978-3-319-61425-0_51

23. Graesser, A.C.: Conversations with AutoTutor help students learn. Int. J. Artif. Intell. Educ. **26**, 124–132 (2016)

24. Graesser, A.C., Forsyth, C., Lehman, B.: Two heads are better than one: learning from agents in conversational trialogues. Teach. Coll. Rec. **119**, 1–20 (2017)

25. Graesser, A.C., Person, N.K.: Question asking during tutoring. Am. Educ. Res. J. **31**, 104–137 (1994)

26. Graesser, A.C., Person, N.K., Harter, D.: The tutoring research group: teaching tactics and dialogue in AutoTutor. Int. J. Artif. Intell. Educ. **12**, 257–279 (2001)

27. Greer, J., McCalla, G. (eds.): Student Models: The Key to Individualized Educational Systems. Springer, New York (1994)

28. Gunning, D.: Explainable Artificial Intelligence (XAI). Defense Advanced Research Projects Agency (DARPA) (2017)

29. Hao, J., Zapata-Rivera, D., Graesser, A.C., Cai, Z., Hu, X., Goldberg, B.: Towards an intelligent tutor for teamwork: responding to human sentiments. In: Sottilare, R., Graesser, A., Hu, X., Sinatra, A.M. (eds.) Design Recommendations for Intelligent Tutoring Systems: Volume 6 - Team Tutoring, pp. 151–160. Army Research Laboratory, Orlando (2018). ISBN 978-0-9977257-4-2

30. Hsiao, I.H., Brusilovsky, P.: Guiding and motivating students through open social student modeling: lessons learned. Teach. Coll. Rec. **119**(3), 1–42 (2017)

31. Kay, J., Zapata-Rivera, D., Conati, C.: The GIFT of scrutable learner models: why and how. In: Sinatra, R.A.M., Graesser, A.C., Hu, X., Goldberg, B., Hampton, A.J. (eds.) Design Recommendations for Intelligent Tutoring Systems: Volume 8 – Data Visualization, pp. 25–40. U.S. Army CCDC - Soldier Center, Orlando (2020)

32. Katz, S., Albacete, P., Chounta, I.-A., Jordan, P., McLaren, B.M., Zapata-Rivera, D.: Linking dialogue with student modelling to create an adaptive tutoring system for conceptual physics. Int. J. Artif. Intell. Educ. **31**(3), 397–445 (2021). https://doi.org/10.1007/s40593-020-00226-y

33. Kerly, A., Ellis, R., Bull, S.: CALMsystem: a conversational agent for learner modelling. In: Ellis, R., Allen, T., Petridis, M. (eds.) Applications and Innovations in Intelligent Systems XV, pp. 89–102. Springer, London (2008). https://doi.org/10.1007/978-1-84800-086-5_7
34. Latané, B., Williams, K., Harkins, S.: Many hands make light the work: the causes and consequences of social loafing. J. Pers. Soc. Psychol. **37**(6), 822–832 (1979)
35. Long, Y., Aleven, V.: Enhancing learning outcomes through self-regulated learning support with an open learner model. User Model. User-Adap. Inter. **27**(1), 55–88 (2017)
36. Lopez, A.A., Guzman-Orth, D., Zapata-Rivera, D., Forsyth, C.M., Luce, C.: Examining the accuracy of a conversation-based assessment in interpreting English learners' written responses (Research Report No. RR-21-03). Educational Testing Service (2021).https://doi.org/10.1002/ets2.12315
37. Loukina, A., Madnani, N., Zechner, K.: The many dimensions of algorithmic fairness in educational applications. In: Proceedings of the Workshop on Innovative Use of NLP for Building Educational Applications, Florence, Italy, pp. 1–10 (2019)
38. Mehrabi, N., Morstatter, F., Saxena, N., Lerman, K., Galstyan, A.: A survey on bias and fairness in machine learning. CoRR, abs/1908.09635 (2019). http://arxiv.org/abs/1908.09635
39. Millis, K., Forsyth, C., Butler, H., Wallace, P., Graesser, A., Halpern, D.: Operation ARIES!: a serious game for teaching scientific inquiry. In: Ma, M., Oikonomou, A., Jain, L.C. (eds.) Serious Games and Edutainment Applications, pp. 169–195. Springer, London (2011). https://doi.org/10.1007/978-1-4471-2161-9_10
40. Mislevy, R.J., Almond, R.G., Lukas, J.F.: A brief introduction to evidence-centered design. ETS Res. Rep. Ser. **2003**(1), i-29 (2003)
41. Mislevy, R.J., Riconscente, M.M.: Evidence-centered assessment design. In: Handbook of Test Development, pp. 75–104. Routledge (2011)
42. Mitrovic, A.: Fifteen years of constraint-based tutors: what we have achieved and where we are going. User Model. User-Adap. Inter. **22**(1–2), 39–72 (2012). https://doi.org/10.1007/s11257-011-9105-9
43. Mitrovic, A., Martin, B., Suraweera, P.: Intelligent tutors for all: constraint-based modeling methodology, systems and authoring. IEEE Intell. Syst. **22**, 38–45 (2007)
44. Pavlik, P.I., Brawner, K., Olney, A., Mitrovic, A.: A review of student models used in intelligent tutoring systems. In: Sollitare, R.Z., Graesser, A.C., Hu, X., Holden, H. (eds.) Design Recommendations for Intelligent Tutoring Systems: Volume 1 - Learner Modeling, pp. 39–68. U.S. Army Research, Orlando (2013)
45. Rudin, C.: Stop explaining black box machine learning models for high stakes decisions and use interpretable models instead. Nat. Mach. Intell. **1**, 206–215 (2019)
46. Rosé, C.P., McLaughlin, E.A., Liu, R., Koedinger, K.R.: Explanatory learner models: why machine learning (alone) is not the answer. Br. J. Edu. Technol. **50**(6), 2943–2958 (2019)
47. Schaldenbrand, P., et al.: Computer-supported human mentoring for personalized and equitable math learning. In: Roll, I., McNamara, D., Sosnovsky, S., Luckin, R., Dimitrova, V. (eds.) AIED 2021. LNCS (LNAI), vol. 12749, pp. 308–313. Springer, Cham (2021). https://doi.org/10.1007/978-3-030-78270-2_55
48. Shahrour, G., Bull, S.: Interaction preferences and learning in an inspectable learner model for language. In: Artificial Intelligence in Education, pp. 659–661. IOS Press (2009)
49. Shute, V.J.: SMART: student modeling approach for responsive tutoring. User Model. User-Adap. Inter. **5**, 1–44 (1995). https://doi.org/10.1007/BF01101800
50. Shute, V.J., Zapata-Rivera, D.: Adaptive educational systems. In: Durlach, P. (ed.) Adaptive Technologies for Training and Education, pp. 7–27. Cambridge University Press, New York (2012)
51. Somyürek, S., Brusilovsky, P., Guerra, J.: Supporting knowledge monitoring ability: open learner modeling vs. open social learner modeling. Res. Pract. Technol. Enhanc. Learn. **15**(1), 1–24 (2020). https://doi.org/10.1186/s41039-020-00137-5

52. Sottilare, R.A., Brawner, K.W., Sinatra, A.M., Johnston, J.H.: An Updated Concept for a Generalized Intelligent Framework for Tutoring (GIFT). US Army Research Laboratory, Orlando (2017)

53. Sottilare, R., Barr, A., Robson, R., Hu, X., Graesser, A.: Exploring the opportunities and benefits of standards for adaptive instructional systems (AISs). In: Proceedings of the Adaptive Instructional Systems Workshop in the Industry Track of the 14th International Intelligent Tutoring Systems, pp. 49–53 (2018)

54. Tang, L.M., Kay, J.: Scaffolding for an OLM for long-term physical activity goals. In: Proceedings of the 26th Conference on User Modeling, Adaptation and Personalization, pp. 147–156 (2018)

55. Thomson, D., Mitrovic, A.: Preliminary evaluation of a negotiable student model in a constraint-based ITS. Res. Pract. Technol. Enhanc. Learn. **5**(01), 19–33 (2010)

56. VanLehn, K., Graesser, A.C., Jackson, G.T., Jordan, P., Olney, A., Rose, C.P.: When are tutorial dialogues more effective than reading? Cogn. Sci. **3**, 3–62 (2007)

57. Vincent-Lancrin, S., van der Vlies, R.: Trustworthy artificial intelligence (AI) in education: promises and challenges. OECD Education Working Papers, No. 218, OECD Publishing, Paris (2020). https://doi.org/10.1787/a6c90fa9-en

58. Zapata-Rivera, D.: Open student modeling research and its connections to educational assessment. Int. J. Artif. Intell. Educ. **31**(3), 380–396 (2020). https://doi.org/10.1007/s40593-020-00206-2

59. Zapata-Rivera, D., Brawner, K., Jackson, G.T., Katz, I.R.: Reusing evidence in assessment and intelligent tutors. In: Sottilare, R., Graesser, A., Hu, X., Goodwin, G. (eds.) Design Recommendations for Intelligent Tutoring Systems: Volume 5 - Assessment Methods, pp. 125–136. U.S. Army Research Laboratory, Orlando (2017). ISBN 978-0-9893923-9-6

60. Zapata-Rivera, J.D., Greer, J.: Interacting with Bayesian student models. Int. J. Artif. Intell. Educ. **14**(2), 127–163 (2004)

61. Zapata-Rivera, D., Greer, J.E.: Exploring various guidance mechanisms to support interaction with inspectable learner models. In: Cerri, S.A., Gouardères, G., Paraguaçu, F. (eds.) ITS 2002. LNCS, vol. 2363, pp. 442–452. Springer, Heidelberg (2002). https://doi.org/10.1007/3-540-47987-2_47

62. Zapata-Rivera, D., Jackson, T., Katz, I.R.: Authoring conversation-based assessment scenarios. In: Sottilare, R.A., Graesser, A.C., Hu, X., Brawner, K. (eds.) Design Recommendations for Intelligent Tutoring Systems Volume 3: Authoring Tools and Expert Modeling Techniques, pp. 169–178. U.S. Army Research Laboratory (2015)

63. Zapata-Rivera, D., Jackson, T., Liu, L., Bertling, M., Vezzu, M., Katz, I.: Assessing science inquiry skills using trialogues. In: Trausan-Matu, S., Boyer, K.E., Crosby, M., Panourgia, K. (eds.) ITS 2014. LNCS, vol. 8474, pp. 625–626. Springer, Cham (2014). https://doi.org/10.1007/978-3-319-07221-0_84

64. Zapata-Rivera, D., Hansen, E., Shute, V.J., Underwood, J.S., Bauer, M.: Evidence-based approach to interacting with open student models. Int. J. Artif. Intell. Educ. **17**(3), 273–303 (2007)

65. Zapata-Rivera, D., Lehman, B., Sparks, J.R.: Learner modeling in the context of caring assessments. In: Sottilare, R.A., Schwarz, J. (eds.) HCII 2020. LNCS, vol. 12214, pp. 422–431. Springer, Cham (2020). https://doi.org/10.1007/978-3-030-50788-6_31

66. Zapata-Rivera, D., Liu, L., Chen, L., Hao, J., von Davier, A.A.: Assessing science inquiry skills in an immersive, conversation-based scenario. In: Kei Daniel, B. (ed.) Big Data and Learning Analytics in Higher Education, pp. 237–252. Springer, Cham (2017). https://doi.org/10.1007/978-3-319-06520-5_14

67. Zapata-Rivera, D., Liu, L., Katz, I.R., Vezzu, M.: Exploring the use of game elements in the development of innovative assessment tasks for science. Cogn. Technol. **18**(1), 43–50 (2013)

Adaptation Design to Individual Learners and Teams

Adaptively Adding Cards to a Flashcard Deck Improves Learning Compared to Adaptively Dropping Cards Regardless of Cognitive Ability

Lisa Durrance Blalock(✉) [ID]

University of West Florida, Pensacola, FL 32514, USA
lblalock@uwf.edu

Abstract. Flashcards are a popular study tool, however learner decisions can lower their effectiveness. One such decision is whether or not to drop a concept from study. Using objective mastery criteria that adaptively determine when to add or drop an item from study based on performance may improve learning outcomes in flashcard-based tasks. The effectiveness of adaptive flashcard-based learning may also vary based on the cognitive ability of the learner. The current study examined the impact of adaptive mastery instructional strategies on learning butterfly species and whether or not the impact of adaptive mastery varies by cognitive ability. Three learning conditions were compared: a No Add/Drop group (all items remain in the deck throughout study), a Mastery Drop group (start with all items, then drop after an item is mastered), and a Mastery Add group (start with three items, add items once mastered). A pre-post-transfer test design was used both immediately after training and one week later. Participants also completed the symmetry span task and a change detection task to evaluate cognitive ability. Results show the worst overall immediate pre-post learning gains in the Mastery Drop condition compared to the Mastery Add and No Add/Drop conditions which showed similar learning gains. This pattern went away when looking at delayed pre-post learning gains. Cognitive ability did not have any impact on learning performance, suggesting that similar strategies work equally well across all levels of cognitive ability. These results suggest adaptively adding cards is better than dropping them, though if there are no time constraints, leaving all concepts in the deck leads to the best overall learning in the short term.

Keywords: Adaptive training · Mastery · Flashcards · Cognitive ability · Working memory capacity · Visual working memory

1 Introduction

1.1 Overview

Learners frequently use flashcards for study [1–4] and they can help promote long-term retention of information when used effectively [1–3, 5]. However, many students do not use flashcards effectively due to poor metacognitive awareness and a lack of explicit instruction on how to best use them to promote long-term learning [3, 6–8]. For example,

R. A. Sottilare and J. Schwarz (Eds.): HCII 2022, LNCS 13332, pp. 87–103, 2022.
https://doi.org/10.1007/978-3-031-05887-5_7

students often drop flashcards from study too soon, sometimes after only one correct recall [5, 7, 9, 10]. Using objective mastery criteria that adaptively determine when to add or drop an item from study based on learner performance may help improve learning outcomes in flashcard-based tasks [5]. Additionally, while adaptive mastery is likely to benefit all learners, it may have the greatest benefits for novices or those with lower cognitive ability by tailoring learning to their level. The current study examines the impact of different adaptive mastery instructional strategies on learning and whether or not the effectiveness of adaptive mastery varies by cognitive ability.

1.2 Flashcard-Based Learning

Flashcard-based learning engages several effective study strategies such as spacing practice over time [11] and self-testing or retrieval practice [12]. The spacing effect refers to improved long-term retention when studying is spaced out over time compared to massed practice [11, 13, 14]. When used effectively, flashcards can promote spaced learning when learners study a deck multiple times across a learning interval (e.g., course, semester, or learning unit). Flashcards can also promote retrieval practice when learners use flashcards to test their recall of concepts. Retrieval practice, or self-testing, improves learning compared to restudying, a well-documented phenomenon known as the testing effect [12, 15–17]. Testing is particularly beneficial for long-term retention [12].

The built-in ability of flashcards to incorporate spacing and testing makes them an effective and popular study strategy for learners, particularly for high-performing learners [2, 3]. For example, Hartwig and Dunlosky [2] surveyed college students on their learning habits and their GPA. They found high achieving students engaged in self-testing and restudying significantly more often than lower-achieving students. However, recent research suggests the effectiveness of flashcards can vary depending on several factors.

For example, the size of the deck impacts the effectiveness of learning from flashcards. The number of items in a flashcard deck impacts the overall spacing of concepts during study, meaning decisions on how many items to study at once have a direct impact on spacing within a study session. A larger deck of cards inherently means more items will be studied between each view of a concept, thereby increasing spacing, while studying several smaller decks of cards will reduce spacing between items. Kornell [19] examined this question using GRE word pairs. Across three experiments, participants studied word pairs in either one large stack (within-spacing) or over several smaller stacks (massed-practice) either within one study session or over several days of study. Within-spacing led to significantly better recall one day later across all studies, with 90% of participants benefiting from the larger study decks [19]. However, this benefit was in contrast to learner perceptions: 72% believed they learned more with the smaller, massed decks. Kornell noted that the negative impact of smaller deck sizes on learning is likely greater in real-world learning environments where students may drop flashcards from study when they feel those concepts have been mastered, reducing the deck size even further.

Indeed, students report dropping cards from study to make better use of study time [20, 21]. However, students often overestimate how much they have learned [22] and thus make suboptimal decisions on when to drop an item from study. Several studies have shown participants will drop a card from study after only one correct recall [9, 10, 20] which is not enough to build long-term knowledge [23–25]. Thus, learners fail to recognize the benefits of repeated and spaced practice [20, 21].

An alternative approach to one large deck versus several small decks or dropping cards from study would be to increase the deck size gradually as learners master concepts. Kornell [19] suggests this strategy may be an effective way to increase spacing of concepts while avoiding overloading concepts at the start of study. However, no research has directly examined this question, as most work focuses on when to drop items from study rather than when to add items. Increasing deck size may be particularly beneficial for struggling learners, by reducing cognitive load at early learning stages when the intrinsic load is highest [26–28]. The current paper examines if adaptively increasing the deck size is as effective as starting out with a large deck and determines how effective this approach is across different learners.

1.3 Adaptive Mastery-Based Approaches to Flashcard-Based Learning

The research is clear that learners make many sub-optimal decisions when using flashcards [9, 10, 20–22]. How can we improve flashcards-based learning to overcome these pitfalls? One approach is to provide explicit instructions on how to use flashcards. However, attempts to provide instruction on effective strategies have found it very difficult to overcome these metacognitive illusions, even when learners are shown their own results that contradict their assumptions [29]. An alternative to explicit instruction is implementing adaptive instructional strategies which could help overcome these metacognitive pitfalls and improve the effectiveness of flashcards as study tools [5]. Adaptive systems tailor learning based on learner performance and can improve learning outcomes and efficiency [30, 31]. However, these adaptive systems often focus on training complex skills such as periscope operation [31] with relatively less focus on incorporating adaptive interventions in simpler learning paradigms. Currently, only a handful of studies have examined adaptive instructional strategies in flashcard-based learning, though the results thus far have been promising.

One adaptive technique that has been explored in this literature is using objective mastery criteria to drop flashcards as a way to address students' limitations with metacognitive monitoring. For example, an item would be dropped after a learner correctly answered three out of the last four trials within a certain time frame as the adaptive mastery criteria [32, 33]. While using objective mastery criteria has significant advantages over learner-controlled dropping [5], the literature on the benefits of adaptive mastery-based learning is somewhat mixed.

Whitmer and colleagues [5] directly compared the pre- to post-test learning gains of three groups that varied based on the type of dropping while learning to identify armored vehicles with flashcards while keeping the sequencing random for all conditions. The groups included a Mastery Drop condition (i.e., flashcards were dropped based on objective mastery criteria using accuracy and reaction time), a No Mastery condition (i.e., flashcards were not dropped during learning), and a learner control condition (i.e., learners decided when to drop a card). Participants' post-test performance was assessed 30 min after learning and again 48 h later. Whitmer and colleagues found that the Mastery Drop group showed less forgetting after a 48-h delay, whereas the No Mastery group had a significant rate of forgetting across the two days. However, there was no significant difference between the Mastery Drop and No Mastery conditions at delayed recall (i.e., 48-h retention interval) or in the number of training trials completed, time spent training, or number of mastered items. Additionally, the learner-controlled condition had the worst performance across all testing. Thus, there may be some benefits to delayed pre-post gain scores when adaptively dropping items from study, but many questions remain unanswered as to the most effective use of adaptive mastery.

The theoretical context for adaptively varying the flashcard deck is primarily based on Cognitive Load Theory (CLT), which argues cognitive load during learning is based on three primary load sources: intrinsic, germane, and extraneous load [27, 28]. Intrinsic load is the inherent difficulty of the task at hand, germane load is the cognitive processing involved in a task, and extraneous load is any irrelevant or environmental load that distracts from the learning task. The goal of instruction is to maximize germane load by eliminating extraneous load and reducing intrinsic load as much as possible. For novice learners, intrinsic load is very high in the early stages of learning. By reducing the number of flashcard concepts at the start of learning, the intrinsic load is reduced which allows for more germane learning. As they learn, concepts can then be added, making the task more challenging without overloading the intrinsic load of the task. Thus, novice learners start with fewer concepts to learn at the start, then add more as they gain proficiency.

The concept of varying a learning task to reduce intrinsic load and support germane processing is also supported by the large literature on desirable difficulties. Desirable difficulties are instructional strategies that make learning more challenging during study but ultimately lead to better retrieval [6, 34, 35]. This includes varying study conditions, spacing study over time, testing instead of restudying material, and interleaving concepts during study [6, 34]. The goal is to increase the difficulty during study enough to promote learning and engagement without hindering comprehension. However, as noted above, learners often make suboptimal study choices [6, 9, 20–22]. Implementing these desirable difficulties in an adaptive way using objective mastery criteria may be the best way to overcome these metacognitive illusions and promote germane learning.

However, CLT and the desirable difficulties literature do make somewhat different predictions on how to best adapt the flashcard deck. Both would advocate for increasing difficulty with learning, however, the desirable difficulties literature would suggest dropping well-learned items from study would benefit learning by focusing on the more difficult items for that learner while CLT would argue adding cards is best. Thus, while both CLT and the desirable difficulties literature support adaptive mastery in principle, the ideal implementation of adaptive mastery is not clear.

Thus, while there is a theoretical basis for adaptive mastery to improve learning, the current research on the topic has left many unanswered questions. Given that adaptive mastery has the potential to dramatically impact spacing of items during learning by changing the size of the deck, by adding or removing items from the deck, it remains an important area for future research. However, adaptive mastery may not be the best choice for all learners and/or it may provide greater benefits for lower aptitude learners. Thus, in order to fully understand the impact of adaptive mastery on learning, we must also evaluate individual differences in cognitive ability.

1.4   Cognitive Ability and Flashcard-Based Learning

The effectiveness of learning interventions may vary by learner characteristics. Most of the research on individual differences in flashcard learning has focused on working memory capacity (WMC). WMC is the ability to maintain and manipulate information in one's memory during a short period of time [36–40]. WMC has a limited capacity and this capacity has shown to be an important individual difference in completing complex cognitive tasks [41–43].

Research on WMC and flashcard-based learning has been somewhat mixed. Agarwal and colleagues [44] looked at the role of WMC in learning general-knowledge facts in a flashcard-style task. Learners low in WMC showed a greater benefit of retrieval practice in a delayed retention task compared to those high in WMC. Their data suggests the effectiveness of various instructional strategies varies depending on critical cognitive ability metrics like WMC.

Wang and colleagues [45], however, examined how WMC may moderate the impact of interleaving flashcards (i.e., studying all paintings by one artist at a time or mixing up paintings by multiple artists) when learning painting styles. Participants completed the task under single and dual-task conditions to determine if the interleaving effect relies on WMC. Interleaving items led to significantly better classification of novel paintings compared to blocking items under both single and dual-task conditions, suggesting the interleaving benefit does not rely heavily on WM resources. While high WMC individuals performed better overall, capacity did not interact with the interleaving condition, suggesting interleaving was equally beneficial for high and low WMC learners. Therefore, it may be the case that additional instructional strategies other than spacing during retrieval practice with flashcards help low WMC individuals reach the same performance levels as high WMC individuals.

One individual difference that has not been examined in the flashcard learning literature is visual working memory capacity (VWM-C). Similar to WMC, visual working memory is a capacity-limited system dedicated to the storage and manipulation of visual information [46, 47]. VWM-C is typically measured using a color change detection task, in which participants determine if a colored square changes from study to test. Performance on this task is then used to calculate the capacity metric k, which represents the number of items currently available in visual working memory [46, 48].

Visual working memory plays a critical role in everyday visual processing, including guiding saccadic eye movements [49, 50]. Additionally, there is evidence that VWM-C is uniquely predictive of higher-order cognitive processes like fluid intelligence [40, 51, 52]. Given the importance of visual processing in learning visual items like the butterfly species used in the current study, it is possible that VWM-C plays an important role that is distinct from the contributions of WMC. The current study examines the impact of both WMC and V-WMC on learning in a flashcard task.

1.5 Present Study

The current study examines the impact of different adaptive mastery instructional strategies on learning and whether or not the impact of adaptive mastery varies by cognitive ability. To evaluate the impact of adaptive mastery, I compared three mastery conditions: a Mastery Add condition in which flashcards were added to study as concepts were mastered; a Mastery Drop condition in which flashcards were dropped from study as concepts were mastered; and a No Add/Drop control condition in which there were no changes to the flashcard deck. Participants also completed two measures of cognitive ability, a WMC span task (symmetry span) and a change detection task to evaluate visual working memory capacity (V-WMC). Based on CLT and prior research on mastery-based learning, I expected to see the best short- and long-term learning in the Mastery Add condition, particularly for individuals with low cognitive ability.

2 Method

2.1 Participants and Design

A total of 122 undergraduate psychology students (median age = 19; 27 males, 84 females, 2 prefer not to answer; 9 participants with missing data due to computer error) participated in the study for partial course credit. Of the original 122 participants, 12 were dropped from further analysis due to chance or below chance performance on the training task, 9 were not included due to missing session 1 data as a result of a computer error, and 2 were dropped for completing significantly more training trials during learning (2 SD above the mean), resulting in 99 participants for the current analysis. Participants were randomly assigned to one of three training conditions: Mastery Add ($n = 32$), Mastery Drop ($n = 36$), and No Add/Drop ($n = 31$).

2.2 Materials

Stimuli. Participants learned to identify six butterfly species: Argus, Cabbage White, American Copper, African Grass Blue, Pipevine, and Spicebush (see Fig. 1). The stimuli were real-life images of butterflies, with eight images of each species. The images showed the butterflies in real-world environments from different angles and in different states (e.g., wings open versus wings closed). A ninth image of each butterfly was reserved for use during the transfer test. Additionally, 48 images were used as distractor images during training. These images were matched to each species such that each trained species had a set of eight images that were used as distractors. The distractors were selected based on similarity in color and shape so to the trained species.

Fig. 1. Example images for each trained butterfly species. (Color figure online)

Pre, Post, and Transfer Tests. The same format was used for the pre-, post-, and transfer tests. For each test, participants saw an image of a butterfly along with a list of species. They were instructed to select the correct species for each image then click submit to move to the next image. There were six questions for each test. Images for the pre- and post-tests came from the set of trained images and the same images were used across all participants. Images for the transfer test were novel images of each species and the same images were used for all participants. Participants did not receive any feedback on any tests. The same tests were used for both immediate and delayed recall and transfer.

Training Task. During training, participants saw the name of a butterfly species along with four images of butterflies. They were instructed to select the image that matched the species name (see Fig. 2 Panel A). Participants received feedback after making their choice. Correct answers were outlined in green while incorrect answers were outlined in red. If a participant chose the incorrect image, they would see both the incorrect response in red and the correct response in green (see Fig. 2 Panel B). The total number of trials varied based on participant performance and condition.

Fig. 2. Example training trial. Panel A shows the start of each training trial. Panel B shows the feedback provided following an incorrect response. The butterfly highlighted in green was the correct answer. (Color figure online)

WMC Span Task. The automated symmetry span task was used to evaluate WMC [53, 54]. During the task, participants first determined if an image was vertically symmetrical then they saw a 4×4 grid with one square shaded in red shown for 850 ms. Depending on the set size of the trial, they would view 2–5 images and locations before recalling the location of the squares in the order they were presented. Participants used the mouse to click on the locations on a blank 4×4 grid. When a location was selected, it would be shaded red with a number in the center of the selected location to indicate the recalled order. If participants could not remember a location in the sequence they were instructed to click on a 'BLANK' button. Participants completed 12 total trials, three trials at each set size. The trials were automated, such that if a participant did not respond to the symmetry judgment within 5 s, the trial automatically advanced to the location and the trial would be counted as incorrect. Symmetry span scores were calculated using the all-or-nothing method [37, 53].

VWM Capacity Task. A color change detection task was used to evaluate VWM capacity [55]. For each trial, participants saw four unique colored squares on a grey background for 100 ms followed by a 900 ms retention interval. Then, a second array of colored squares was shown with one square highlighted by a magenta box. Participants had to respond with the S key if the highlighted square was the same color as before or the D key if it was a different color. The colors used were red, black, green, blue, violet, yellow, and white. Participants completed 60 change detection trials. V-WMC was calculated using the single-probe formula for k [46, 48].

2.3 Procedure

Due to COVID-19 related restrictions, 67 of the participants completed the experiment online using Inquisit 6 Web [56]. The remaining 32 participants completed the study in a lab setting in groups of two using Inquisit 6 Lab [56]. The procedure was the same for both groups with the exception of the informed consent process. Online participants provided online informed consent by clicking agree to continue to the study. In-person participants signed a paper consent form. The study took place over two sessions spaced 7 days apart.

Session 1 Procedure. In session one participants completed the informed consent followed by the pre-test. They were instructed to select the butterfly species that matched the picture then click the submit button to move to the next trial. After all six pre-test questions, participants read instructions for the training task. They were instructed to select the butterfly image that matched the species name at the top of the screen. Depending on what condition they were in, they were also told that butterflies would be added or dropped after they correctly identified an item four times in a row.

During training, in the No Add/Drop condition, all items remained in the flashcard deck throughout training for 30 min. In the Mastery Add condition, two species were selected at random at the start of training. Once one species was answered correctly four times in a row, a new species was added to study (selected at random). This continued until all six species were mastered or they reached 30 min of training at which point training ended. In the Mastery Drop condition, participants started with all six species. Once a species was answered correctly four times in a row, that species was dropped from study. This continued until all species were mastered or they reached 30 min of training. Feedback was given in all conditions.

Following training, participants completed a short anagram task where they had 10 s to unscramble a word (30 words shown in total; [57]). This task took approximately

10 min. After the anagrams task, participants completed the post-test followed by the transfer test. Finally, they provided their age and gender and were given a reminder to return for the second session. The total time for the first session was about 45–50 min.

Session 2 Procedure. Seven days after session 1, participants returned for session 2. In session 2, participants first completed the delayed post-test and delayed transfer test. Next, they completed the symmetry span task and change detection tasks. The symmetry span task and the change detection tasks took roughly 10–15 min each to complete. Finally, participants were provided with the debriefing statement. The total time for the second session was about 20–30 min.

3 Results

To evaluate learning I analyzed both immediate and delayed pre-post learning gains, transfer test scores, and efficiency scores (see Table 1 for descriptive statistics). Pre-post learning gain scores were calculated by subtracting the post-test score from the pre-test score, then dividing by 1-pre-test score. Learning efficiency was calculated by dividing post-test scores by the total number of trials completed. To examine the impact of cognitive ability on learning performance, a median split was used to separate participants into high and low groups for both WMC (symmetry span score; median = 22.5, $M = 22.5$, $SD = 8.32$) and V-WMC (single-probe k; median = 2.77, $M = 2.31$, $SD = 1.16$).

Table 1. Means and standard deviations (in parentheses) for immediate and delayed gain scores, transfer test scores, and efficiency scores.

	Mastery Add	Mastery Drop	No Add/Drop
Immediate pre-post gain scores	71% (22%)	51% (35%)	81% (23%)
Delayed pre-post gain scores	42% (38%)	35% (51%)	49% (50%)
Immediate transfer test scores	63% (22%)	63% (22%)	64% (18%)
Delayed transfer test scores	63% (17%)	61% (19%)	65% (28%)
Immediate efficiency scores	0.43 (.21)	0.63 (.46)	0.43 (.29)
Delayed efficiency scores	0.29 (.30)	0.51 (.59)	0.24 (.27)

3.1 Online Versus In-Person Session Comparison

To determine if completing the experiment in person or online had a significant impact on performance, independent samples t-tests were conducted on each dependent variable listed above. There were no significant differences between the online and in-person groups on immediate or delayed pre-post gain scores [immediate: $t(97) = .72$, $p = .48$,

$d = .15$; delayed: $t(68) = 1.65, p = .10, d = .41$], immediate or delayed transfer test scores [immediate: $t(97) = -.31, p = .76, d = -.07$; delayed: $t(68) = -1.60, p = .12, d = -.40$], or immediate or delayed efficiency scores [immediate: $t(97) = -1.22, p = .22 d = -.26$; delayed: $t(68) = .43, p = .68, d = .11$]. Given the high similarity, the groups were analyzed together to evaluate the impact of adaptive mastery on learning.

3.2 Learning Performance

To evaluate immediate and delayed learning for both high and low cognitive ability, a 3 (adaptive mastery: Mastery Add, Mastery Drop, and No Add/Drop) × 2 (WMC: high vs. low) × 2 (V-WMC: high vs. low) between-subjects MANOVA was conducted with immediate pre-post gain scores, immediate transfer test scores, immediate efficiency scores, delayed pre-post gain scores, delayed transfer test scores, and delayed efficiency scores as dependent variables. There was a significant effect of adaptive mastery condition on performance [$F(12, 100) = 5.41$, Wilks' Lambda $= .37, p < .001$]. There was no significant effect of WMC [$F < 1$] or VWM-C [$F < 1$] on performance. No interactions between mastery condition and either cognitive ability measure reached significance [adaptive mastery × WMC: $F < 1$; adaptive mastery × VWM-C: $F(12, 100) = 1.17$, Wilks' Lambda $= .77, p = .31$; WMC × VWM-C: $F < 1$; adaptive mastery × WMC × VWM-C: $F < 1$]. Thus, while adaptive mastery significantly impacted learning performance, no cognitive ability measures had any significant impacts.

To further examine the impact of adaptive mastery on performance, univariate ANOVAs were conducted for each dependent variable. Adaptive mastery conditions had a significant impact on immediate pre-post gain scores [$F(2, 96) = 9.56, p < .001$] and immediate efficiency scores [$F(2, 96) = 4.05, p < .05$]. There was no significant impact of adaptive mastery on delayed pre-post gain scores ($Fs < 1$), delayed efficiency scores [$F(2, 67) = 2.81, p = .07$], or immediate or delayed transfer test scores ($Fs < 1$).

Post-hoc tests were conducted on immediate pre-post gain scores using the Tukey highly significant difference test. The Mastery Drop condition ($M = 52\%, SD = 35\%$) showed significantly lower learning gains compared to the Mastery Add [$M = 71\%, SD = 22\%$; $t(96) = 2.90, p < .05$] and the No Add/Drop [$M = 81\%$; $SD = 23\%$; $t(96) = -4.25, p < .001$] conditions (see Fig. 3). There was no significant difference between the Mastery Add and No Add/Drop conditions [$t(96) = -1.33, p = .38$]. Thus, adaptively dropping cards from study based on mastery led to the worst overall immediate learning gains but adaptively adding cards was not significantly different from a random presentation of cards. These group differences went away for the delayed pre-post gain scores, suggesting that the benefits of adaptively adding cards or not adapting the deck are short-term.

Post-hoc tests were conducted on immediate efficiency scores using the Tukey highly significant difference test. The Mastery Drop condition showed significantly better efficiency ($M = .63; SD = .46$) compared to both the Mastery Add ($M = .43; SD = .21$; $t(96) = -2.42, p < .05$) and the No Add/Drop [$M = .43, SD = .29; t(96) = 2.45, p < .05$] conditions (see Fig. 4). There was no significant difference in efficiency between the Mastery Add and No Add/Drop groups [$t(96) = .05, p = .99$). The greater efficiency in the Mastery Drop group is likely due to the lower number of trials completed in that condition. A one-way ANOVA on the total number of learning trials showed a significant

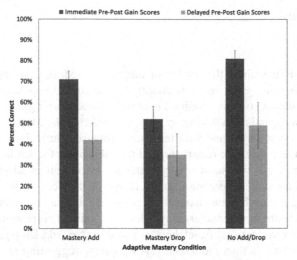

Fig. 3. Immediate and delayed pre-post gain scores for each adaptive mastery condition.

effect of mastery condition [$F(2, 96) = 25.87, p < .001$], with the Mastery Drop condition ($M = 92.8, SD = 33.1$) completing significantly fewer trials than both the Mastery Add [$M = 190; SD = 71.1; t(96) = 4.54, p < .001$] and No Add/Drop [$M = 245; SD = 135; t(96) = 4.54, p < .001$] conditions. The No Add\Drop condition also completed significantly more trials compared to the Mastery Add condition [$t(96) = -2.48, p < .05$]. However, the low number of trials completed in the Mastery Drop condition led to significantly worse learning overall, showing significant costs to better efficiency.

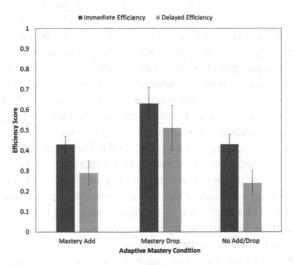

Fig. 4. Immediate and delayed efficiency scores for each adaptive mastery condition.

4 Discussion

4.1 Overview

The current study examined the impact of adaptively adding or dropping flashcards from study on learning gains and how adaptive instructional approaches may vary in effectiveness based on cognitive ability. I compared three learning conditions: Mastery Add, Mastery Drop, and a No Add/Drop group. I predicted the Mastery Add condition would show the best short- and long-term performance, particularly for individuals with low cognitive ability. The results showed partial support for this hypothesis in that adaptively adding cards was substantially better for learning than adaptively dropping cards from study for immediate learning gains. However, the No Add/Drop group showed the best performance both immediately after study and after a one-week delay, though this advantage was not significant for delayed learning gains. This advantage for the non-adaptive condition was not significant, however, and the No Add/Drop group completed significantly more trials than the Mastery Add group, suggesting there may still be an advantage for adaptively adding cards when there are time constraints for learning. Additionally, I found no impact of cognitive ability on performance in this task, a point that will be further discussed below.

4.2 When Should Flashcard-Based Learning Be Adaptive?

One clear conclusion from the current study is to avoid dropping items from study when learning from flashcards. The Mastery Drop condition showed the lowest overall performance both immediately after training and one week later. The current data show that even when using objective mastery criteria to overcome metacognitive pitfalls [3–5, 10, 22], dropping items from study leads to lower learning gains. This is consistent with research showing learners benefit from overlearning [20, 21]; by continually reviewing concepts learners benefit from additional retrieval practice. While the Mastery Drop condition showed the greatest efficiency, that is likely because participants completed so few trials. Performance greatly suffered as a result of the efficiency.

Beyond that, the data become somewhat less definitive. Given the best performance was found in the non-adaptive condition, one could conclude the best approach is to not adapt the learning deck in a flashcard-based learning task. Leaving all concepts in the deck and using a random order seems to maximize the potential of flashcards to incorporate principles of spacing and testing [1–3, 5]. This approach also has the benefit of simplicity. One advantage of flashcards from a practical standpoint is that they do not require advanced technology or a large number of resources in order to use them. As long as learners are given explicit instructions to not drop items from study and they study items in a large flashcard deck to maximize spacing [19], it is likely they will maximize their learning without adaptive interventions.

However, the cost to the non-adaptive approach is participants in the No Add/Drop group completed significantly more learning trials compared to the adaptive conditions. Additionally, performance in the No Add/Drop condition was not significantly better than performance in the Mastery Add condition. That is, learners in the Mastery Add condition showed comparable learning outcomes in fewer trials when compared to the

No Add/Drop group. Thus, adaptively adding cards may be better in terms of efficiency if there are time constraints on learning and there are technology resources available to implement this adaptive strategy.

As we see in the Mastery Drop condition, however, there is a point where efficiency is no longer a useful learning outcome. Additionally, in a real-world learning environment when learners have more control over their learning, prior research shows that it can be difficult for learners to overcome metacognitive illusions and make better learning decisions [29]. Yan and colleagues [29] showed even when explicitly instructed on how to best approach learning they still make suboptimal decisions during study. Future research should compare adaptive and non-adaptive interventions to improve learner decision-making.

Thus, when determining whether or not to include adaptive mastery in flashcards-based learning, instructors and learners should consider a few factors. First, all things being equal using a non-adaptive approach to flashcard-based learning will likely yield good learning outcomes assuming learners keep all items in the deck for study. Second, if there are any time constraints on learning the material, adaptively adding cards is likely the best approach as it provides a good balance of learning gains and efficiency. Third, based on prior research, instructors should provide explicit instruction on how to maximize the use of flashcards. Finally, if an adaptive approach is used, adding cards is significantly better than dropping cards.

4.3   Cognitive Ability and Flashcard-Based Learning

The current study found no evidence that cognitive ability impacted performance in any of the adaptive mastery conditions, suggesting that the same adaptive instructional strategy for flashcard-based learning is equally effective across all levels of cognitive ability. It is possible that the flashcard task used in the current study did not have a sufficiently high enough cognitive load to elicit individual differences in performance. Specifically, participants only trained on six butterfly species, making the task relatively easy. By adding more categories or by interleaving multiple learning domains during learning, the cognitive load would presumably increase, making cognitive ability more predictive of performance. However, as noted above, it may be that adaptive instructional strategies are not necessary for flashcard-based learning tasks and that a full flashcard deck is the most straightforward and effective strategy for all learners regardless of cognitive ability.

Additionally, in real-world learning environments, participants with lower cognitive ability may make more suboptimal learning decisions such as dropping a card from study too soon [5, 9, 10, 20, 22]. Prior work has shown allowing participants control over when to drop a concept from study leads to poor learning outcomes [5]. In the current study, high and low cognitive ability participants may have been equalized simply by not allowing participants to modify the learning deck. Future work can examine the impact of flashcard difficulty and learner control may vary depending on cognitive ability.

4.4 Conclusion

This paper used an adaptive mastery approach to flashcard-based learning and evaluated whether or not the ideal adaptive instructional strategy varied depending on cognitive ability. I showed that both adaptively adding flashcards to study as well as leaving the flashcard deck unchanged lead to better learning gains compared to adaptively dropping cards from study. Cognitive ability had no impact on learning outcomes, suggesting similar instructional strategies work equally well across all learners, at least when learners have no control over the flashcard deck. Overall the results suggest non-adaptive approaches to learning from flashcards may be the simplest choice unless there are time constraints. In situations where time is limited, adaptively adding cards to study will lead to better immediate learning gains compared to adaptively dropping cards. The data clearly show dropping flashcards from study is not optimal for learning. Learners and instructors alike should keep that in mind when they use or recommend flashcards for study.

Acknowledgments. Many thanks to Sean Chancellor, Morgan Kelley, and Crystal Meyer for their assistance with data collection and to Sadan Yagci for his assistance with programming.

References

1. Golding, J.M., Wasarhaley, N.E., Fletcher, B.: The use of flashcards in an introduction to psychology class. Teach. Psychol. **39**, 199–202 (2012). https://doi.org/10.1177/009862831 2450436
2. Hartwig, M.K., Dunlosky, J.: Study strategies of college students: are self-testing and scheduling related to achievement? Psychon. Bull. Rev. **19**, 126–134 (2012). https://doi.org/10.3758/s13423-011-0181-y
3. Karpicke, J.D., Butler, A.C., Roediger, H.L.: Metacognitive strategies in student learning: do students practise retrieval when they study on their own? Memory **17**, 471–479 (2009). https://doi.org/10.1080/09658210802647009
4. Kornell, N., Bjork, R.A.: The promise and perils of self-regulated study. Psychon. Bull. Rev. **14**, 219–224 (2007). https://doi.org/10.3758/bf03194055
5. Whitmer, D.E., Johnson, C.I., Marraffino, M.D., Pharmer, R.L., Blalock, L.D.: A Mastery Approach to Flashcard-Based Adaptive Training. In: Sottilare, R.A., Schwarz, J. (eds.) HCII 2020. LNCS, vol. 12214, pp. 555–568. Springer, Cham (2020). https://doi.org/10.1007/978-3-030-50788-6_41
6. Bjork, R.A., Dunlosky, J., Kornell, N.: Self-regulated learning: beliefs, techniques, and illusions. Annu. Rev. Psychol. **64**, 417–444 (2013). https://doi.org/10.1146/annurev-psych-113 011-143823
7. Kornell, N., Bjork, R.A.: Learning concepts and categories: is spacing the "enemy of induction"? Psychol. Sci. **19**, 585–592 (2008). https://doi.org/10.1111/j.1467-9280.2008.021 27.x
8. Senzaki, S., Hackathorn, J., Appleby, D.C., Gurung, R.A.R.: Reinventing flashcards to increase student learning. Psychol. Learn. Teach. **16**, 353–368 (2017). https://doi.org/10.1177/1475725717719771
9. Karpicke, J.D.: Metacognitive control and strategy selection: deciding to practice retrieval during learning. J. Exp. Psychol. Gen. **138**, 469–486 (2009)

10. Miyatsu, T., Nguyen, K., McDaniel, M.A.: Five popular study strategies: their pitfalls and optimal implementations. Perspect. Psychol. Sci. **13**, 390–407 (2018). https://doi.org/10.1177/1745691617710510

11. Cepeda, N.J., Pashler, H., Vul, E., Wixted, J.T., Rohrer, D.: Distributed practice in verbal recall tasks: a review and quantitative synthesis. Psychol. Bull. **132**, 354–380 (2006). https://doi.org/10.1037/0033-2909.132.3.354

12. Roediger, H.L., Karpicke, J.D.: The power of testing memory: basic research and implications for educational practice. Perspect. Psychol. Sci. **1**, 181–210 (2006). https://doi.org/10.1111/j.1745-6916.2006.00012.x

13. Cepeda, N.J., Coburn, N., Rohrer, D., Wixted, J.T., Mozer, M.C., Pashler, H.: Optimizing distributed practice: theoretical analysis and practical implications. Exp. Psychol. **56**, 236–246 (2009). https://doi.org/10.1027/1618-3169.56.4.236

14. Janiszewski, C., Noel, H., Sawyer, A.G.: A meta-analysis of the spacing effect in verbal learning: implications for research on advertising repetition and consumer memory. J. Consum. Res. **30**, 138–149 (2003). https://doi.org/10.1086/374692

15. Johnson, C.I., Mayer, R.E.: A testing effect with multimedia learning. J. Educ. Psychol. **101**, 621–629 (2009). https://doi.org/10.1037/a0015183

16. Karpicke, J.D., Roediger, H.L.: Repeated retrieval during learning is the key to long-term retention. J. Mem. Lang. **57**, 151–162 (2007). https://doi.org/10.1016/j.jml.2006.09.004

17. Rowland, C.A.: The effect of testing versus study on retention: a meta-analytic review of the testing effect. Psychol. Bull. **140**, 1432–1463 (2014). https://doi.org/10.1037/a0037559

18. Roediger, H.L., Karpicke, J.D.: Test-enhanced learning: taking memory tests improves long-term retention. Psychol. Sci. **17**, 249–255 (2006). https://doi.org/10.1111/j.1467-9280.2006.01693.x

19. Kornell, N.: Optimising learning using flashcards: spacing is more effective than cramming. Appl. Cog. Psychol. **23**, 1297–1317 (2009). https://doi.org/10.1002/acp.1537

20. Kornell, N., Bjork, R.A.: Optimising self-regulated study: the benefits—and costs—of dropping flashcards. Memory **16**, 125–136 (2008). https://doi.org/10.1080/09658210701763899

21. Wissman, K.T., Rawson, K.A., Pyc, M.A.: How and when do students use flashcards? Memory **20**, 568–579 (2012). https://doi.org/10.1080/09658211.2012.687052

22. Pyc, M.A., Rawson, K.A., Aschenbrenner, A.J.: Metacognitive monitoring during criterion learning: when and why are judgments accurate? Mem. Cognit. **42**(6), 886–897 (2014). https://doi.org/10.3758/s13421-014-0403-4

23. Pyc, M.A., Rawson, K.A.: Examining the efficiency of schedules of distributed retrieval practice. Mem. Cogn. **35**, 1917–1927 (2007). https://doi.org/10.3758/bf03192925

24. Pyc, M.A., Rawson, K.A.: Testing the retrieval effort hypothesis: does greater difficulty correctly recalling information lead to higher levels of memory? J. Mem. Lang. **60**, 437–447 (2009). https://doi.org/10.1016/j.jml.2009.01.004

25. Pyc, M.A., Rawson, K.A.: Costs and benefits of dropout schedules of test–restudy practice: implications for student learning. Appl. Cog. Psychol. **25**, 87–95 (2011). https://doi.org/10.1002/acp.1646

26. Van Merriënboer, J.J.G., Kester, L., Paas, F.: Teaching complex rather than simple tasks: balancing intrinsic and germane load to enhance transfer of learning. Appl. Cog. Psychol. **20**, 343–352 (2006). https://doi.org/10.1002/acp.1250

27. Sweller, J.: Cognitive load theory, learning difficulty and instructional design. Learn. Instr. **4**, 295–312 (1994). https://doi.org/10.1016/0959-4752(94)90003-5

28. Sweller, J.: Element interactivity and intrinsic, extraneous and germane cognitive load. Educ. Psychol. Rev. **22**, 123–138 (2010). https://doi.org/10.1007/s10648-010-9128-5

29. Yan, V.X., Bjork, E.L., Bjork, R.A.: On the difficulty of mending metacognitive illusions: a priori theories, fluency effects, and misattributions of the interleaving benefit. J. Exp. Psychol. Gen. **145**, 918–933 (2016). https://doi.org/10.1037/xge0000177

30. Durlach, P.J., Ray, J.M.: Designing Adaptive Instructional Environments: Insights from Empirical Evidence. Technical Report, U.S. Army Research Institute for the Behavioral and Social Sciences (2011)

31. Landsberg, C.R., Astwood, R.S., Jr., Van Buskirk, W.L., Townsend, L.N., Steinhauser, N.B., Mercado, A.D.: Review of adaptive training system techniques. Mil. Psychol. **24**, 96–113 (2012). https://doi.org/10.1080/08995605.2012.672903

32. Mettler, E., Burke, T., Massey, C.M., Kellman, P.J.: Comparing adaptive and random spacing schedules during learning to mastery criteria. In: Proceedings of the 42nd Annual Conference of the Cognitive Science Society, pp. 773–779. Cognitive Sciences Society, Virtual (2020)

33. Mettler, E., Kellman, P.J.: Adaptive response-time-based category sequencing in perceptual learning. Vision. Res. **99**, 111–123 (2014). https://doi.org/10.1016/j.visres.2013.12.009

34. Bjork, E.L., Bjork, R.A.: Making things hard on yourself, but in a good way: creating desirable difficulties to enhance learning. In: Gernsbacher, M.A., Pew, R.W., Hough, L.M., Pomerantz, J.R. (eds.) Psychology And the Real World: Essays Illustrating Fundamental Contributions to Society, pp. 56–64. Worth Publishers, New York (2011)

35. Bjork, R.A.: Memory and metamemory considerations in the training of human beings. In: Metcalfe, J., Shimamura, A. (eds.) Metacognition: Knowing About Knowing, pp. 185–205. MIT Press, Cambridge (1994)

36. Baddeley, A.D., Hitch, G.: Working memory. In: Psychology of Learning and Motivation, vol. 8, pp. 47–89. Academic Press (1974)

37. Conway, A.R.A., Kane, M.J., Bunting, M.F., Hambrick, D.Z., Wilhelm, O., Engle, R.W.: Working memory span tasks: a methodological review and user's guide. Psychon. Bull. Rev. **12**, 769–786 (2005). https://doi.org/10.3758/bf03196772

38. Engle, R.W.: Working memory capacity as executive attention. Curr. Dir. Psychol. Sci. **11**, 19–23 (2002). https://doi.org/10.1111/1467-8721.00160

39. Engle, R.W.: Working memory and executive attention: a revisit. Perspect. Psychol. Sci. **13**, 190–193 (2018). https://doi.org/10.1177/1745691617720478

40. Kane, M., Hambrick, Z., Tuholski, S., Wilhelm, O., Payne, T., Engle, R.: The generality of working memory capacity: a latent-variable approach to verbal and visuospatial memory span and reasoning. J. Exp. Psychol. Gen. **133**, 189–217 (2004). https://doi.org/10.1037/0096-3445.133.2.189

41. Engle, R.W., Laughlin, J.E., Tuholski, S.W., Conway, A.R.A.: Working memory, short-term memory, and general fluid intelligence: a latent-variable approach. J. Exp. Psychol. Gen. **128**, 309–331 (1999)

42. Kane, M.J., Bleckley, M.K., Conway, A.R., Engle, R.W.: A controlled-attention view of working-memory capacity. J. Exp. Psychol. Gen. **130**, 169–183 (2001)

43. Klein, K., Boals, A.: The relationship of life event stress and working memory capacity. Appl. Cognit. Psychol. **15**, 565–579 (2001). https://doi.org/10.1002/acp.727

44. Agarwal, P.K., Finley, J.R., Rose, N.S., Roediger, H.L.: Benefits from retrieval practice are greater for students with lower working memory capacity. Memory **25**, 764–771 (2017). https://doi.org/10.1080/09658211.2016.1220579

45. Wang, J., Liu, Z., Xing, Q., Seger, C.A.: The benefit of interleaved presentation in category learning is independent of working memory. Memory **28**, 285–292 (2020). https://doi.org/10.1080/09658211.2019.1705490

46. Cowan, N.: The magical number 4 in short-term memory: a reconsideration of mental storage capacity. Behav. Brain Sci. **24**, 87–114 (2001). https://doi.org/10.1017/S0140525X01003922

47. Luck, S.J., Vogel, E.K.: Visual working memory capacity: from psychophysics and neurobiology to individual differences. Trends Cogn. Sci. **17**, 391–400 (2013). https://doi.org/10.1016/j.tics.2013.06.006

48. Rouder, J.N., Morey, R.D., Morey, C.C., Cowan, N.: How to measure working memory capacity in the change detection paradigm. Psychon. Bull. Rev. **18**(2), 324–330 (2011). https://doi.org/10.3758/s13423-011-0055-3

49. Hollingworth, A., Richard, A.M., Luck, S.L.: Understanding the function of visual short-term memory: transsaccadic memory, object correspondence, and gaze correction. J. Exp. Psychol. Gen. **137**, 163–181 (2008). https://doi.org/10.1037/0096-3445.137.1.163

50. Hollingworth, A., Matsukura, M., Luck, S.J.: Visual working memory modulates rapid eye movements to simple onset targets. Psychol. Sci. **24**, 790–796 (2013). https://doi.org/10.1177/0956797612459767

51. Alloway, T.P., Alloway, R.G.: Investigating the predictive roles of working memory and IQ in academic attainment. J. Exp. Child Psychol. **106**, 20–29 (2010). https://doi.org/10.1016/j.jecp.2009.11.003

52. Fukuda, K., Vogel, E., Mayr, U., Awh, E.: Quantity, not quality: the relationship between fluid intelligence and working memory capacity. Psychon. Bull. Rev. **17**(5), 673–679 (2010). https://doi.org/10.3758/17.5.673

53. Redick, T.S., et al.: Measuring working memory capacity with automated complex span tasks. Eur. J. Psychol. Assess. **28**, 164–171 (2012). https://doi.org/10.1027/1015-5759/a000123

54. Inquisit 6 Automated Symmetry Span Task (ASSPAN) [computer software]. https://www.millisecond.com (2020)

55. Luck, S.J., Vogel, E.K.: The capacity of visual working memory for features and conjunctions. Nature **390**, 279–281 (1997). https://doi.org/10.1038/36846

56. Inquisit 6: https://www.millisecond.com (2021)

57. Inquisit 6: Solving Anagrams. https://www.millisecond.com (2019)

AIS Challenges in Evaluating the Selection of Learner Interventions

Robert A. Sottilare^(✉) (iD)

Soar Technology Inc., Orlando, FL 32817, USA
bob.sottilare@soartech.com

Abstract. This paper identifies challenges related to the optimal selection of adaptive instructional system (AIS) learner interventions. There are major challenges associated with accurately assessing learner performance, deciding when and how to intervene with learner, and evaluating the effectiveness of any intervention (e.g., feedback, support, direction, change in content level of difficulty) delivered during adaptive instruction, and improving intervention selection over time. AISs are computer-based systems that accommodate individual differences and tailor instruction to match learner capabilities to acquire knowledge through a guided learning process. While AISs are highly effective learning tools, they are costly and require very high-level skills to develop them and create complex, effective, and efficient adaptive courses. In some well-defined domains of instruction (e.g., mathematics), AIS developers have focused on the identification of errors and the interventions selected are focused on correcting these errors. Common errors are often identified in a misconception library that associates content and practice required to overcome them. In more complex domains, learner data may be sparse, and the simple identification of errors may not be sufficient to support optimal selection of learner interventions. Under these conditions, the challenge is to identify and address poor learner performance by testing plausible root causes and addressing them directly instead of focusing on errors that may only address symptoms and not root causes of learner performance. This paper identifies several challenge areas, discusses potential solutions, and provides recommendations for future research to overcome difficult challenges associated with machine selection of adaptive instructional interventions.

Keywords: Adaptive instructional systems (AISs) · Learner assessments · Learner performance · Instructional interventions · Learning effect

1 Introduction to Adaptive Instruction

Adaptive instructional systems (AISs) are computer-based systems that accommodate individual differences and tailor instruction to match learner capabilities to acquire knowledge through a guided learning process [1–3] and support education and training experiences in cognitive, psychomotor and collaborative learning domains (Fig. 1). While AISs are highly effective learning tools [4], they are costly to build, and require very high-level skills to create these complex courses. Effective adaptive

© The Author(s), under exclusive license to Springer Nature Switzerland AG 2022
R. A. Sottilare and J. Schwarz (Eds.): HCII 2022, LNCS 13332, pp. 104–112, 2022.
https://doi.org/10.1007/978-3-031-05887-5_8

instructional course that automatically identify learner performance and other states use machine learning and other artificially-intelligent methods to automatically assess learner progress toward a set of learning objectives, and then intervene with the learner and the instructional environment to tailor the teaching experience so it engages the learner, stimulates behaviors leading to skill development, and overcomes gaps in the learner's knowledge relative to expected performance.

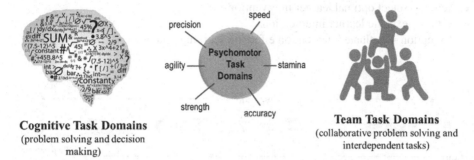

Cognitive Task Domains
(problem solving and decision making)

Team Task Domains
(collaborative problem solving and interdependent tasks)

Fig. 1. AISs support tailored instruction for individuals and team in cognitive, psychomotor and collaborative learning instructional domains.

Interventions generally take two forms: interactions with the learner or changes to the content (e.g., problems, information, scenarios) presented to the learner. Learner interactions may include verbal or textual feedback, support, direction, prompts for additional information, or assessments. Changes to content usually include either real-time modification of the content or the selection of different content. In either case, the goal is to present the learner with content at an appropriate level of difficulty.

Per Vygotsky's zone of proximal development (ZPD) [5], the goal of good instruction is to engage the learner with content that exercises their knowledge and skill at a level of difficulty that is just beyond their current capabilities so new learning occurs, but within a level of difficulty that can be managed with appropriate support from the tutor. The goal of the ZPD is to manage the instructional experience within a lane of difficulty that will neither bore the learner nor cause them to withdraw, give up or be frustrated by the adaptive instructional process that is beyond their capacity. Bjork [6, 7] also describes the need for "desirable difficulties" which are needed to engage the learner, challenge them to move forward with their learning and support their learning goals.

Given this set of complex interventions, it is crucial for the AIS to have an accurate model of the learner's competence in the domain/topic of instruction being taught. In designing AISs, it is crucial for the artificial intelligence (AI) within the system to be transparent and explainable in their assessment of the learner's states and their selection of interventions. In the next section, we describe the tutoring cycle within AISs.

2 Learning Effect Model (LEM)

The LEM (Fig. 2) was first described by Sottilare in 2012 [8] and has subsequently evolved to provide additional details about the tutoring cycle within AISs. Extensive research supports and document the LEM [9–11]. In this section, we describe the LEM process along with AIS components and features and how they lead to effective learning outcomes. The LEM can be broken down into five discrete subprocesses within an intelligent action-perception framework:

- Perception – assess learner domain competency from previous experiences
- Perception – assess learner's performance in real-time
- Action – select optimal learner interventions
- Action – execute learner interventions
- Perception – evaluate intervention effect on learning outcomes

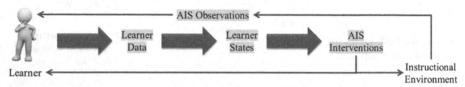

Fig. 2. This simplified depiction of the learning effect model (LEM) highlights AI processes in yellow that operating within AISs. AIS interventions act on both the learner and the instructional environment with the goal of optimizing learner performance.

2.1 Assessment of Learner Domain Competency

In this section, we discuss the process and challenges associated with accurately determining the competency of learners based on their previous experiences. This subprocess uses past achievements stored in a learning record store (or long-term learner model) are used to assess competency in a domain of instruction (e.g., algebra, physics or baseball). The Oxford English Dictionary defines *competency* as "the ability to do something successfully or efficiently" [12]. However, AISs require measures of assessment to define a learner's level of competency (e.g., at, below or above a required standard). Specific measures of domain competency include achievement types, experience duration, experience source information, domain learning and forgetting, and assessment within learning experiences [13].

Learning achievement types are based on the types of experiences used to increase their knowledge. Achievement types include instruction, reading, auditing and deliberate practice. More active experiences like practice (learning by doing) versus auditing (learning through examples) are considered to be more effective in transferring skills for use in work or operational environments. Experience duration distinguishes between achievements that might include a one-hour online algebra course and a four-year degree in mathematics. Information about the source of learning experience help distinguish the quality of the experience. Domain learning and forgetting measures for assessed domain experiences provide quantifiable contributions to domain competency modeling. Retention (memory) of domain information is strengthened during learning experiences as knowledge and skills are regularly applied. However, once the last learning experience ends, forgetting begins as knowledge and skill decay. Finally, the result of formal assessments provides yet another measure of domain competency.

Together, measures of competency may be used to determine a level competency which may also be thought of as a probability of successful domain performance. To record achievements in a consistent way, IEEE under Project 9274 [14] has developed a standard statement that captures measures of achievement in an experience application program interface (xAPI) statement. AIS architectures can use information in LRSs to maintain competency models, initialize learner models, and tailor new learning experiences.

2.2 Assessment of Learner Real-Time Performance

In this section, we discuss the process and challenges in accurately assessing real-time learner performance. In much the same way that competency models are developed and maintained, learner models are also initialized and updated as the learner achieves learning objectives during adaptive instruction. Assessments are tangible measures of learning and may take many forms in training and educational contexts. Formal assessments may include tests or checks on learning that involve problem solving, but could also include checklist of completed actions, successful completion of a scenario in a training simulation, a logic puzzle, completing steps in a process or a decision that results in an optimal outcome. In AIS software architectures such as the Generalized Intelligent Framework for Tutoring (GIFT) [15, 16], performance assessments are categorized as at, above or below expectations where expectations are set based on the learner model at the start of instruction. The learner model is typically initialized by the LRS domain achievements contribution in the form of the competency model. Other systems use raw assessment scores or identify errors to support decisions about the type of learner intervention required. Errors tend to be identified as deviations from a prescribed path. This view limits creativity when learners attempt new solutions that define paths have never been tried, but are still successful in terms of learning outcomes (e.g., solved the problem, completed the mission). Successful outcomes or achievements are documented, transmitted, and stored in the LRS as shown in the sample xAPI format below.

"result": {solving quadratic equations}
"completion": true,
"duration": 1 hour,
"success": true,
"score": 92
"refresher needed": one year after completion

Now that we have discussed the need for an accurate model of the learner's performance, the next step is to discuss the selection process for optimal learner interventions to address identified performance issues (e.g., learner errors or root causes of poor performance).

2.3 Selection of Optimal Learner Interventions

In this section, we discuss the process and challenges in selecting optimal learner interventions. In some well-defined domains of instruction (e.g., mathematics), AIS designers and developers have focused on the identification of errors and interventions selected are related to correcting these errors. Common errors are often identified in a misconception library that associates content and practice required to overcome those errors. While it is good instructional practice to help learners reduce the number and types of errors. Simply identifying errors may not be sufficient for the learner to achieve mastery of the domain. In more complex domains, learner data may be sparse, and measures of assessment may not be sufficient to support optimal selection of learner interventions.

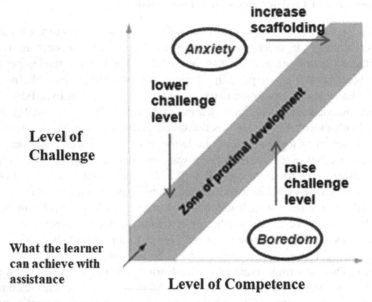

Fig. 3. Zone of Proximal Development not only guides learning experiences that are challenging, but also influences emotions that are critical to efficient learning.

Per Vygotsky [5], a goal for educational technology like AISs is to match the level of skill of the learner to the difficulty of the content (Fig. 3). This assumes a robust model of the learner is available to support interventions prior to start of adaptive instruction. This also assumes that difficulty of the content can be evaluated. For example, it might be desirable to objectively measure the ability of an opposing force in a military game for training so that the AIS can adapt the difficulty of the training scenario events appropriately to match the learner's capabilities.

While some selection criteria for learner interventions might be relatively straightforward in the case of remediating errors, it is often difficult for human tutors to project the impact of their interventions. Experienced human tutors who are experts in the domain of instruction understand common errors made by learners, have knowledge of multiple methods or instructional paths leading to successful outcomes, and deep rapport with their students to facilitate their trust and compliance. This experience along with a deep knowledge of the learner's knowledge gaps helps facilitate the selection of interventions. So how do we transfer this expertise to a computer-based tutor in the form of an AIS?

Just as a human tutor collects information about their students, an AIS facilitating a course can collect information about the impact of their intervention decisions, the conditions of the learner prior to, during, and after the intervention decision, and the conditions of the instruction (e.g., where the learner is in the course content or in a training scenario). If the AIS is used to facilitate an adaptive course several hundred times, it is possible to use AI to build a neural network or Markov Decision Process that can project the impact of various interventions. In addition to projections, the success of content adaptation interventions used to match learn capabilities to the difficulty of the course content (ala Vygotsky's ZPD) are well documented in the literature [17–19].

2.4 Execution of Selected Learner Interventions

This section discusses the process of AIS actions for executing any selected learner interventions. As noted previously, interventions can take many forms, but the most prevalent interventions center on providing feedback to the learner. The delivery of feedback should be provided in a manner that is engaging and is specific enough to action by the learner [20]. Feedback should be timely and sensitive to the needs of the learner. Feedback should refer to a skill or knowledge deficit and encourage learners to continue toward their achievement goals.

The mode of communications with the learner can be verbal or textual and through different sources (e.g., verbal with no visible source, verbal directly from an embodied virtual character within the AIS or verbal directly from an external application). Communication modes paired with virtual character have shown mixed results. Wang & Gratch [21] found that the immediacy of virtual human tutor's non-verbal response had a positive impact on student learning and affect. However, they also found that while virtual human tutors were able to establish rapport with learners, this rapport did not help them achieve better learning results. Several researchers found that learner engagement improved with virtual human tutors and higher engagement provides the potential for greater learning opportunities [22–24]. Given these findings, it seems that responsive virtual humans should be considered when designing AIS intervention strategies.

2.5 Evaluation of Intervention Effect on Learning

Figure 4 outlines a process and software architecture for conducting and evaluating adaptive instruction provided by AISs. A similar testbed architecture has been prototyped under a US Army project called the Learning & Readiness Intelligent Agent Testbed (LARIAT) [25]. LARIAT's goal is slightly different from the framework pictured in Fig. 4. LARIAT's goal is to evaluate the effectiveness of intelligent agents that are used to represent entities in training simulations. Figure 4 pictures a framework that evaluates the performance of agents within an AIS that manage classification and decision processes to optimize learner performance.

The ability to collect large amounts of data about learner interventions, learner states and traits, and a mapping of successful and unsuccessful instructional paths will provide

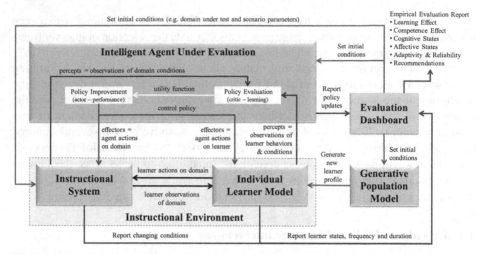

Fig. 4. This evaluation architecture provides an experimental framework for assessing AIS processes for classifying/predicting learner states, selecting optimal interventions, and then executing or delivering those interventions. The architecture supports self-improving principles that assess the effectiveness of current policies and update them as data demonstrates better methods to optimize learning.

the information required to analyze the effect of interventions on learning. A major challenge is understanding all the interactions and impacts of multiple interventions on a single learner under changing conditions. Another challenge is that the impact of learner interventions is clearly dependent upon the accurate classification or prediction of learner performance and other states (e.g., emotions, workload) impacting learning focus and capacity. Finally, a third challenge is to develop low data methods for evaluating new adaptive courses where data is sparse or unreliable.

In addition to evaluation effect, we recommend AIS researchers track the influence of instructional strategies and policies on learning performance and other learning outcomes. We recommend that domain independent instructional strategies like mastery learning (especially in simulated environments) [26], worked examples [27], and error-sensitive feedback [28–30] be re-examined in the light of adaptive instructional applications. Overall, learning effectiveness could be greatly enhanced by designing AISs as self-improving systems [31, 32].

3   Additional Recommendations

The following recommendations are provided to focus on AIS technology gaps beyond the challenges of evaluating the effectiveness of interventions to optimize learner performance. First, the personalization features in AISs should also provide the ability to evaluate opportunities to reduce the time required for a learner to reach a level of proficiency. Second, we recommend research to understand the impact of pairing AISs with other immersive training technologies (e.g., virtual, augmented, and mixed reality (XR) simulations) and engaging game-based training environments. Third, we recommend research be conducted to evaluate opportunities to improve learner engagement during interactions with AISs. Improved realism and responsiveness of virtual humans and their integration with AISs may enable better bonding/rapport with learners leading to improved learning outcomes.

Finally, we recommend research be conducted to evaluate the effect of various types of learner control on adaptive instruction. Areas where learners might be given more control include controlling the attributes of a virtual instructor, controlling the choice of learning peers, initiating and selecting of on-demand learning topics, enabling control over their learner model content and information. Control over the domain and teaching approach, and control over the amount of control the learner has over AIS processes may also have an influence on learning outcomes [33].

References

1. Wang, M.C., Walberg, H.J.: Adaptive instruction and classroom time. Am. Educ. Res. J. **20**(4), 601–626 (1983)
2. Tsai, C.C., Hsu, C.Y.: Adaptive instruction systems and learning. In: Seel, N.M. (ed.) Encyclopedia of the Sciences of Learning. Springer, Boston (2012). https://doi.org/10.1007/978-1-4419-1428-6_1092
3. Sottilare, R., Brawner, K.: Component interaction within the generalized intelligent framework for tutoring (GIFT) as a model for adaptive instructional system standards. In: The Adaptive Instructional System (AIS) Standards Workshop of the 14th International Conference of the Intelligent Tutoring Systems (ITS) Conference, Montreal (2018)
4. VanLehn, K.: The relative effectiveness of human tutoring, intelligent tutoring systems, and other tutoring systems. Educ. Psychol. **46**(4), 197–221 (2011)
5. Vygotsky, L.: Zone of proximal development. Mind in society: the development of higher psychological processes. **5291**, 157 (1987)
6. Bjork, R.A.: Creating Desirable Difficulties to Enhance Learning. Crown House Publishing, Carmarthen (2017)
7. Bjork, R.A., Bjork, E.L.: Desirable difficulties in theory and practice. J. Appl. Res. Mem. Cogn. **9**(4), 475 (2020)
8. Sottilare, R.: Considerations in the development of an ontology for a generalized intelligent framework for tutoring. In: International Defense and Homeland Security Simulation Workshop in Proceedings of the I3M Conference, pp. 19–25 (2012)
9. Sottilare, R., Ragusa, C., Hoffman, M., Goldberg, B.: Characterizing an adaptive tutoring learning effect chain for individual and team tutoring. In: Proceedings of the Interservice/Industry Training Simulation and Education Conference, Orlando (2013)
10. Sottilare, R.: Elements of a learning effect model to support an adaptive instructional framework. In: Generalized Intelligent Framework for Tutoring (GIFT) Users Symposium (GIFTSym4), p. 7 (2016)
11. Sottilare, R.A., Shawn Burke, C., Salas, E., Sinatra, A.M., Johnston, J.H., Gilbert, S.B.: Designing adaptive instruction for teams: a meta-analysis. Int. J. Artif. Intell. Educ. **28**(2), 225–264 (2018)
12. Competency defined (1931). Oxford English Dictionary
13. Sottilare, R.A., Long, R.A., Goldberg, B.S.: Enhancing the experience application program interface (xAPI) to improve domain competency modeling for adaptive instruction. In: Proceedings of the Fourth (2017) ACM Conference on Learning@ Scale, pp. 265–268 (2017)

14. Sottilare, R.: Understanding the AIS problem space. In: Adaptive Instructional System (AIS) Standards Workshop (2019)
15. Sottilare, R.A., Baker, R.S., Graesser, A.C., Lester, J.C.: Special issue on the generalized intelligent framework for tutoring (GIFT): creating a stable and flexible platform for innovations in AIED research. Int. J. Artif. Intell. Educ. **28**(2), 139–151 (2018)
16. Sottilare, R.A., Brawner, K.W., Sinatra, A.M., Johnston, J.H.: An updated concept for a generalized intelligent framework for tutoring (GIFT). GIFTtutoring.org. 1–9 (2017)
17. Gallego-Durán, F.J., Molina-Carmona, R., Llorens-Largo, F.: Measuring the difficulty of activities for adaptive learning. Univ. Access Inf. Soc. **17**(2), 335–348 (2018)
18. Sottilare, R.A., Goodwin, G.A.: Adaptive instructional methods to accelerate learning and enhance learning capacity. In: International Defense and Homeland Security Simulation Workshop of the I3M Conference (2017)
19. Shute, V.J., Zapata-Rivera, D.: Adaptive educational systems. Adapt. Technol. Training Educ. **7**(27), 1–35 (2012)
20. Merrill, D.C., Reiser, B.J., Ranney, M., Trafton, J.G.: Effective tutoring techniques: a comparison of human tutors and intelligent tutoring systems. J. Learn. Sci. **2**(3), 277–305 (1992)
21. Wang, N., Gratch, J.: Can virtual human build rapport and promote learning? In: Artificial Intelligence in Education, pp. 737–739. IOS Press (2009)
22. Ward, W., Cole, R., Bolaños, D., Buchenroth-Martin, C., Svirsky, E., Weston, T.: My science tutor: a conversational multimedia virtual tutor. J. Educ. Psychol. **105**(4), 1115 (2013)
23. Park, S., Kim, C.: Boosting learning-by-teaching in virtual tutoring. Comput. Educ. **82**, 129–140 (2015)
24. Castellano, G., et al.: Towards empathic virtual and robotic tutors. In: Lane, H.C., Yacef, K., Mostow, J., Pavlik, P. (eds) Artificial Intelligence in Education. AIED 2013. LNCS, vol. 7926. Springer, Heidelberg (2013). https://doi.org/10.1007/978-3-642-39112-5_100
25. Brawner, K., Ballinger, C., Sottilare, R.: Evaluating the Effectiveness of Artificially Intelligent Agents. Florida AI Research Society (2022, submitted)
26. McGaghie, W.C., Issenberg, S.B., Barsuk, J.H., Wayne, D.B.: A critical review of simulation-based mastery learning with translational outcomes. Med. Educ. **48**(4), 375–385 (2014)
27. Atkinson, R.K., Derry, S.J., Renkl, A., Wortham, D.: Learning from examples: instructional principles from the worked examples research. Rev. Educ. Res. **70**(2), 181–214 (2000)
28. Durlach, P.J., Ray, J.M.: Designing adaptive instructional environments: insights from empirical evidence (2011)
29. Durlach, P.J., Spain, R.D.: Framework for instructional technology. Adv. Appl. Hum. Model. Simul. **9**, 222–231 (2012)
30. Sottilare, R.A., DeFalco, J.A., Connor, J.: A guide to instructional techniques, strategies and tactics to manage learner affect, engagement, and grit. Des. Recommendations Intell. Tutoring Syst. **2**, 7–33 (2014)
31. Hu, X., Tong, R., Cai, Z., Cockroft, J.L., Kim, J.W.: Self-improvable adaptive instructional systems (SIAISS)–a proposed model. Des. Recommendations Intell. Tutoring Syst. **23**, 11 (2019)
32. Long, Z., Andrasik, F., Liu, K., Hu, X.: Self-improvable, self-improving, and self-improvability adaptive instructional system. In: Pinkwart, N., Liu, S. (eds) Artificial Intelligence Supported Educational Technologies. Advances in Analytics for Learning and Teaching, pp. 77–91. Springer, Cham (2020). https://doi.org/10.1007/978-3-030-41099-5_5
33. Kay, J.: Learner control. User Model. User-Adap. Inter. **11**(1), 111–127 (2001)

Iterative Refinement of an AIS Rewards System

Karen Wang[1]($\boxtimes$), Zhenjun Ma[2], Ryan S. Baker[3] (iD), and Yuanyuan Li[2]

[1] Worcester Polytechnic Institute, 100 Institute Rd, Worcester, MA 01609, USA
karenw2005@gmail.com
[2] Learnta, 1460 Broadway, New York, NY 10036, USA
[3] University of Pennsylvania, 3700 Walnut St., Philadelphia, PA 19104, USA

Abstract. Gamification-based reward systems are a key part of the design of modern adaptive instructional systems and can have substantial impacts on learner choices and engagement. In this paper, we discuss our efforts to engineer the rewards system of Kupei AI, an adaptive instructional system used by elementary and middle school students in afterschool programs to study English and Mathematics. Kupei AI's rewards system was iteratively engineered across four versions to improve student engagement and increase progress, involving changes to how many points were awarded for success in different activities. This paper discusses the design changes and their impacts, reviewing the impacts (both positive and negative) of each generation of re-design. The end result of the design was improved learning and more progress for students. We conclude with a discussion of the implications of these findings for the design of gamification for adaptive instructional systems.

Keywords: Adaptive instructional system · Gamification · Reward system

1 Introduction

Gamification-based reward systems are an increasingly common part of the design of modern adaptive instructional systems and can have substantial impacts on learner choices and engagement [15]. For example, GamiCAD introduced arcade-style bonus missions based on successful performance, leading to faster completion of content and more completion of content, as well as self-report of better engagement [10]. The gold bars given when students master a concept in a Cognitive Tutor have also been found to be a positive incentive for many students [11]. Some research has suggested that not all students find reward systems compelling, but those who do tend to see increased motivation and learning [16]. Much of the work using rewards for gamification builds on an intellectual tradition of behavior management and modification that dates back to early behaviorist work. In that work, reward systems were studied for their impact on behavior and learning [17]. Morford and colleagues [13] note the intellectual contribution that the applied behavior analysis and behavior modification literatures have made to contemporary work on gamification.

However, as been noted since the 1970s, reward systems can impact students in negative fashions as well as positive fashions [9]. Reward systems, if incorrectly designed, can

© The Author(s), under exclusive license to Springer Nature Switzerland AG 2022
R. A. Sottilare and J. Schwarz (Eds.): HCII 2022, LNCS 13332, pp. 113–125, 2022.
https://doi.org/10.1007/978-3-031-05887-5_9

focus learners on short-term outcomes rather than long-term outcomes and lead students to adopt behaviors that maximize the reward rather than the learning it is intended to promote. Furthermore, not all reward systems are even effective at promoting the intended behaviors [12] – the details of the design appear to matter considerably. Reward systems, if poorly designed, can also reduce long-term intrinsic motivation for the learning activity [3]. As such, considerable attention needs to be paid to the design of reward systems in adaptive instructional systems. Design principles for gamification [6] can be useful but, as discussed above, even a seemingly excellent design can have unintended consequences [5, 9]. As such, the methods of learning engineering [5] are needed to ensure that designs achieve their intended goals. In particular, ongoing monitored iterative design [8] is needed, where the developer repeatedly modifies the system and tests the consequences of those modifications.

In this paper, we present the monitored iterative design of a gamified rewards system for the Kupei AI adaptive instructional system, an AI-driven learning system tutoring elementary (3rd–6th graders), middle (7th–9th), and high (9th–12th) schoolers in English and Math in China. Across four iterations, a point system was implemented in order to encourage students to master concepts and improve their learning progress. The point system was intended to give students an incentive to work on specific concepts until they reached mastery, using the system's various features appropriately and efficiently. However, students developed strategies targeted towards earning points with the maximum efficiency that were not optimal for learning. For instance, certain design choices led students to put more effort into curricular areas where it was easier to rapidly master topics. Through multiple design iterations, we were able to design a system that guided students towards more appropriate system usage.

In this paper, we present the story of the iterative design of this system, presenting each iteration in design, and empirically investigating its results on student usage behaviors and learning outcomes. In presenting this narrative, we consider broader themes around how to design effective gamification systems, distilling what we learned at each phase of redesign.

2 Methods

2.1 Platform

Kupei AI is an AI-driven learning system tutoring elementary (3rd–6th graders), middle (7th–9th), and high (9th–12th) schoolers in English and Math in China. Courses are divided up by grade level and then by units then concepts. The units are correlated with what students are learning in school. Each unit comprises a list of concepts, with the subjects the learning system recommends for the student at the top (Fig. 1).

The intended design is for students to begin a unit with a diagnostic test, a short quiz, although they have the choice to directly go into practice (Fig. 1). If students decided to skip the test and go straight to practice, then a message window popped up that encouraged the student to take a diagnostic test first. The diagnostic test covers multiple concepts to determine which concepts the student should work on within the unit. Since concepts are connected through a knowledge graph, if a student achieves advanced mastery on a concept, then the learning system infers that the student has also mastered the concept's prerequisites. Following the diagnostic test, students have the option to practice. Practice is divided into problem sets of 5 items on the same concept – however, if a student achieves advanced mastery (defined below) after the first 3 items in a problem set, the system stops the student's work on the problem set automatically.

Once finished with either the practice or test, students are met with a concluding screen showing concepts where they achieved basic mastery and those that were not mastered (Fig. 2). The learning system categorizes student proficiency on concepts students have worked on into three levels: unmastered, basic mastery, and advanced mastery. Using Bayesian Knowledge Tracing [3], the system estimates student proficiency in real-time. If a student has under an 80% probability of knowing a skill, the concept is unmastered; if the probability is between 80% (a cut-off used by many commercial systems) and 95% (the original cut-off in [3]), then the concept is labeled as basic mastered; if the concept mastery level is above 95% then it is labeled as advanced mastery. By design, advanced mastery can only be earned through practice while basic mastery can be earned through both practice and tests.

The original system (referred to below as version V0) did not incorporate gamification, but starting April 29th, 2021, the Kupei AI team released and began the process of iterating a point system. In its first version (referred to below as version V1), the point system gave students one point for each basic mastered concept. Students were able to click on a little treasure chest next to a summary of the specific learning concept they mastered to receive the point (Fig. 3). Starting with V1, after completing a practice or test, students can click open treasure chests next to each concept they gained basic mastery on and earn a point. Students were also ranked against their classmates by the number of points they have, and this was shown in a leaderboard available to students to view after finishing a test or practice set.

Fig. 1. Student unit learning screen. The large orange button says "test" and the red button at the bottom right says "practice." In this unit, there are 15 concepts to master. The progress bar next to the practice button shows mastery progress for each concept. (Color figure online)

Fig. 2. The results screen after a diagnostic test. The left circle shows the student received a score of 15/100. The right circle shows the number of concepts mastered in the unit, which is 2/15. The bottom row of numbers from 1 to 7 are the questions the student answered in the diagnostic test: red means they answered the question incorrectly and green means they answered correctly. (Color figure online)

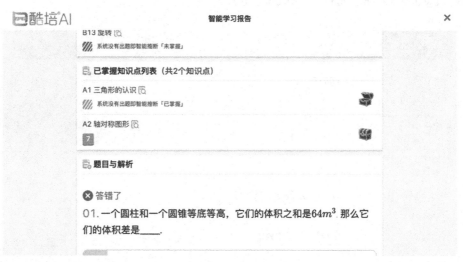

Fig. 3. Students scroll to the questions they answered correctly and click on the treasure chest next to each learning concept to receive their points.

2.2 Data Collection

The entire student user base of Kupei, composed of 5726 students, was tracked from April 22nd to June 17th. From April 22 to 28, the students used the baseline version without the point system (V0). From April 29 to May 8, the students used the first version of the point system (V1). On May 9 and June 1, respectively, iterated versions V2 and V3 of the point system were introduced.

Across these iterations, the students worked on 27785 concepts, making a total of 1,393,529 actions. During this time, students achieved basic mastery on 29.7% of the concepts they attempted and achieved advanced mastery on 41.2% of the concepts they attempted. Although the system notifies the student when they achieve basic mastery, students generally did not stop at basic mastery, and went on to advanced mastery, leading to a higher proportion of advanced mastery.

2.3 Overview of Design Iterations

When the first version V1 of the point system first launched on April 29th, its primary purpose was to make learning more engaging by giving students a reward for earning basic mastery on a concept. By giving a point for each concept mastered, and showing the class's scores on a leaderboard, the intention was to incentivize students to compete [14] to earn a large number of points to surpass their classmates on the leaderboard. However, though the student's goal would be to perform well and compete, they would be learning more concepts and making more progress within the system to do so.

Before the point system, Version 0 (V0), students were unable to earn points. The first version of the point system (V1) gave the student a point the first time they earned basic mastery on a specific learning concept, and introduced the leaderboard. Version V1 gave a single point for mastering any concept, regardless of whether the concept was in English or Math.

As will be discussed below, Math concepts generally take longer to complete in the system than English concepts. As a result, students who wanted to be at the top of the leaderboard shifted to working mostly on concepts in English. Version V2 attempted to correct for this unanticipated incentive by making basic mastery in English worth one point and mastery in Math worth three points. This change re-balances work on the two subject areas.

Having addressed this limitation, the design team noticed that students were now much more likely to skip the diagnostic test. The diagnostic test is designed to determine which concepts in the unit the student needs to work on, avoiding allocating student time to concepts they already know. However, by skipping the test, students can master concepts they already know, gaining points without learning. Version V3, therefore, made it possible to earn three points for achieving Math basic mastery and one point for achieving English basic mastery in practice and on tests.

In the following section, we go into greater detail on the impact of each of these design changes on student behavior, learning, and learning efficiency.

3 Analysis and Results

3.1 V0 to V1 Design Change Impact

V0 represented the first version of the system, before the introduction of the point system. We added a point system and leaderboard to V0, creating V1. In V1, students could earn points when they achieved basic mastery of concepts while practicing. English and Math were both worth one point for each basic mastered concept.

First, we can compare students between V0 and V1 in terms of their mastery rates. We consider the basic mastery rate, or the number of concepts a student achieved basic mastery on divided by the total number of concepts they worked on. We also consider advanced mastery rate, the number of concepts a student achieved advanced mastery on, divided by the total number of concepts they worked on. In doing so, we only look at students who used the system during both the V0 and V1 periods and conduct a paired comparison, as only a very small number of students quit or started the system right at the switchover date.

In V0, students had a basic mastery rate of 62.5% skewness $= -0.563$. In V1, students had a basic mastery rate of 64.1%, a statistically significantly faster basic mastery rate, $t(3683) = -3.3367$, $p < 0.001$, skewness $= -0.654$. In V0, students had an advanced mastery rate of 43.1%. In V1, students had an advanced mastery rate of 45.4%, a statistically significantly faster mastery rate, $t(3368) = -4.2585$, $p < 0.001$, skewness $= 0.229$ V0, 0.080 V1. All values of skewness for mastery rate were in the moderately skewed or approximately symmetric ranges, justifying the use of parametric tests.

From V0 to V1, the time students took to achieve basic mastery on each concept decreased by a median of 35.7 s in English, $V = 498656$, $p < 0.001$, and went up by a median of 7.3 s which is statistically significant, $V = 1283079$, $p < 0.001$, in Math.

However, the proportion of time spent on English versus Mathematics shifted between the two versions. In V0, 47.403% of students' completed concepts were English and 52.6% were Math. This changed considerably in V1, where students spent 59.5% of their completed concepts were English, and 40.5% of their completed concepts were Math. The increase in the proportion of time spent on English from V0 to V1 was statistically significant, $t(779) = -10.614$, $p < 0.001$, skewness $= 0.0692$ V0, -0.441 V1. The decrease in the proportion of time spent on Math from V0 to V1 was also statistically significant, $t(779) = 10.614$, $p < 0.001$.

This difference in time spent led to a difference in the rate of basic mastery between the two content domains. The basic and advanced rates in each subject increased from V0 to V1. In V0, students' median English basic mastery rate was 75%. In V1, it was 74%, not a statistically significant difference (Wilcoxon used due to high skewness), $V = 285187$, p-value $= 0.1878$, skewness $= -0.911$ V0, -0.945 V1. Math basic mastery rate in V1 went up as well. Students had a basic mastery rate of 59.4% in V0 which went up to 60.8% in V1, a statistically significant increase, $t(2208) = 2.28$, $p = 0.023$, skewness $= -0.411$ V0, -0.511V1. Additionally, the advanced mastery rate in English statistically significantly increased from 49.7% in V0 to 52.6% in V1, $t(1159) = -3.2294$, $p = 0.001$, skewness $= -0.091$ V0, -0.222 V1. The advanced mastery in math increased from 39.6% in V0 to 41.6% in V1, $t(2208) = -2.9198$, $p = 0.004$, skewness $= -0.411$ V0, -0.511 V1. Both mastery rate increases were significant.

Another shift found was that students were much more likely to skip the diagnostic tests in V1 than in V0. In V0, 49.1% of student items were on practice and 50.9% of student items were on diagnostic tests. In V1, 64.0% of student items were on practice and 36.0% of student items were on diagnostic tests. Students were statistically significantly more likely to practice in V1 than V0, by 14.9%, $t(871) = 16.15$, p-value < 0.001, skewness $= 0.0279$ V0, -0.507 V1.

3.2 V1 to V2 Design Change

To try to re-balance the proportion of time spent in English and Math, we changed the points given for mastering a concept in Math from 1 point to 3 points (version V2). As Fig. 6 shows, this led to a gradual shift back to Math. Students completed more Math concepts in V2 than in V1, leading to the overall proportion of concepts completed being approximately equal for English and Math in V2. The proportion of English completed by students in V1 was 55.3% and the proportion of math completed was 44.73878%. In V2, English concepts completed significantly decreased to 50.7%, $t(1414) = 5.92$, $p < 0.001$ and Math significantly increased to 49.3%, $t(1414) = -5.92$, $p < 0.001$, skewness $= -0.250$ V1, -0.177 V2.

The overall proportion of practice continued to significantly increase, from 56.5% in V1 to 66.5% in V2, $t(1552) = 15.02$, $p < 0.001$, skewness $= -0.218$ V1, -0.694 V2, and the proportion of tests significantly decreased from 43.5% to 33.5% in V2, $t(1552) = -15.02$, $p < 0.001$, skewness $= 0.218$ V1, 0.694 V2. As students could only earn points in practice, the increase in the proportion of practice is seen in both English and Math, with the change in Math being especially large. The proportion of English practice completed significantly increased from 66.7% in V1 to 72.7% in V2, $t(730) = 7.58$, $p < 0.001$, skewness $= -0.541$V1, -0.767 V2. In addition, Math practice significantly increased from 45.0% in V1 to 59.6% in V2, $t(923) = 16.60$, $p < 0.001$, skewness $= 0.180$ V1, -0.439 V2.

However, since students completed more practice and each practice set targets one learning concept at a time, this led to a statistically significant increase in the time taken to master both Math and English concepts. In Math, the median time taken to master a concept increased from 734.67 s in V1 to 813.97 s in V2, $V = 1165568$, $p < 0.001$. In English, the median time taken to master a concept increased from a median of 347.45 s in V1 to 377.6437 s in V2, also statistically significant, $V = 328174$, p-value < 0.001.

Despite changes in student learning behavior, the only significant change for both basic and advanced mastery rate was a decrease in the English advanced mastery rate. Math basic mastery rate was 60.0% in V1 and 60.4% in V2, which was not a statistically significant difference, $t(3141) = 0.93$, $p = 0.350$, skewness $= -0.460$ V1, -0.460 V2. Although, the median English basic mastery rate was 71% in V1 and 73% in V2, was a statistically significant increase, $V = 892435$, p-value $= 0.001628$, skewness $= -0.907$ V1, -0.824 V2. Advanced mastery for Math was 37.98% in V1 and 37.99% in V2, $t(3141) = -0.021366$, $p = 0.983$, skewness $= 0.411$ V1, 0.399 V2, but significantly decreased for English from 49.5% in V1 to 47.7% in V2, $t(1903) = 2.89$, $p = 0.004$, skewness $= -0.151$ V1, -0.011 V2.

3.3 V2 to V3 Design Change

The increase in practice in V1 and V2 implied that students were no longer using the diagnostic test as often to focus their time on the skills they needed to learn. We therefore re-designed V3 to give students a point for demonstrating mastery of a concept within the diagnostic test. In V2, students completed the majority of the concepts they encountered through practice, with 64.6% of concepts completed being practice and 35.4% being test. After V3 launched, this ratio switched, with 46.4% of concepts completed in practice and 53.6% in tests. Overall, the proportion of completed concepts in practice significantly decreased from 64.6% in V2 to 46.4% in V3, $t(2489) = 30.84$, $p < 0.001$, skewness $= -0.521$ V2, 0.157 V3, from V2 to V3, and tests significantly increased from 35.4% in V2 to 53.6% in V3, $t(2489) = -30.84$, $p < 0.001$. As shown in Figs. 4 and 5, this applied to both English and Math concepts. Math testing increased from 42.2% in V2 to 58.9% in V3, $t(1712) = 24.02$, $p < 0.001$, skewness $= 0.302$ V2, -0.329 V3 compared to Math practice. English testing increased from 27.4% in V2 to 46.5% in V3, $t(1167) = 23.44$, $p < 0.001$, skewness $= -0.818$ V2, -0.142 V3, compared to English practice. As shown in Fig. 7, as students completed more diagnostic tests in V3, they were also able to achieve basic mastery on concepts faster. For English, the median time spent to achieve basic mastery decreased by 12.4% from 372.4 s in V2 to 326.2 s in V3, $V = 2184932$, $p < 0.001$, and in Math, the median time spent decreased by 27.5% from 773.2 s in V2 to 560.3 s in V3, $V = 6007553$, $p < 0.001$.

Although basic mastery rate did not significantly change for Math from V2 to V3, it did significantly decrease from 67.8% in V2 to 64.7% in V3 for English, $t(2317) = 12.674$, $p < 0.001$, skewness $= -0.812$ V2, -0.721 V3. In addition, advanced mastery significantly decreased for both subjects with a decrease of 2.6% in Math, $V = 3733852$, $p < 0.001$, and 7.1% in English, $t(2317) = 12.67$, $p < 0.001$, skewness $= 0.035$ V2, 0.391 V3.

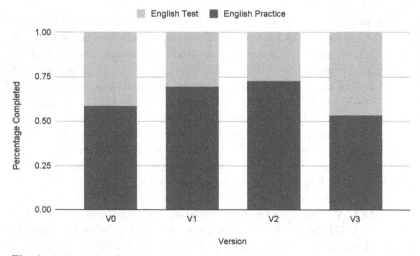

Fig. 4. Student completed English concept proportion test and practice per version.

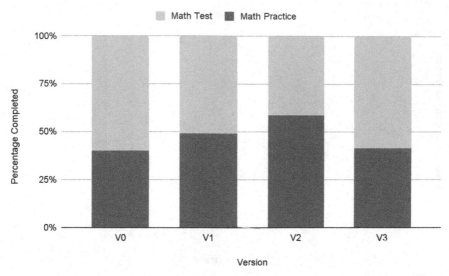

Fig. 5. Student completed Math concept proportion test and practice per version.

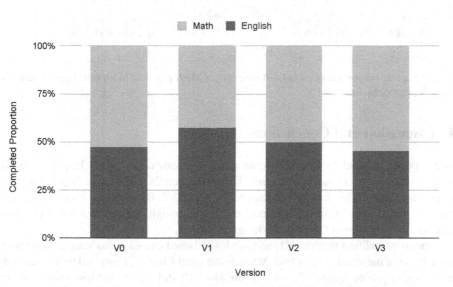

Fig. 6. Student completed concepts proportions of Math and English.

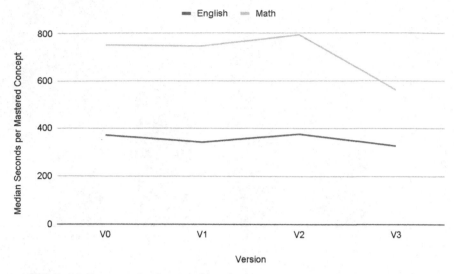

Fig. 7. Median seconds taken to achieve basic mastery on one concept per version.

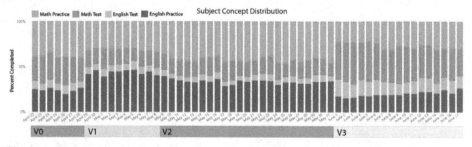

Fig. 8. Student proportion of completed concepts in Math practice, Math test, English test, and English practice by day.

4 Discussion and Conclusions

In this paper, we add gamification to an adaptive instruction system, Kupei AI, in the form of a point system and leaderboard. This paper details the iterative improvement of that point system through three versions, V1, V2, and V3. By adding a rewards system to this AIS, it was possible to shift student behavior, although there were some unexpected consequences in some of the earlier design iterations.

The first modified version, V1, added a leaderboard and gave students a single point for achieving basic mastery in both Mathematics and English. From V0 to V1, learning efficiency improved, especially for English. Not only did they spend less time to master a learning concept for English, decreasing by 35.7 s, but their mastery rate, which is their total number of mastered concepts over their total completed concepts, increased for both English and Math by 2.1% and 1.4% respectively. In addition, student learning behavior changed from V0 to V1 as students focused on completing more English and more practice. Students were incentivized to complete tasks that earned them points quickly. So, since students could only earn points in practice, not test, they completed an increase in practice. Also, English took less time to master English than Math so students mastered more English concepts than Math. Despite students being more motivated to master concepts quickly to earn points, they focused on earning points the quicker way through English practice rather than through Math.

After seeing students switch to working much more on English than Math, V2 was implemented to balance the number of concepts students spent learning. Since it was easier to earn points quickly in English, V2 balanced things by giving students more points per concept mastered in Math. As such, the proportion of English worked on reduced, bringing English and Math back into greater balance. This change affected student mastery performance, with English basic mastery rate increasing by 2%. However, students spent more time to master a concept. This occurred because students completed an even greater proportion of practice compared to tests, meaning that students were practicing concepts in full that they would have bypassed through taking the diagnostic test. If students mastered a concept in the diagnostic test, they would not earn points for it, so they skipped straight to practice to earn points. This shift from test to practice between V1 and V2 may have been a continued trend from V0 to V1, or it may have been that the greater number of points available was a stronger inducement to complete practice rather than tests.

Doing only practice is not the most effective way to learn in Kupei AI, as the diagnostic tests help students focus on concepts they need to improve rather than going through all concepts. This is especially important for Math. To address this issue and encourage students to complete the diagnostic tests, V3 made it possible for students to earn mastery points through demonstrating mastery within tests. Quickly, students began to complete more tests and less practice, as the proportion of practice completed compared to tests decreased by 18.2%. Testing increased considerably in both subjects. As a result, the time spent mastering concepts decreased significantly, with English decreasing 46 s and Math decreasing 213 s. However, The amount of Basic mastery went down in English and advanced mastery decreased in both Math and English, by 2.6% and 7.1% respectively. The decrease in English basic mastery rate was likely due to students spending more time on Math diagnostic tests. The decrease in advanced mastery was most likely because students can only achieve advanced mastery through practice, not tests. Since achieving advanced mastery did not reward students with any points, once a student mastered a concept through tests, there was no incentive for them to go back and practice that concept further to earn advanced mastery.

Although there has been considerable design progress from V0 to V3, more progress remains possible. In a sense, the design of gamification and rewards systems can be an extended journey, as each change has a mixture of positive and negative (typically unintended) consequences. In a future version of a system like this nature, it would probably make sense to also introduce rewards for achieving advanced mastery. Even if this results in some over-practice, there is likely some advantage to this in terms of preparation for learning future concepts [7] and retention of knowledge over time [1]. Further changes will become necessary, as each change shifts the overall pattern of student behavior.

In reflecting not just on the implications of our design but also on the implications of our approach to design, we can see that there are several advantages (and of course some limitations) to the approach we have developed. Switching rapidly between designs enables rapid iteration towards a better design, but it also means that the impacts of changes may bleed together. Longer periods of each version would make differences between versions clearer but requires more time to hone in on a better system. In this analysis, we switched all students to the next version at the same time and compared time periods. An alternate (more common) approach would have been to conduct A/B tests. However, there would be significant possible issues if we randomly assigned at the student level (i.e. students would have different reward structures in the same learning center, creating effects such as resentful demoralization [2]). Therefore, we would need to assign students to condition by learning center, creating challenges in the comparability of students between conditions. One of the largest reasons to use an A/B test rather than compare across time periods is the possibility that previous versions impact current behavior – however, Fig. 8 shows the relatively rapid response of students to design changes. In general, the need to iterate design based on the effects of the current intervention makes it less feasible to conduct many-condition A/B/n studies, slowing the degree to which assignment of students to condition could speed design progress in this case. Overall, rapid iteration of the type used in this study has some limitations, as students who use multiple versions may respond to later versions differently than students who start with those later versions. This risk reduces the degree to which we can infer causality and generalizability for our changes, but the result is a system that produces better results. Clear and unquestionable evidence on what caused the impact can only be produced through a randomized comparison on entirely new students, but is that always the primary goal? Ultimately, the effort needed for an unconfounded randomized study may be better spent on an efficacy study comparing to a control condition.

Further research may also want to look for the impact of these design iterations on other aspects of student behavior. For instance, since Kupei AI's learning system has many different grade levels available for students to work on, students may be working on problems beneath their grade level just to earn points quickly. Relatedly, it may be worth looking into the degree to which the changes observed across versions differs for younger and older students using the platform. It is also possible that these design changes may be influencing the frequency and form of gaming the system (Baker et al. 2004) that is occurring in the system.

Overall, over the course of these modifications, we were able to rapidly iterate Kupei AI's points system, removing the unintended consequences emerging in earlier versions and increasing the speed with which students mastered concepts through the system. This higher efficiency creates several opportunities, including the ability to cover a greater proportion of the year's content in a limited amount of after-school time and focus student time better on the content they need to learn. A next step will be to look at whether this improved distribution of time leads to greater learning gains on an external examination of knowledge, and the impacts of the re-design on student retention in the afterschool program. Overall, by allocating student time more effectively, we can lead to more efficient learning and in the long term, hopefully, better outcomes for learners.

Acknowledgements. We would like to thank the developers of the Kupei AI system and the students, teachers, and parents who used the system, for their participation in this research.

References

1. Cen, H., Koedinger, K.R., Junker, B.: Is over practice necessary?-improving learning efficiency with the cognitive tutor through educational data mining. In: Proceedings of the International Conference on Artificial Intelligence and Education, pp. 551–558. IOS Press (2007)
2. Cook, T.D., Campbell, D.T.: Quasi-Experimentation: Design and Analysis Issues for Field Settings. Houghton Mifflin, Boston (1979)
3. Corbett, A.T., Anderson, J.R.: Knowledge tracing: modeling the acquisition of procedural knowledge. User Model. User-Adapt. Interact. **4**(4), 253–278 (1995)
4. Deci, E.L., Koestner, R., Ryan, R.M.: A meta-analytic review of experiments examining the effects of extrinsic rewards on intrinsic motivation. Psychol. Bull. **125**, 627–668 (1999)
5. Dede, C., Richards, J., Saxberg, B. (eds.): Learning Engineering for Online Education: Theoretical Contexts and Design-Based Examples. Routledge (2018)
6. Dicheva, D., Dichev, C., Agre, G., Angelova, G.: Gamification in education: a systematic mapping study. J. Educ. Technol. Soc. **18**(3), 75–88 (2015)
7. Koedinger, K.R., Corbett, A.T., Perfetti, C.: The knowledge-learning-instruction framework: bridging the science-practice chasm to enhance robust student learning. Cogn. Sci. **36**(5), 757–798 (2012)
8. Koedinger, K.R., Sueker, E.L.F.: Monitored design of an effective learning environment for algebraic problem solving. Technical report CMUHCII-14-102. Carnegie Mellon University (2014)
9. Lepper, M.R., Greene, D.: The Hidden Costs of Reward: New Perspectives on the Psychology of Human Motivation. Psychology Press (2015)
10. Li, W., Grossman, T., Fitzmaurice, G.: GamiCAD: a gamified tutorial system for first time autocad users. In: Proceedings of the 25th Annual ACM Symposium on User Interface Software and Technology, pp. 103–112. Association for Computing Machinery (2012)
11. Long, Y., Aleven, V.: Students' understanding of their student model. In: Biswas, G., Bull, Susan, Kay, J., Mitrovic, A. (eds.) AIED 2011. LNCS (LNAI), vol. 6738, pp. 179–186. Springer, Heidelberg (2011). https://doi.org/10.1007/978-3-642-21869-9_25
12. Long, Y., Aleven, V.: Gamification of joint student/system control over problem selection in a linear equation tutor. In: Trausan-Matu, S., Boyer, K.E., Crosby, M., Panourgia, K. (eds.) Intelligent Tutoring Systems. ITS 2014. Lecture Notes in Computer Science, vol. 8474, pp. 378–387. Springer, Cham (2014). https://doi.org/10.1007/978-3-319-07221-0_47
13. Morford, Z.H., Witts, B.N., Killingsworth, K.J., Alavosius, M.P.: Gamification: the intersection between behavior analysis and game design technologies. Behav. Anal. **37**(1), 25–40 (2014)
14. Muñoz-Merino, P.J., Molina, M.F., Muñoz-Organero, M., Kloos, C.D.: An adaptive and innovative question-driven competition-based intelligent tutoring system for learning. Expert Syst. Appl. **39**(8), 6932–6948 (2012)
15. Shortt, M., Tilak, S., Kuznetcova, I., Martens, B., Akinkuolie, B.: Gamification in mobile-assisted language learning: a systematic review of Duolingo literature from public release of 2012 to early 2020. Comput. Assist. Lang. Learn. 1–38 (2021)
16. Tahir, F., Mitrovic, A., Sotardi, V.: Investigating the effects of gamifying SQL-tutor. In: Proceedings of the International Conference on Computers in Education (2020)
17. Terrell Jr., G., Kennedy, W.A.: Discrimination learning and transposition in children as a function of the nature of the reward. J. Exp. Psychol. **53**(4), 257 (1957)

Examining Two Adaptive Sequencing Approaches for Flashcard Learning: The Tradeoff Between Training Efficiency and Long-Term Retention

Daphne E. Whitmer[✉], Cheryl I. Johnson, and Matthew D. Marraffino

Naval Air Warfare Center Training Systems Division, Orlando, FL 32826, USA
daphne.whitmer@gmail.com

Abstract. In an effort to modernize training, the United States Navy and Marine Corps have placed an emphasis on identifying effective, learner-centric instructional methods. One avenue is to apply individualized training techniques, such as adaptive sequencing to flashcard-based study, a popular tool used for independent study. Therefore, the goal of this research was to compare two adaptive sequencing methods to identify the most efficient and effective approach for long-term retention. Participants learned armored vehicle identification in one of three conditions, 1) Adaptive Response-Time Based-Sequencing (ARTS), which uses accuracy and reaction time to prioritize flashcards adaptively, 2) Leitner, which uses accuracy to create decks and prioritize flashcards adaptively, and 3) Random, which was the control condition that sequenced the flashcards randomly. We found no differences between the three conditions in terms of learning efficiency and delayed learning gains. However, we found that participants in the Leitner condition completed training significantly faster those in the other conditions. After controlling for in-training measures, we found participants in the Leitner condition had the lowest delayed learning gains and there were no significant differences between the ARTS and Random conditions. These data suggest that the "efficiency" associated with the Leitner condition translated to the worst long-term retention. Additionally, neither adaptive approach outperformed the random sequencing condition, suggesting that random sequencing may provide its own spacing due to the number of cards in the deck. More research is needed to determine whether adaptive sequencing provides added value as an adaptive approach for single-session flashcard study.

Keywords: Adaptive training · Flashcards · Spacing effect · Testing effect · Long-term retention · Learning efficiency

1 Introduction

The United States Navy and United States Marine Corps (USMC) have an increased interest in improving and modernizing current training methods for "21st century learning" by making them more learner-centered (Berger 2019; Gilday 2019). As part of their

© The Author(s), under exclusive license to Springer Nature Switzerland AG 2022
R. A. Sottilare and J. Schwarz (Eds.): HCII 2022, LNCS 13332, pp. 126–139, 2022.
https://doi.org/10.1007/978-3-031-05887-5_10

professional military training and education, Service members are required to memorize information such as ranks and insignia, map symbols, unit locations, and vehicle identification, just to name a few. Therefore, the present study focused on exploring individualized training methods to improve rote learning for Service members to align with these 21st century learning goals. In particular, we focused on merging flashcard-based learning with adaptive training techniques to improve rote learning outcomes and meet the needs of modern military training and education. Among students and instructors alike, flashcards are a popular study tool to accomplish independent rote memorization on basic topics (Hartwig and Dunlosky 2012; Karpicke, Butler, and Roediger 2009; Kornell and Bjork 2007). There is a need to explore individualized approaches for these types of digital-aged resources for students, as students are responsible for studying outside of course instruction. One way to individualize flashcard learning is to implement adaptive training techniques. Previous work has demonstrated that implementing mastery learning adaptively can aid consistent long-term retention compared to non-adaptive and learner controlled techniques (Whitmer et al. 2020). In the present research, the goal was to compare adaptive sequencing approaches for flashcard learning to promote long-term retention. In this paper, we discuss the literature behind flashcard learning and adaptive training, and we report an experiment comparing two different adaptive sequencing approaches to random sequencing.

1.1 Retrieval Practice with Flashcards

The extensive flashcard learning literature has demonstrated that flashcards help students during knowledge acquisition, but the way students engage with the materials can have a large impact on learning. In general, students use flashcards to prepare for exams by memorizing the information, testing themselves on the information, and assessing their proficiency (Hartwig and Dunlosky 2012; Karpicke, Butler, and Roediger 2009; Kornell and Bjork 2007; Wissman, Rawson and Pyc 2012). The way students use flashcards to self-test is related to the classic finding in the literature called the testing effect, which states that testing oneself is more beneficial to long-term retention than just studying or re-reading the material. Karpicke and Roediger (2008) showed that repeated testing had large and impactful effects on retention compared to restudying. In this study, 80% of the repeatedly tested foreign language vocabulary words were recalled correctly one week after training, compared to the 30% of the words that were restudied. This finding, along with many related studies (e.g., Carrier and Pashler 1992; Roediger and Butler 2011; Roediger and Karpicke 2006; Toppino and Cohen 2009), has provided evidence that self-testing with flashcards and practicing the memory retrieval process (otherwise known as retrieval practice) can have positive effects on long-term retention.

A technique within the flashcard learning literature that is often considered is the time interval between flashcards. The spacing effect is a robust finding that suggests that increasing the time interval or spacing out multiple study sessions over longer timespans is more beneficial to long-term retention than "cramming" the information once or twice before an exam (Bahrick, Bahrick, Bahrick and Bahrick 1993; Cepeda et al. 2006; Ebbinghaus 1885, 1964; Glenberg and Lehmann 1980; Kornell 2009; Paivio 1974; Shaughnessy, Zimmerman and Underwood 1974). However, research shows that students prefer to cram their studying (Kornell and Bjork 2008a), which only aids short-term memory with a rapid rate of forgetting (Kornell 2009). Students do not implement spacing effectively on their own, perhaps due to poor metacognitive strategies (Bjork, Dunlosky and Kornell 2013), less concern over long-term retention, or they have to make the most of limited study time (Sommer 1990). A way to address some of these concerns and meet these limitations is to build flashcard-learning systems that space out or sequence the flashcards on behalf of the student to increase the efficiency of retrieval practice while maximizing long-term learning gains in shorter-term study periods.

1.2 Adaptive Training and Adaptive Flashcards

Adaptive training, or training that is tailored to a trainee's performance during training (Landsberg et al. 2011), is a learner-centered approach that has been shown to produce positive learning outcomes in several different military domains (e.g., Landsberg et al. 2012; Marraffino et al. 2019, 2021; Van Buskirk et al. 2019; Whitmer et al. 2020, 2021) and non-military domains (e.g., Peirce and Wade 2010; Romero et al. 2006; VanLehn et al. 2005). One example of using adaptive training strategies with flashcards for a military domain includes using mastery criteria to determine when to drop cards from further study. In an experiment by Whitmer and colleagues (2020), participants learned to identify armored vehicles and their retention was measured two days after training. There were three training conditions: an adaptive condition where mastering a flashcard based on pre-determined criteria dropped it from the deck, a non-adaptive condition where no dropping occurred, and a learner-controlled condition where the learner was in control of dropping the flashcards. Those in the adaptive condition had the most consistent retention on a delayed test two days later compared to the non-adaptive and learner-controlled conditions, and the learner-controlled condition had the worst overall performance.

Another learner-centered approach to adaptive flashcard learning is to customize the order of the flashcards using the individual learner's performance. Adaptively sequencing the flashcards can make the time interval of the next flashcard presentation optimal based on how the student is performing. This approach reshuffles the deck using the learner's last response. Past research in the flashcard learning literature has suggested that adaptively sequencing the flashcards, or increasing the time interval between correctly answered flashcards (i.e., spacing effect), can lead to better learning outcomes than randomly sequencing flashcards or having fixed learning intervals (Kornell and Bjork 2008a; Mettler and Kellman 2014; Mettler, Massey and Kellman 2016). This approach reshuffles incorrect flashcards to the "top" of the deck, giving learners more time with the concepts that need more practice, and the easier items to the "bottom" of the deck. Recently, Whitmer and colleagues (2021) examined the use of adaptive flashcards in a USMC automotive maintenance course, where the flashcards were both adaptively sequenced and adaptively dropped based on mastery criteria. Their results showed a positive relationship between usage of the adaptive flashcards and course performance. Additionally, they showed that a cohort of students who used the adaptive flashcards had better exam performance than a previous cohort that did not use the flashcards. However, this study did not manipulate adaptive instructional techniques and there is a need to examine the unique contribution of different approaches to adaptive sequencing on long-term learning outcomes.

1.3   Two Adaptive Sequencing Approaches

Adaptive Response-Time Based-Sequencing (ARTS). In the present research, we explored two potential adaptive sequencing techniques in the literature that aligned with the spacing effect. First, the ARTS algorithm uses accuracy and reaction time to assign priority values to individual flashcards to determine the order of the flashcards. Specifically, ARTS increases the time between correct flashcards by prioritizing incorrect flashcards, then correct flashcards with slow reaction times, and correct flashcards with fast reaction times last (Mettler, Massey and Kellman 2011). When students take longer to respond to a flashcard, it may suggest that they may have difficulty retrieving the information or are less confident about their answer (e.g., Shiffrin and Schneider 1977; Sternberg 1969). The ARTS approach uses reaction time to distinguish between correctly answered flashcards, which may be useful to inform sequencing algorithms and provide a more thorough picture into a student's retrieval processes, shedding light on concepts with which a student needs more practice.

Some researchers propose that ARTS can lead to better long-term learning outcomes than random or fixed sequencing (Mettler and Kellman 2014; Mettler et al. 2016, 2020). In an exemplary study, Mettler and Kellman (2014; Experiment 1) compared a random sequencing condition to an ARTS condition using butterfly categories as the flashcard materials. They found that the ARTS condition was more efficient (i.e., calculated by Post-test score/ #Trials) than the Random condition. Although not statistically significant, results suggested the Random condition to be more accurate on the delayed post-test one week later than the ARTS condition. However, these findings are challenging to interpret because dropping was implemented differently for each condition in addition to sequencing. The ARTS condition dropped cards based on whether the category was mastered (i.e., three cards would be dropped when a category was mastered, because three species were presented per category), whereas the Random sequencing condition dropped individual cards when mastered. This difference in how cards were dropped by condition may explain why the Random condition completed significantly more trials than the ARTS condition, as opposed to the sequencing algorithm itself contributing to these differences in learning efficiency.

Additionally, Mettler and colleagues (2020) conducted two experiments comparing ARTS to random sequencing using a geography task as the flashcard learning materials. In the first experiment, they compared ARTS with mastery criteria to drop flashcards from future study to a condition that included random sequencing and no dropping (but training could terminate early if all flashcards were mastered before 45 min had elapsed). They found that ARTS was significantly more efficient (i.e., calculated by $[[(\text{Post-test} - \text{Pre-test})/ \text{\#Trials}] * \text{\#Flashcards}]$) than the Random condition, and the Random condition took twice as many trials to end training than the ARTS condition. However, the Random condition had a significantly larger accuracy improvement from the pre-test to delayed post-test than the ARTS condition. Nonetheless, the interpretation of Experiment 1 (Mettler et al. 2020) is limited because it is unclear whether adaptive sequencing played a role in improving training efficiency due to the differences in how the mastery criteria were implemented to end training. In other words, it is not surprising that the Random condition completed so many more trials than the ARTS condition because flashcards were not dropped, which is a part of the efficiency score calculation. In Experiment 2, they compared ARTS with mastery criteria to drop flashcards (same as Experiment 1) to a condition that had random sequencing with the same mastery criteria to drop flashcards during training. They found that the ARTS condition was significantly more efficient than the Random condition. Again, they found that the Random condition completed significantly more trials than the ARTS condition, but it was about 50 trials more, on average. Unfortunately, the authors did not report inferential or descriptive statistics, but state no differences were found between conditions for accuracy improvement, which contribute to the challenge in interpreting these results. Based on the studies reviewed (Mettler and Kellman 2014; Mettler et al. 2020), it is unclear whether ARTS provides an efficiency or long-term retention advantage compared to random sequencing based on the limitations of these experiments.

Leitner. A second adaptive sequencing approach that is consistent with the spacing effect is the Leitner method. The Leitner algorithm increases the time between correct flashcards by prioritizing incorrect flashcards and creating decks based on number of correct recalls (Leitner 1972). As flashcards are correctly recalled, multiple decks get created such that a learner might have a deck of flashcards that have yet to be recalled correctly, a deck of flashcards that have been correctly recalled once, and another deck of flashcards that have been correctly recalled twice. In this instance, flashcards with two correct recalls are not prioritized until incorrect flashcards and flashcards with one correct recall all reach two correct recalls. Using the individual learner's performance, flashcards that a learner is struggling with are seen more often than well-known flashcards. The Leitner approach is often encouraged by teachers and written about in popular-press as an example of the spacing effect (Gromada 2021; Whelan 2019), however, empirical investigations into this technique are limited. There are some papers in the literature that suggest they implemented the Leitner algorithm into their work (e.g., Pham et al. 2016; Schuetze 2015), but comparisons to other forms of sequencing or spacing are lacking. Additionally, the Leitner algorithm is built-in to many commercial flashcard software programs such as MemoryLifter.com and early versions of Duolingo (Settles 2016). More research is needed to investigate the efficacy of the Leitner algorithm on long-term retention and compare it with algorithms that are more complex.

1.4 Current Research and Hypotheses

The goal of this research was to compare adaptive sequencing approaches for flashcard learning to determine the superior adaptive technique for learning efficiency and long-term retention. The research questions were whether 1) ARTS had a larger benefit for efficiency and long-term retention than the Leitner method, and 2) whether both adaptive methods would outperform a control condition with random sequencing.

We hypothesized that ARTS would have the best learning outcomes (i.e., delayed learning gains and learning efficiency), because ARTS uses both accuracy and reaction time to prioritize flashcards whereas Leitner only uses accuracy. ARTS uses reaction time to distinguish between correctly answered flashcards, which may help discern concepts with which students need more practice and therefore, prioritize them sooner. Additionally, we expected that both adaptive sequencing conditions would have higher learning outcomes than the control condition that used random sequencing.

2 Method

2.1 Participants and Design

A total of 84 participants were recruited to participate in this two-session experiment from a local large state university. Six participants did not return for the second session of the study. Because we were interested in effects on long-term retention, this left 78 participants in the final sample for analysis. The average age of participants was 21 years old ($SD = 2.76$) and the sample was half women ($n = 39$) and half men ($n = 39$). Participants were compensated $30 for participating in Session 1 and $30 for Session 2 of the experiment.

Participants were assigned randomly to one of three between-subjects conditions: Leitner ($n = 25$), ARTS ($n = 28$), or Random ($n = 25$), which varied based on how flashcards were sequenced.

2.2 Materials

Training Materials. The training stimuli were 18 models of armored vehicles taken with permission from the Army Model Exchange. Fifteen of those models were learning objectives that were assessed on the pre-test and the post-test. The additional 3 vehicles were used as filler items during training to ensure adequate spacing between the flashcards and were not assessed on the pre-test or post-test. The same isometric view used in Whitmer et al. (2020) was used in the present experiment in order to show the top, side, and front view of each vehicle (see Fig. 1, left panel).

Testbed. The testbed we developed presented all instructions and phases of the experiment to the participant: pre-test, familiarization phase, training, and post-test. On the pre-test and post-tests (which were identical but items were presented in a random order), participants were presented with a vehicle model and had to select the name from a list of all 15 possible vehicles. There were two parts to the familiarization phase. First, participants were given a short tutorial on distinguishing features of the vehicles (e.g., hull, turret, weapons, and suspension) to guide participants during learning. Second, each vehicle was presented twice with its name for 5 s in a random order. During training, each flashcard showed a vehicle model and four possible vehicle names. Participants were instructed to select the name of the vehicle as quickly and accurately as possible, and based on the participant's performance the spacing of flashcards differed by condition. Participants received feedback after every trial. After an incorrect response, participants saw the vehicle they incorrectly chose in a side-by-side comparison with the target. Likewise, the button of the incorrect option changed to red and the correct option changed to green. For correct responses, the button of the correct option changed to green (see Fig. 1, right panel).

Fig. 1. Example flashcard (left) with feedback for an incorrect response (right).

2.3 Procedure

The present experiment is a subset of a larger study. Only the procedure pertinent to the research questions are discussed here. The experiment was a two-session study, with the second session occurring one week after the first session. After consenting to participate during the first session, participants completed a pre-test that assessed their ability to identify 15 armored vehicles by selecting the name from a list. Next, participants completed the familiarization phase, involving a short tutorial and passively reviewing the flashcards. Then, participants completed the training which varied by their condition assignment with a maximum training time of 25 min.

Those in the *ARTS* condition had the flashcards adaptively sequenced to their performance. Specifically, the algorithm uses the accuracy and reaction time of the previous trial to assign a priority value to each flashcard. Incorrect flashcards are prioritized first and seen more often, followed by correct flashcards with slow reaction times, and correct flashcards with fast reaction times are least prioritized. The priority values are constantly updated as more and more interactions with the flashcards occur (see Mettler et al. 2016 for more details on the algorithm). A mastery criterion of four consecutive recalls within six seconds per trial was used to drop flashcards from the deck (based on previous work, Whitmer et al. 2020). Training ended if participants mastered all 15 flashcards or the maximum time had elapsed.

Those in the *Leitner* condition also had the flashcards adaptively sequenced to their performance. However, in contrast to the ARTS algorithm, the Leitner algorithm only uses accuracy to prioritize incorrect flashcards over correct flashcards. Additionally, the Leitner algorithm creates decks based on number of correct recalls. For instance, flashcards with one correct recall are not prioritized until incorrect flashcards all reach one correct recall. Similar to the ARTS condition, a mastery criterion of four consecutive recalls was used to drop flashcards from the deck. However, unlike the ARTS condition, reaction time was not used in the mastery criterion because it is not traditionally used by this method. Training ended if participants mastered all 15 flashcards or the maximum time had elapsed.

Those in the *Random* condition were presented with flashcards in a random sequence until the maximum time of 25 min had elapsed or all 15 flashcards had been mastered and dropped from the deck using the criterion of four consecutive recalls within six seconds per trial (the same as the ARTS condition).

After training, participants completed surveys for approximately ten minutes. At the end of Session 1, participants completed a post-test. Altogether, Session 1 took approximately two hours to complete. Participants returned for Session 2 one week later, where participants completed the delayed post-test and additional measures for approximately one hour. Lastly, participants were debriefed and thanked for their participation.

3 Results

We first compared conditions on their delayed gain score to test our research questions related to differences between conditions' long-term retention. We computed a delayed gain score by examining how much participants learned and retained one week after training by accounting for how much they could have improved from their pre-test score ([Delayed Post-Test Score – Pre-Test Score]/[100% – Pre-Test Score]), which is a standard measure in the literature (see Hake 1998; Marraffino et al. 2019; Whitmer et al. 2020). We conducted a one-way ANOVA and found no differences between the conditions on their delayed gain score, $F(2,75) = 1.34$, $p = .27$, $\eta_p^2 = .03$. Next, we compared conditions on their learning efficiency to test whether different sequencing approaches led to improvements in the number of trials invested to learn and retain the material. We computed an efficiency score for each participant by taking the delayed gain score and dividing by the number of trials completed. We conducted a one-way ANOVA and found no differences between the conditions on their learning efficiency score, $F(2,75) = 0.56$, $p = .57$, $\eta_p^2 = .02$. It should be noted that we also computed efficiency as defined by Mettler and colleagues (2014, 2020) and did not find a significant difference between conditions.

Due to the lack of differences between conditions in learning efficiency and delayed learning gains, we sought to explore how each condition completed training. Next, we compared conditions based on in-training variables (i.e., time in training, number of trials, and number of cards mastered specific to condition) by conducting one-way ANOVAs. Each of the omnibus ANOVAs was statistically significant: time in training ($F(2,75) = 7.40$, $p = .001$, $\eta_p^2 = .17$), number of trials ($F(2,75) = 5.16$, $p = .008$, $\eta_p^2 = .12$), and number of cards mastered ($F(2,75) = 5.43$, $p = .006$, $\eta_p^2 = .13$). Post-hocs showed that those in the Leitner condition completed training significantly faster in terms of time in training (i.e., 5 min faster) than the ARTS ($p < .001$, $d = 1.04$) and Random ($p = .02$, $d = 0.67$) conditions, but there was no difference between the ARTS and Random conditions ($p = .20$, $d = 0.37$). Likewise, those in the Leitner condition completed training significantly faster in terms of total number of trials (i.e., 20 fewer trials) than the ARTS ($p = .04$, $d = 0.57$) and Random ($p = .002$, $d = 0.87$) conditions, but there was no difference between the ARTS and Random conditions ($p = .30$, $d = 0.30$). Based on the mastery criteria specific to the training condition, those in the Leitner condition mastered significantly more cards than those in the ARTS condition ($p = .002$, $d = 0.91$), but no other significant differences emerged based on number of cards mastered. We also examined how many flashcards participants mastered in the Leitner condition using the same mastery criteria as the other conditions (i.e., including reaction time). Those in the Leitner condition mastered significantly fewer items ($M = 6.54$, $SD = 4.28$) than the Random condition ($p = .03$, $d = 0.79$), when including reaction time into the mastery criteria. Descriptive statistics are reported in Table 1.

Table 1. Descriptive statistics of in-training measures by condition.

	Time in training (min)		Total number of trials		Number of flashcards mastered specific to condition	
Condition	*M*	*SD*	*M*	*SD*	*M*	*SD*
ARTS	22.66	4.01	140.52	30.41	7.56	6.38
Leitner	17.47	4.10	121.43	33.08	12.86	5.35
Random	21.00	5.09	150.56	40.82	10.60	5.84

Next, we aimed to explore how controlling for these in-training measures may explain differences in long-term retention. We conducted an ANCOVA to examine how the training conditions differed in delayed learning gains and used the three in-training measures as covariates. The overall model was statistically significant, $F(5,72) = 13.98$, $p < .001$. Each covariate significantly predicted delayed learning gains: training time ($p = .03$), number of trials ($p < .001$), and number mastered ($p < .001$). The post-hocs showed that after controlling for the in-training measures, those in the Leitner condition had significantly lower delayed gain scores than the ARTS ($p < .001$, $d = 1.05$) and Random ($p < .001$, $d = 1.23$) conditions, and there was no difference between the ARTS and Random ($p = .54$, $d = 0.18$) conditions (see Fig. 2).

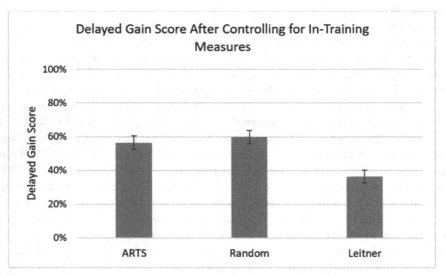

Fig. 2. Differences between conditions' long-term retention using in-training measures as covariates. Error bars denote standard error.

4 Discussion

In this experiment, we investigated two research questions: whether adaptively sequencing flashcards would be more efficient and produce higher delayed learning gains than randomly sequenced flashcards, and whether ARTS would be a better sequencing approach than the Leitner method for these learning outcomes. Contrary to our predictions, we found no differences between the three conditions on learning efficiency or delayed learning gains. Therefore, we decided to explore in-training variables more closely. We observed that those in the Leitner condition completed training much faster than those in the ARTS and Random conditions, in terms of number of trials, time to complete training, and number of mastered flashcards. After controlling for these in-training measures, we found that those in the Leitner condition had the lowest delayed learning gain score, compared to those in the ARTS and Random conditions.

Specifically, we found no differences between the ARTS and Random conditions across the in-training measures and learning outcomes. These results are in contrast with Mettler and colleagues (2014, 2020) who reported an efficiency advantage for ARTS over Random sequencing. However, in those experiments, the ARTS and Random conditions were not compared equitably. Mettler et al. (2014) implemented dropping flashcards differently in the ARTS and Random sequencing conditions, and Mettler et al. (2020; Experiment 1) used mastery criteria to end training differently in these two conditions. In the present study, we attempted to address this issue by matching dropping criteria between the ARTS and Random conditions, and we found no benefits for adaptive sequencing on its own.

Furthermore, we compared two adaptive sequencing approaches, ARTS (which uses accuracy and reaction time to prioritize the order of cards) and the Leitner method (which uses just accuracy to create decks and prioritize the order of cards) on learning outcomes. To our knowledge, this is the first empirical paper that has directly compared the Leitner method to another adaptive sequencing approach. The data suggest that the speed at which those in the Leitner condition completed training led to the lowest delayed learning gain scores. It should be noted that reaction time was not used in the mastery criteria for the Leitner condition because it is not typically utilized by this method. However, we analyzed the number of mastered cards in the Leitner condition using the mastery criteria applied to the other two conditions, which revealed fewer mastered cards for those in the Leitner condition. Multiple ways of analyzing the data suggest that the less stringent mastery criteria used by the Leitner method allowed participants to finish training earlier after completing fewer trials, and perhaps prior to a flashcard being meaningfully "mastered" and encoded into long-term memory. In other words, the Leitner method seems to have a tradeoff between the efficiency to complete the training and long-term retention.

4.1 Implications and Limitations

Overall, all three conditions demonstrated positive learning gains after a one-week retention interval, suggesting that flashcard-based study using adaptive training techniques is effective for promoting retention of the material (Carrier and Pashler, 1992; Roediger and Butler 2011; Roediger and Karpicke 2006; Toppino and Cohen 2009). It is notable that all three conditions utilized performance-based mastery criteria for dropping flash-cards from future study, and this technique has been demonstrated to be effective across a variety of studies (e.g., Kornell and Bjork 2008b; Pyc and Rawson 2011; Vaughn and Rawson 2011; Whitmer et al. 2020). Although there were no significant differences on learning gains by sequencing condition, our data suggest that other sequencing methods are preferable to the Leitner method when taking into account in-training performance. However, this may be due to the more lenient mastery criteria used relative to the other conditions investigated during this experiment. From a theoretical perspective, the present results are inconclusive as to whether adaptive sequencing is superior to random sequencing. Although other researchers have found positive results for ARTS over random sequencing (Mettler et al. 2014, 2020), randomly sequenced flashcards may provide their own spacing due to the number of cards in the deck. A randomly shuffled deck could incidentally space flashcards in a way that leads to a spacing effect due to chance.

One limitation of the present experiment is that we only examined two adaptive sequencing approaches and there are potentially other ways to order the flashcards adaptively. It should be noted that the spacing effect is typically found when multiple study sessions are spread out over a period of several days (Cepeda et al. 2006). In contrast, the current experiment examined two adaptive sequencing algorithms, ARTS and Leitner, which used spacing intervals within a single study session that are aligned with the spacing effect principle, as discussed by Mettler and colleagues for ARTS (2011, 2014, 2016, 2020) and Pham (2016) and Schuetze (2014) for the Leitner method. These one-shot training approaches may be more akin to "cramming" (i.e., studying once or twice before an exam), and perhaps these adaptive spacing methods may be more effective across multiple training sessions. To conclude, it might be the case that applying effective mastery criteria should be the focus in order to modernize and deliver learner-centered training methods to the USMC (e.g., Whitmer et al. 2020). Given the tradeoffs between time-in-training and retention, removing cards when they are meaningfully encoded may help strike a balance between long-term retention and efficiency.

4.2 Future Research

The findings in this experiment provide opportunities to expand on adaptive training approaches in flashcard-based study. For example, our study compared only two different types of adaptive sequencing. Future research should explore other types of sequencing and mastery criteria combinations that may lead to improvements in training efficiency and long-term retention. Additional experiments should compare these approaches in situations when multiple learning opportunities are available, such as a semester-long course, to examine if they aid long-term retention. Additionally, the current experiment examined a single content area - armored vehicle identification. Future research should compare adaptive approaches across multiple domains to determine whether learning outcomes are reliable when different content areas are used. Lastly, although the Leitner method does not traditionally use reaction time, future research should explore implementing reaction time as part of its mastery criteria for dropout (as it should increase the number of flashcards completed) to examine how this change affects learning outcomes in comparison to other adaptive and non-adaptive sequencing approaches.

4.3   Conclusion

The data suggest that the increased efficiency associated with the Leitner condition resulted in the worst long-term retention compared to other sequencing conditions. Additionally, neither adaptive sequencing approach, ARTS or Leitner, outperformed the Random condition, contrary to previous findings. Although adaptive flashcard techniques do hold promise for improving learning gains via 21st century learning approaches, more research is needed to ascertain the benefits of an adaptively sequenced flashcard deck in one-shot learning opportunities.

Acknowledgments. We gratefully acknowledge Dr. Peter Squire and the Office of Naval Research who sponsored this work (Funding Doc# N0001421WX00349). Presentation of this material does not constitute or imply its endorsement, recommendation, or favoring by the U.S. Navy or the Department of Defense (DoD). The opinions of the authors expressed herein do not necessarily state or reflect those of the U.S. Navy of DoD.

References

Bahrick, H.P., Bahrick, L.E., Bahrick, A.S., Bahrick, P.E.: Maintenance of foreign language vocabulary and the spacing effect. Psychol. Sci. **4**(5), 316–321 (1993)

Berger, D.H.: 38th Commandant's Planning Guidance (SSIC No. 05000 General Admin & Management) (2019). https://www.marines.mil/News/Publications/MCPEL/Electronic-Library-Display/Article/1907265/38th-commandants-planning-guidance-cpg/

Bjork, R.A., Dunlosky, J., Kornell, N.: Self-regulated learning: beliefs, techniques, and illusions. Annu. Rev. Psychol. **64**, 417–444 (2013)

Carrier, M., Pashler, H.: The influence of retrieval on retention. Mem. Cognit. **20**(6), 633–642 (1992). https://doi.org/10.3758/BF03202713

Cepeda, N.J., Pashler, H., Vul, E., Wixted, J.T., Rohrer, D.: Distributed practice in verbal recall tasks: a review and quantitative synthesis. Psychol. Bull. **132**(3), 354–380 (2006)

Ebbinghaus, H.: Über das Gedächtnis: Untersuchungen zur Experimentellen Psychologie (About Memory: Studies on Experimental Psychology). Duncker & Humblot (1885)

Ebbinghaus, H.: Memory: a contribution to experimental psychology. In: Ruger, H.A., Bussenius, C.E., Hiligard, E.R. (ets.) Trans. Dover Publications. (Original work published in 1885) (1964)

Gilday, M.M.: Fragmentary order: a design for maintaining maritime superiority (2019). https://www.navy.mil/DesktopModules/ArtithecleCS/Print.aspx?PortalId=1&ModuleId=685&Articcle=2237608

Glenberg, A.M., Lehmann, T.S.: Spacing repetitions over 1 week. Mem. Cognit. **8**(6), 528–538 (1980). https://doi.org/10.3758/BF03213772

Gromada, J.: The Leitner system: how does it work? Mindedge (2021) https://www.mindedge.com/learning-science/the-leitner-system-how-does-it-work/

Hake, R.R.: Interactive-engagement versus traditional methods: a six-thousand-student survey of mechanics test data for introductory physics courses. Am. J. Phys. **66**(1), 64–74 (1998)

Hartwig, M.K., Dunlosky, J.: Study strategies of college students: Are self-testing and scheduling related to achievement? Psychon. Bull. Rev. **19**(1), 126–134 (2012). https://doi.org/10.3758/s13423-011-0181-y

Karpicke, J.D., Butler, A.C., Roediger, H.L., III.: Metacognitive strategies in student learning: do students practise retrieval when they study on their own? Memory **17**(4), 471–479 (2009)

Karpicke, J.D., Roediger, H.L., III.: The critical importance of retrieval for learning. Science **319**(5865), 966–968 (2008)

Kornell, N.: Optimising learning using flashcards: spacing is more effective than cramming. Appl. Cogn. Psychol. **23**(9), 1297–1317 (2009)

Kornell, N., Bjork, R.A.: The promise and perils of self-regulated study. Psychon. Bull. Rev. **14**(2), 219–224 (2007). https://doi.org/10.3758/BF03194055

Kornell, N., Bjork, R.A.: Learning concepts and categories: Is spacing the "enemy of induction"? Psychol. Sci. **19**(6), 585–592 (2008a)

Kornell, N., Bjork, R.A.: Optimising self-regulated study: the benefits—and costs—of dropping flashcards. Memory **16**(2), 125–136 (2008b)

Landsberg, C.R., Astwood, R.S., Jr., Van Buskirk, W.L., Townsend, L.N., Steinhauser, N.B., Mercado, A.D.: Review of adaptive training system techniques. Mil. Psychol. **24**(2), 96–113 (2012)

Landsberg, C.R., Van Buskirk, W.L., Astwood, R.S., Mercado, A.D., Aakre, A.J.: Adaptive training considerations for simulation-based training systems (Special report 2010–001). Defense Technical Information Center (DTIC) (2011)

Leitner, S.: So lernt man lernen. AngewandteLernpsychologie – ein Weg zum Erfolg. Herder (1972)

Marraffino, M.D., Johnson, C.I., Whitmer, D.E., Steinhauser, N.B., Clement, A.: Advise when ready for game plan: Adaptive training for JTACs. In: Proceedings of the Interservice/Industry Training, Simulation and Education Conference (2019)

Marraffino, M.D., Schroeder, B.L., Fraulini, N.W., Van Buskirk, W.L., Johnson, C.I.: Adapting training in real time: an empirical test of adaptive difficulty schedules. Mil. Psychol. **33**(3), 136–151 (2021)

Mettler, E., Burke, T., Massey, C.M., Kellman, P.J.: Comparing adaptive and random spacing schedules during learning to mastery criteria. In: Proceedings of the 42nd Annual Conference of the Cognitive Science Society, pp. 773–779 (2020)

Mettler, E., Kellman, P.J.: Adaptive response-time-based category sequencing in perceptual learning. Vision. Res. **99**, 111–123 (2014)

Mettler, E., Massey, C.M., Kellman, P.J.: Improving adaptive learning technology through the use of response times. In: Proceedings of the 33rd Annual Conference of the Cognitive Science Society, pp. 2532–2537 (2011)

Mettler, E., Massey, C.M., Kellman, P.J.: A comparison of adaptive and fixed schedules of practice. J. Exp. Psychol. Gen. **145**(7), 897–917 (2016)

Paivio, A.: Spacing of repetitions in the incidental and intentional free recall of pictures and words. J. Verbal Learn. Verbal Behav. **13**(5), 497–511 (1974)

Peirce, N., Wade, V.: Personalised learning for casual games: The "Language Trap" online language learning game. In: Meyer, B. (Ed.) Proceedings of the 4th European Conference on Games Based Learning (ECGBL 2010), pp. 306–315. Academic Publishing (2010)

Pham, X.L., Chen, G.D., Nguyen, T.H., Hwang, W.Y.: Card-based design combined with spaced repetition: a new interface for displaying learning elements and improving active recall. Comput. Educ. **98**, 142–156 (2016)

Pyc, M.A., Rawson, K.A.: Costs and benefits of dropout schedules of test–restudy practice: Implications for student learning. Appl. Cogn. Psychol. **25**(1), 87–95 (2011)

Roediger, H.L., III., Butler, A.C.: The critical role of retrieval practice in long-term retention. Trends Cogn. Sci. **15**(1), 20–27 (2011)

Roediger, H.L., III., Karpicke, J.D.: Test-enhanced learning: Taking memory tests improves long-term retention. Psychol. Sci. **17**(3), 249–255 (2006)

Romero, C., Ventura, S., Gibaja, E.L., Hervás, C., Romero, F.: Web-based adaptive training simulator system for cardiac life support. Artif. Intell. Med. **38**(1), 67–78 (2006)

Schuetze, U.: Spacing techniques in second language vocabulary acquisition: Short-term gains vs. long-term memory. Lang. Teach. Res. **19**(1), 28–42 (2015)

Settles, B.: How we learn how you learn. Duolingo Blog (2016). https://blog.duolingo.com/how-we-learn-how-you-learn/

Shaughnessy, J.J., Zimmerman, J., Underwood, B.J.: The spacing effect in the learning of word pairs. Mem. Cognit. **2**(4), 742–774 (1974). https://doi.org/10.3758/BF03198150

Shiffrin, R.M., Schneider, W.: Controlled and automatic human information processing: II. Perceptual learning, automatic attending and a general theory. Psychol. Rev. **84**(2), 127–190 (1977)

Sommer, W.G.: Procrastination and cramming: how adept students ace the system. J. Am. Coll. Health **39**, 5–10 (1990)

Sternberg, S.: Memory-scanning: mental processes revealed by reaction-time experiments. Am. Sci. **57**(4), 421–457 (1969)

Toppino, T.C., Cohen, M.S.: The testing effect and the retention interval: questions and answers. Exp. Psychol. **56**(4), 252–257 (2009)

Van Buskirk, W.L., Fraulini, N.W., Schroeder, B.L., Johnson, C.I., Marraffino, M.D.: Application of theory to the development of an adaptive training system for a submarine electronic warfare task. In: Sottilare, R.A., Schwarz, J. (eds.) HCII 2019. LNCS, vol. 11597, pp. 352–362. Springer, Cham (2019). https://doi.org/10.1007/978-3-030-22341-0_28

VanLehn, K., et al.: The Andes physics tutoring system: lessons learned. Int. J. Artif. Intell. Educ. **15**(3), 147–204 (2005)

Vaughn, K.E., Rawson, K.A.: Diagnosing criterion-level effects on memory: What aspects of memory are enhanced by repeated retrieval? Psychol. Sci. **22**(9), 1127–1131 (2011)

Whelan, J.: Using the Leitner system to improve your study. Medium (2019) https://jessewhelan.medium.com/using-the-leitner-system-to-improve-your-study-d5edafae7f0

Whitmer, D.E., Johnson, C.I., Marraffino, M.D., Hovorka, J.: Using adaptive flashcards for automotive maintenance training in the wild. In: Sottilare, R.A., Schwarz, J. (eds.) HCII 2021. LNCS, vol. 12792, pp. 466–480. Springer, Cham (2021). https://doi.org/10.1007/978-3-030-77857-6_33

Whitmer, D.E., Johnson, C.I., Marraffino, M.D., Pharmer, R.L., Blalock, L.D.: A mastery approach to flashcard-based adaptive training. In: Sottilare, R.A., Schwarz, J. (eds.) HCII 2020. LNCS, vol. 12214, pp. 555–568. Springer, Cham (2020). https://doi.org/10.1007/978-3-030-50788-6_41

Wissman, K.T., Rawson, K.A., Pyc, M.A.: How and when do students use flashcards? Memory **20**(6), 568–579 (2012)

Design and Development of Adaptive Instructional Systems

Development of AIS Using Simulated Learners, Bayesian Networks and Knowledge Elicitation Methods

Bruno Emond[4]([✉])[iD], Jennifer Smith[1][iD], Mashrura Musharraf[3][iD],
Reza Zeinali Torbati[1][iD], Randy Billard[2][iD], Joshua Barnes[4][iD],
and Brian Veitch[1][iD]

[1] Memorial University of Newfoundland, St. John's, Canada
{jennifersmith,rzt313,bveitch}@mun.ca
[2] Virtual Marine, Paradise, Canada
randy.billard@virtualmarine.ca
[3] Aalto University, Espoo, Finland
mashrura.musharraf@aalto.fi
[4] National Research Council Canada, Ottawa, Canada
{bruno.emond,joshua.barnes}@nrc-cnrc.gc.ca

Abstract. The development of adaptive instructional systems (AIS) is an iterative process where both empirical data on human performance and learning, and experimentation using computer simulations can play a role. The paper presents our current efforts to advance adaptive instructional system technology conceived as self-improvement systems [37]. The paper describes our methodological approach for informing the design and implementation of adaptive instructional systems by conducting concurrent research activities using 1) Bayesian networks for modelling learning processes, 2) knowledge elicitation of expert instructors, and 3) simulated learners and tutors to explore AIS system design options. Each activity fulfills separate but complementary objectives. Bayesian networks modelling of learners' performance provides the means to implement predictions of learners' performance, and selection of adaptive learning content. Knowledge elicitation methods are fundamental in understanding human capabilities and limitations in the context of AIS systems design that support and regulate the cognitive demands of the learner and instructor. Simulated learner and tutor interactions enable the specification of detailed cognitive process models of learning and instructions.

Keywords: Simulated learners · Bayesian networks · Expert knowledge elicitation · Tutoring strategies · Marine operations

This project was supported in part by collaborative research funding from the National Research Council of Canada's Artificial Intelligence for Logistics Program.

1   Introduction

Adaptive instructional systems (AISs) are a class of training systems including intelligent tutoring systems (ITSs), intelligent mentors or recommender systems, and intelligent instructional media [35]. From an ITS perspective, the development of AISs is an iterative process involving adjustments to domain knowledge, pedagogical strategies, learner models, and user interface through a systematic empirical validation based on human learner data. In principle, adaptive instructional systems can be configured for experimentation to support the evaluation of learner/instructional/domain models and hypotheses testing [36]. This paper presents our current efforts to advance adaptive instructional system technology conceived as self-improvement systems [37]. Our approach consists of informing the design and implementation of adaptive instructional systems by conducting concurrent research activities using 1) Bayesian networks for modelling learning processes, 2) knowledge elicitation of expert instructors, and 3) simulated learners and tutors to explore AIS system design options.

More specifically, we seek to improve maritime operations in Canadian waters using innovative scientific and engineering methods by enhancing operators' competencies using training simulators. The main objective is to develop a set of tools to verify and optimize the design of adaptive instructional systems. The anticipated benefits of the project are: 1) an increase in knowledge of the cognitive demands and skill acquisition related to sea ice management in the context of freight transportation and small vessel emergency operations; 2) a cost reduction of training scenario development by testing alternative learning scenarios and instructional designs prior to empirical validation with human participants; 3) an increase in training efficiency using optimized scenarios and adaptive instructions.

The paper is divided in four sections. The first section defines the problem statement, the second section outlines the benefits of Bayesian networks for AIS and discusses its application in optimizing simulator training with adaptive and personalized training. A third section presents knowledge elicitation methods for expert navigation instructors, in order to provide guidance for the development of AIS tutoring policies. A fourth section presents an overview of an AIS simulation platform to explore how simulated learners and tutors system can learn from each other. The last section summarizes the key elements of the paper.

2   Problem Statement

Training simulators are essential in areas such as aviation, maritime navigation, medicine, emergency responses, and the military. Many elements of training simulations are of value for minimizing risks for trainees and equipment, allowing for repeated skills practice, and developing adequate responses to rare but dangerous conditions. In addition, training simulation provides a means to collect novel data on human performance and learning in situations that are otherwise prohibitive due to risk [5]. Simulation based assessments provide a safe means to practice and acquire data, though the data collected is often new and cannot be easily compared to historic data sets. Various data analysis methods have been shown to offer good models of content (item difficulty) and learner knowledge components. On the basis of these analyses, training simulations can be optimized by sequencing training scenarios (problems and tasks), and estimating a learner's domain knowledge and skills mastery.

However, mastery diagnostic and sequencing of training tasks can be enhanced with refined tutoring instructions. Capturing human tutoring expertise is key for the design of domain-specific AIS. Expert knowledge elicitation methods are fundamental in understanding human capabilities and limitations at an information processing level. The detailed knowledge gathered from experts on a particular task or set of tasks is necessary to design AIS systems that support and regulate the cognitive demands of the learner and instructor.

Even when there is a large data set to support learner modelling, the most common situation in the design and development of adaptive instructional system, is that the system is to address a new training need and that there is likely very limited prior user testing data. The lack of testing makes it difficult to ensure that lessons and guidance from design recommendations and prior studies in other domains have been effectively applied in the training application [44]. Ultimately, optimized tutoring strategies should be determined through empirical investigation, but the space of AIS design options is large [10] and it is difficult to validate by human performance and learning data, which is often sparse. Synthetic data generated by simulated students could be one approach to provide a broader range of learner behaviours to explore adaptive system states' reinforcement learning policies.

The current paper presents our initial efforts to advance adaptive instructional system technology conceived as self-improvement systems [37]. Our approach consists of informing the design and implementation of adaptive instructional systems by conducting concurrent research activities using 1) Bayesian networks for modelling learning processes, 2) knowledge elicitation of expert instructors, and 3) simulated learners and tutors to explore AIS system design options. The objective is to optimize instructional strategies given variations in learner models, or improvement in learners' performances. The multidisciplinary approach combines Bayesian modelling, expert knowledge elicitation methods, and cognitive modelling.

3   Bayesian Networks for Learner and Tasks Models

Significant purposes of AIS include predicting a learner's need and selecting adaptive learning content, adaptive tests and exercises, and adaptive hints or recommendations [28]. This makes Bayesian networks (BN) the most suitable and most frequently used artificially intelligent (AI) techniques in adaptive learning systems [20]. This section outlines the benefits of BN and discusses its application in optimizing simulator training by introducing adaptivity and personalization.

BNs are probabilistic models that are constituted by a qualitative and a quantitative part. The qualitative part is composed of a directed acyclic graphical structure that represents the interactions and causal relationships among a set of random variables using nodes and links [30]. The quantitative part consists of marginal probabilities of the independent variables (also called parent nodes) and conditional probability tables (CPT) of dependent variables (also called the child nodes). A CPT specifies the probability of a dependent variable for all state combinations of the variables on which it is directly dependent [29].

In the context of adaptive learning systems, BN is often used to drive the diagnosis in a student model. During the diagnosis, the cognitive state (latent variable) of a student is inferred by his/her interaction (observable variable) with the learning system. Compared to other diagnostic approaches, BNs have the advantages of allowing the transparent measurement of students' knowledge at different granularity levels, simplifying the specification of model parameters, and incorporating the possibility of lucky guesses and unintentional errors or slips [21,27]. These advantages withstand for training simulators making BN a prominent choice to model trainee diagnosis, trainee performance, and trainee competence. In these BNs, the state of latent variables such as knowledge and competence are inferred from the states of observable variables such as tasks performed correctly or incorrectly in a simulated scenario. For example, Billard et al. [6] used data from expert inputs and simulators to build a BN of competence of lifeboat coxswains during maneuvering at slow speeds. In this study, the BN was informed by expert predictions to create initial CPTs, which improved the model diagnostic capabilities compared to models that were only informed by a sparse data set. A sample BN adopted from Billard et al. [6] is shown in Fig. 1.

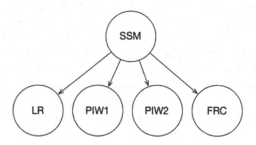

Fig. 1. A sample BN for SSM assessment scenario [6].

As shown in Fig. 1, the BN is able to describe the relationships and interactions i) among the different tasks in a slow speed manoeuvring (SSM) assessment scenario, and ii) among competence and the different tasks in the same scenario. As a trainee performs in a SSM scenario, evidence is collected for four different tasks including stopping next to a life raft for inspection (LR), picking up two persons in the water (PIW1, PIW2), and stopping next to a Fast Rescue Craft (FRC) for transfer of personnel. The trainee either completes the tasks successfully or they fail in one or more tasks. Depending on the states of the observable variables LR, PIW1, PIW2, and FRC, the state of trainee competence in slow speed manoeuvring is inferred. Backward inference can then be applied to diagnose trainees' strengths and weaknesses, and tailor training curriculum to individual needs [27]. Thus, BNs applications in training simulators can improve training outcomes through trainee assessment and optimizing instructional design [5]. It is possible to extend the traditional BN to dynamic models so that the trainee's learning progression through multiple attempts at the same task can be captured, which is common in conventional simulator training. By monitoring students' changing knowledge states while practising a skill, instructions can be individualized to enable students to achieve mastery within a practical time frame [9]. BNs can also be applied to evaluate and optimize simulator training through comparing student performance under adaptive and non-adaptive training. BNs can also be used to build models of expert and novice performance in scenarios that can only be explored safely through simulator training, including emergency scenarios involving rough weather. BNs can be used to evaluate the behaviours that result in the highest probability of task completion, and to study limitations in human performance.

4 Expert Tutors Knowledge Elicitation

Knowledge elicitation is commonly a qualitative approach to gathering expert domain knowledge for intelligent systems and involves interviews, observations, and task analysis [8,11]. These methods are fundamental in understanding human capabilities and limitations at an information processing level. The detailed knowledge gathered from experts on a particular task or set of tasks is necessary to design AIS systems that support and regulate the cognitive demands of the learner and instructor. This section outlines the knowledge elicitation methodology (i.e., semi-structured interviews and observations in simulated settings) and the data gathered from expert navigation instructors [34].

4.1 Semi-structured Interviews

In this case, semi-structured interviews involve asking experts detailed questions on how they would complete a particular task (i.e., maneuvering a lifeboat at slow speeds or navigation through an ice field) followed by probing questions to understand the factors that influence the expert's decision-making. Semi-structured interviews were used by [34] to elicit knowledge related to how a navigation task is performed by an experienced operator and what advice and feedback should be provided to students performing the task in a simulator. This information can be used to inform the domain, student, and pedagogical models of an AIS.

From the perspective of developing expert tutors, future interviews with experts will focus on how instructors teach navigation skills, provide exercises to practice, and deliver corrective feedback during simulator training. Specifically, tabletop exercises and student performance critiques will be used to elicit instructor advice and feedback in specific contexts. Tabletop exercises involve describing a situation and asking an expert instructor to explain how they would approach a particular task and discussing their thought process. Specific interview questions will ask instructors how they adapt their instructions and feedback for individual learners. Critiques of student performance involve watching video examples, diagnosing the level of the student, rating the student's performance, and indicating when the instructor should provide corrective feedback. The videos are paused at various points in the scenario to ask targeted questions on when the instructor would intervene and if so, what specific instruction and feedback they would provide. The probing questions will be informed by existing pedagogical frameworks (e.g., simulation-based mastery learning) [26] and motor learning theory (i.e., the timing of feedback, feedback providing knowledge of performance or results) as outlined in Table 1 [7,33]. The proposed knowledge elicitation interviews can be used to develop the instructional model (i.e., inform how skills are taught and identify the teaching frameworks/styles) and to develop a sense of the variability in these strategies.

4.2 Simulation Exercises

Empirical data on expert performance in a simulated setting can also be used to gather information related to how a navigation task is performed and how these skills are taught. The information collected from navigation experts in a series of experiments [34,39,40] can be used to guide the development of AIS and its tutoring policies. The experiments were conducted using a ship bridge simulator that consisted of a 360-degree panoramic projection screen surrounding a simplified bridge console. An Anchor Handling Tug Supply (AHTS) vessel that is common for offshore sea ice management support operations was modelled in the simulator for these exercises.

The first experiment [40] was designed to investigate the influence of bridge officer experience on the effectiveness of ice management operations in a simulated setting. A total of 18 experienced bridge officers and 18 cadets training to become seafarers participated in the experiment. Each participant was tasked with completing ice management scenarios that varied in severity (e.g., precautionary or emergency conditions in 4-tenths or 7-tenths initial ice concentrations). Details of the experiment are provided in [40].

The second experiment [39] was designed to measure the effect of simulator training on cadet performance. A total of 35 seafaring cadets participated in the experiment, forming two groups that received different amounts of training. The cadets were trained to a target performance level in three ice management techniques (e.g., pushing, prop-wash, and leeway) and subsequently tested on their ability to manage ice in the same emergency scenario using the simulator. Details of the experiment are provided in [39].

The published results of [39,40] can be used to inform the domain and student models of the AIS. Specifically, data from [40] encompasses the different ways experienced seafarers performed the ice management task and how untrained novice cadets approached the same task. Similarly, data from [39] provided a range in performance data from trained cadets. This information can be used in the development of the student models (i.e., identifying what information students pay attention to and how training affects their performance).

The third experiment [34] was a knowledge elicitation pilot study designed to gather more domain knowledge related to how experienced seafarers perform ice management tasks. The pilot experiment involved four experienced seafarers who were interviewed using tabletop exercises and student examples to capture the contextual information related to ice management operations. The seafarers were also asked to complete the same exercises in the simulator to capture the dynamic aspects of the ice management operations. The information gathered from the pilot helped inform how experienced seafarers approached and executed the ice management scenarios (i.e., what factors were considered when forming a strategy, what ice management techniques were most appropriate for the conditions, and what information the experienced seafarers paid attention to for each ice management technique). Details of the experiment are provided in [34].

4.3   Data to Inform and Optimize Export Tutors

The information collected from the semi-structured interviews can be used to classify the teaching style of the expert instructors and inform the pedagogical model of the AIS. Further, collecting empirical data on how expert tutors would contextually instruct, evaluate, deliver feedback, and interact with students in a simulated setting can be used to constrain the variability of the AIS tutoring strategies. For instance, empirical data from the simulation experiments [34, 39, 40] can be used to optimize the tutoring strategies (outlined in Table 1) and the variability in the cadets' performance data [39, 40] constrains the development of the student learner computational models. Similarly, the interview and empirical data on expert instructors [34, 40] constrains the variability of tutoring strategies, which will help to scaffold the instructional strategies in a self-adapted system.

5   Simulated Learners and Tutors for Exploring AIS Design Options

Simulated learners are computational models of learners [45]. They have been recognized to play various roles in training and learning environments. Among these roles are efficiently author instruction at scale, evaluate pedagogical effectiveness of instruction, test theories of how humans learn [19], the possibility of providing an environment to support teacher's practice (teachable agents), embed simulated learners as part of the learning environment, and to provide an environment for exploring and testing learning system design issues [24, 44]. More recently, Wray has brought forward the use of simulated learners as a software verification method "to attempt to understand, prior to full-scale development, the potential benefits of adaptive algorithms and the requirements they impose on students and instructors" [45]. The Apprentice Learner Architecture [23] follows a similar approach where simulated learners can be combined with novel interaction designs to provide model transparency, input flexibility, and problem solving control to achieve greater model completeness in less time than existing authoring methods [42]. Other modelling applications of simulated learners include modelling: learning sequences [25], the role of time in learning [17], the design of AIS for self-directed longer-term learners [22], adaptive remediation in online training [38], and differential error types between human and simulated learners [43].

5.1   ACT-R Cognitive Architecture

The current approach to the simulation of learners and tutors is based on the ACT-R cognitive architecture [1], which provides a range of empirically validated mechanisms to model human memory performance and learning. The project builds on prior work in training simulations [34,39,40], and cognitive modelling [12–16]. Figure 2 provides a system overview of the simulation components using the ACT-R cognitive architecture for modelling learner's performance and learning of training scenario tasks. The ACT-R cognitive models can also be expressed as analytic likelihood functions using statistical concepts such as serial vs. parallel process, convolution, minimum/maximum processing time, and mixtures [18].

Central to the method is the ACT-R cognitive modules for perception (visual and aural), motor control (manual and vocal), and their coordination in task execution through procedural, and declarative memory as well as goal and problem state representations. The general approach is to encode a minimal set of perceptual, and motor linked to interface devices affordances and let ACT-R learn the knowledge and skills required to execute the training scenario tasks by attending to the tutor feedback (declarative information and rewards), and applying some of its learning mechanisms (base-level learning, associative-learning, production compilation, and utility learning).

Three learning mechanisms are currently used to simulate the interaction between a simulated learner and a simulated tutor. These mechanisms are base-level learning of declarative memories, utility learning of procedural knowledge (reinforcement learning), and production compilation for learning to skip steps and learning from instructions (from declarative to procedural knowledge). Equations 1 and 2 define two learning mechanisms used by the ACT-R cognitive architecture.

Equation 1 is the memory base-level learning equation and reflects the activation of declarative memory chunks as a function of the number of times a chunk is referred to and the time since it was last referenced. In addition, the equation takes to constant parameters for the decay rate and the basic initial activation level of memory chunks. The activation level of memory chunks is an important value of memory chunks which determines retrieval time and possible retrieval failure in case the activation level is not above some threshold.

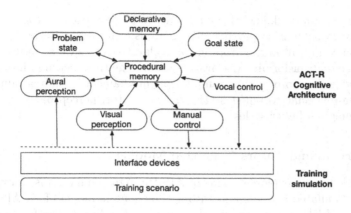

Fig. 2. ACT-R cognitive architecture.

$$B_i = ln(\sum_{j=1}^{n} t_j^{-d}) - \beta_i \qquad (1)$$

where:
n = the number of presentation for chunk i.
t_j = time since the jth presentation.
d = decay parameter.
β_i = constant offset parameter.

Equation 2 specifies how the utility of a production is augmented or reduced as a function of the positive or negative reward values. The utility quantity of a production will determine which production is selected in the situation where two or more productions can fire. The production with the highest utility will be selected. The equation also has a learning rate parameter, and includes an initial utility value assigned to a production.

$$U_i(n) = U_i(n-1) + \alpha[R_i(n) - U_i(n-1)] \qquad (2)$$

where:
α = the learning rate.
$R_i(n)$ = effective reward value given to production i on its nth usage.
$U_i(0)$ = initial utility value or a production.

In addition to these two learning mechanisms, the simulated learner and tutor make use of the production compilation. Production compilation works by attempting to compose two productions that fire in sequence into one new production. It allows a problem to be solved with fewer productions over time and therefore to be performed faster. By reducing the number of steps, production compilation can also model learning from instructions. Instructions get encoded initially as declarative knowledge which needs to be retrieved from memory to guide problem solving or psychomotor actions. Through practice, production compilation can result in the drop-out of declarative retrieval as part of the task performance by transforming declarative knowledge (instructions) into procedural knowledge. The production compilation mechanism is built around sets of rules that are specific to buffer styles such as motor, perceptual, retrieval, and goal and imaginal buffer styles.

5.2   Learners and Tutors Cognitive Models

One possible way to explore designs of AIS through simulation is to experiment with both simulated learners and tutors. The generic model of an AIS put forward by the IEEE working group [2] is presented in Fig. 3. The figure indicates that not only the instructional system adapts to the learner by providing feedback and selecting learning tasks at an adequate difficulty level, but the adaptive instructional engine is also subject to improvement by evaluating and adjusting its adaptive policies. At the core of AIS is the basic functionality to measure and assess the level of proficiency of a learner in order to adapt the training and learning experience. There is an inherent complexity and dynamics in adaptive instructional systems where the interaction among the learner(s), the artificially intelligent AIS that guides the learning experience, and the environment forms a complex adaptive system [37].

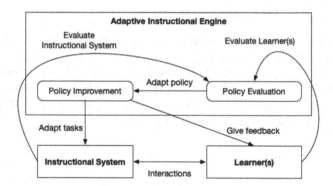

Fig. 3. Generic adaptive instructional system. Adapted from [2].

The goal of the simulated learner is to execute a task with success (learn from the consequence of actions), and the goal of the tutor is to provide the best adapted instructions to a learner (learn from the consequence of instructions). Both the learner and the tutor are made of the same cognitive architecture but they differ in terms of their knowledge, and possible different views of the training simulation. Figure 4 shows that both a simulated learner and tutor are implemented using the ACT-R cognitive architecture. However, the learner and the tutor might have different interfaces to the training scenario. A simulated learner and tutor are two separate models running concurrently, each having different productions and declarative knowledge. This interactive simulation system is very similar to the modular reinforcement learning framework for tutorial planning [32].

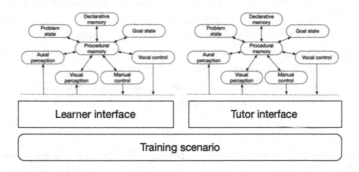

Fig. 4. Interactions between simulated learner and tutor.

One interest in the current approach to modelling both the learner and tutor with the ACT-R cognitive architecture is that it provides a solid foundation for understanding perceptual and cognitive processes in mixed combinations such as human learner/AI tutor and human learner/human tutor/AI tutor. The combined simulation of learners and tutors covers the range of action options for the learner that are related to the execution of a task, while the range of actions for the tutor amounts to the form and timing of providing instructions. Table 1 gives an example of a tutor pedagogical choices. Effects of tutoring strategies and instructions can also be explored through the use of cognitive learner models during skill acquisition [3,4,31,41].

6 Summary

The development of adaptive instructional systems (AIS) is an iterative process involving adjustments to domain knowledge, pedagogical strategies, learner models, and user interface through a systematic empirical validation based on human learner data. The current paper presents our current efforts to advance adaptive instructional system technology conceived as self-improvement systems [37]. Our approach consists of informing the design and implementation of adaptive instructional systems by conducting concurrent research activities using 1) Bayesian networks for modelling learning processes, 2) knowledge elicitation of expert instructors, and 3) simulated learners and tutors to explore AIS system design options. More specifically, we seek to improve maritime operations in Canadian waters using innovative scientific and engineering methods by enhancing operators' competencies using training simulators.

Table 1. Properties related to pedagogical sequences. From Cockburn et al. [7].

Property types	Examples
Temporal	Feedback is provided either concurrently, immediately after, or delayed from a learner action
Aggregation	Feedback is provided independently for each discrete action or accumulated or a sequence of actions (after action review)
Modality	Instructions and feedback can be presented as text, speech synthesis, video, or statistics in tabular format
Performance	Feedback is provided in terms of deviation to an ideal sequence with no reference to its outcome or results
Results	Feedback is provided about the outcome of the action such as success or failure in relation to a desired outcome

Bayesian networks applications in training simulators can improve training outcomes through trainee assessment and optimizing instructional design [5]. It is possible to extend the traditional BN to dynamic models so that trainee's learning progression through multiple attempts at the same task can be captured, which is common in conventional simulator training. By monitoring students' changing knowledge states while practising a skill, instructions can be individualized to enable students to achieve mastery within a practical time frame [9]. BNs can also be applied to evaluate and optimize simulator training through comparing student performance under adaptive and non-adaptive training. Simulation based assessments and BNs can also be used to model and to study human performance in applications where data is sparse, including emergency scenarios.

Knowledge elicitation methods are fundamental in understanding human capabilities and limitations at an information processing level. The detailed knowledge gathered from experts on a particular task or set of tasks is necessary to design AIS systems that support and regulate the cognitive demands of the learner and instructor. The main methods are 1) semi-structured interviews which involve asking experts detailed questions on how they would complete a particular task followed by probing questions to understand the factors that influence the expert's decision-making; 2) simulation exercises provide empirical data on expert performance in a simulated setting and also information related to how a navigation task is performed and how these skills are taught. The information collected from the semi-structured interviews can be used to classify the teaching style of the expert instructors and inform the pedagogical model of the AIS. Further, collecting empirical data on how expert tutors would contextually instruct, evaluate, deliver feedback, and interact with students in a simulated setting can be used to constrain the variability of the AIS tutoring strategies.

Simulated learners are computational models of learners [45]. They have been recognized to play various roles in training and learning environments. The current approach to the simulation of learners and tutors is based on the ACT-R cognitive architecture [1], which provides a range of empirically validated mechanisms to model human memory performance and learning. Three learning mechanisms are currently used to simulate the interaction between a simulated learner and a simulated tutor: 1) base-level learning of declarative memories, 2) utility learning of procedural knowledge (reinforcement learning), and 3) production compilation for learning to skip steps and learning from instructions (from declarative to procedural knowledge). The goal of the simulated learner is to execute a task with success (learn from the consequence of actions), and the goal of the tutor is to provide the best adapted instructions to a learner (learn from the consequence of instructions). Both the learner and the tutor are made of the same cognitive architecture but they differ in terms of their knowledge, and possible different views of the training simulation. One interest in the current approach to modelling both the learner and tutor to with the ACT-R cognitive architecture is that it provides a solid foundation for understanding perceptual and cognitive processes in mixed combinations such as human learner/AI tutor and human learner/human tutor/AI tutor.

References

1. ACT-R research group (2002). http://act-r.psy.cmu.edu
2. Adaptive Instructional Systems (AIS) working group: P2247.1 (2019). https://site.ieee.org/sagroups-2247-1/
3. Anderson, J., Betts, S., Bothell, D., Hope, R.M., Lebiere, C.: Three aspects of skill acquisition, June 2018. https://doi.org/10.31234/osf.io/rh6zt, https://psyarxiv.com/rh6zt
4. Anderson, J.R., Betts, S., Bothell, D., Lebiere, C.: Discovering skill. Cogn. Psychol. **129**, 101410 (2021). https://doi.org/10.1016/j.cogpsych.2021.101410, https://www.sciencedirect.com/science/article/pii/S0010028521000335
5. Billard, R., Musharraf, M., Veitch, B., Smith, J.: Using Bayesian methods and simulator data to model lifeboat coxswain performance. WMU Journal of Maritime Affairs 19(3), 295–312 (09 2020). https://doi.org/10.1007/s13437-020-00204-0
6. Billard, R., Smith, J., Masharraf, M., Veitch, B.: Using Bayesian Networks to Model Competence of Lifeboat Coxswains. TransNav, Int. J. Mar. Navig. Saf. Sea Transp. **14**(3), 585–594 (2020). https://doi.org/10.12716/1001.14.03.09, http://www.transnav.eu/Article_Using_Bayesian_Networks_to_Model_Billard,55,1039.html
7. Cockburn, A., Gutwin, C., Scarr, J., Malacria, S.: Supporting novice to expert transitions in user interfaces. ACM Comput. Surv. **47**(2), 1–36 (01 2015). https://doi.org/10.1145/2659796, https://dl.acm.org/doi/10.1145/2659796
8. Cooke, N.J.: Varieties of knowledge elicitation techniques. Int. J. Hum.-Comput. Stud. **41**(6), 801–849 (1994). https://doi.org/10.1006/ijhc.1994.1083, https://www.sciencedirect.com/science/article/pii/S1071581984710834
9. Corbett, A.T., Anderson, J.R.: Knowledge tracing: modeling the acquisition of procedural knowledge. User Model. User-Adap. Inter. 4(4), 253–278 (1995). https://doi.org/10.1007/BF01099821
10. Domeshek, E., Ramachandran, S., Jensen, R., Ludwig, J., Ong, J., Stottler, D.: Lessons from building diverse Adaptive Instructional Systems (AIS). In: Sottilare, R.A., Schwarz, J. (eds.) HCII 2019. LNCS, vol. 11597, pp. 62–75. Springer, Cham (2019). https://doi.org/10.1007/978-3-030-22341-0_6

11. Durso, R., Nickerson, R., Dumais, S., Lewandowsky, S., Perfect, T.: Handbook of Applied Cognition. Wiley, Hoboken (2007)
12. Emond, B.: WN-LEXICAL: an ACT-R module built from the WordNet lexical database. In: Seventh International Conference on Cognitive Modeling, pp. 359–360. Trieste, Italy (2006)
13. Emond, B., Comeau, G.: Cognitive modelling of early music reading skill acquisition for piano: a comparison of the Middle-C and intervallic methods. Cogn. Syst. Res. **24**, 26–34 (2013). https://doi.org/10.1016/j.cogsys.2012.12.007
14. Emond, B., Vinson, N.G.: Modelling simple ship conning tasks. In: 15th Meeting of the International Conference on Cognitive Modelling, pp. 42–44. Coventry, UK (2017)
15. Emond, B., West, R.R.L.: Cyberpsychology: A Human-Interaction Perspective Based on Cognitive Modeling. Cyberpsychology Behav. **6**(5), 527–536 (2003). https://doi.org/10.1089/109493103769710550
16. Emond, B., West, R.L.: Using cognitive modelling simulations for user interface design decisions. In: Orchard, B., Yang, C., Ali, M. (eds.) IEA/AIE 2004. LNCS (LNAI), vol. 3029, pp. 305–314. Springer, Heidelberg (2004). https://doi.org/10.1007/978-3-540-24677-0_32
17. Essa, A., Mojarad, S.: Does time matter in learning? A computer simulation of Carroll's model of learning. In: Sottilare, R.A., Schwarz, J. (eds.) HCII 2020. LNCS, vol. 12214, pp. 458–474. Springer, Cham (2020). https://doi.org/10.1007/978-3-030-50788-6_34
18. Fisher, C.R., Houpt, J.W., Gunzelmann, G.: Fundamental tools for developing likelihood functions within ACT-R. J. Math. Psychol. **107**, 102636 (2022). https://doi.org/10.1016/j.jmp.2021.102636, https://linkinghub.elsevier.com/retrieve/pii/S0022249621000997
19. Harpstead, E., MacLellan, C.J., Weitekamp, D., Koedinger, K.R.: The use simulated learners in adaptive education. In: AIAED-19: AI+Adaptive Education, pp. 1–3. Beijing, China (2019)
20. Kabudi, T., Pappas, I., Olsen, D.H.: AI-enabled adaptive learning systems: A systematic mapping of the literature. Comput. Educ. Artif. Intell. **2**, 100017 (2021). https://doi.org/10.1016/j.caeai.2021.100017, https://linkinghub.elsevier.com/retrieve/pii/S2666920X21000114
21. Käser, T., Klingler, S., Schwing, A.G., Gross, M.: Dynamic Bayesian networks for student modeling. IEEE Trans. Learn. Technol. **10**(4), 450–462 (2017). https://doi.org/10.1109/TLT.2017.2689017
22. Lelei, D.E.K., McCalla, G.: How to use simulation in the design and evaluation of learning environments with self-directed longer-term learners. In: Penstein Rosé, C., et al. (eds.) AIED 2018. LNCS (LNAI), vol. 10947, pp. 253–266. Springer, Cham (2018). https://doi.org/10.1007/978-3-319-93843-1_19
23. MacLellan, C.J., Koedinger, K.R.: Domain-general tutor authoring with apprentice learner models. Int. J. Artif. Intell. Educ. **32**, 76–117 (2020). https://doi.org/10.1007/s40593-020-00214-2
24. McCalla, G., Champaign, J.: Simulated learners. IEEE Intell. Syst. **28**(4), 67–71 (2013). https://doi.org/10.1109/MIS.2013.116
25. McEneaney, J.E.: Simulation-based evaluation of learning sequences for instructional technologies. Instr. Sci. **44**(1), 87–106 (2016). https://doi.org/10.1007/s11251-016-9369-x
26. McGaghie, W., Issenberg, S., Petrusa, E., Scalese, R.: Effect of practice on standardised learning outcomes in simulation-based medical education. Med. Educ. **40**(8), 792–797 (2006). https://doi.org/10.1111/j.1365-2929.2006.02528.x

27. Millán, E., Pérez-de-la Cruz, J.L.: A Bayesian Diagnostic Algorithm for Student Modeling and its Evaluation. User Model. User-Adap. Inter. **12**(2), 281–330 (2002). https://doi.org/10.1023/A:1015027822614
28. Mousavinasab, E., Zarifsanaiey, N., Kalhori, S.R.N., Rakhshan, M., Keikha, L., Saeedi, M.G.: Intelligent tutoring systems: a systematic review of characteristics, applications, and evaluation methods. Interact. Learn. Environ. **29**(1), 142–163 (2021). https://doi.org/10.1080/10494820.2018.1558257
29. Musharraf, M., Smith, J., Khan, F., Veitch, B., MacKinnon, S.: Assessing offshore emergency evacuation behavior in a virtual environment using a Bayesian network approach. Reliab. Eng. Syst. Saf. **152**, 28–37 (2016). https://doi.org/10.1016/j.ress.2016.02.001, https://www.sciencedirect.com/science/article/pii/S0951832016000399
30. Pearl, J.: Probabilistic reasoning in intelligent systems. Elsevier (1988). https://doi.org/10.1016/C2009-0-27609-4, https://linkinghub.elsevier.com/retrieve/pii/C20090276094
31. Ritter, F.E., Yeh, M.K.C., Yan, Y., Siu, K.C., Oleynikov, D.: Effects of varied surgical simulation training schedules on motor-skill acquisition. Surg. Innovation **27**(1), 68–80 (2020). https://doi.org/10.1177/1553350619881591
32. Rowe, J., Pokorny, B., Goldberg, B., Mott, B., Lester, J.: Toward simulated students for reinforcement learning-driven tutorial planning in gift. In: Sottilare, R. (Ed.) Proceedings of 5th Annual GIFT Users Symposium. Orlando, FL (2017)
33. Schmidt, R., Lee, T., Winstein, C., Wulf, G., Zelaznik, H.: Motor Control and Learning: A Behavioral Emphasis, 6th Edn. Human Kinetics (2019)
34. Smith, J., Yazdanpanah, F., Thistle, R., Musharraf, M., Veitch, B.: Capturing expert knowledge to inform decision support technology for marine operations. J. Mar. Sci. Eng. **8**(9) (2020). https://doi.org/10.3390/JMSE8090689
35. Sottilare, R., Knowles, A., Goodell, J.: Representing functional relationships of adaptive instructional systems in a conceptual model. In: Sottilare, R.A., Schwarz, J. (eds.) HCII 2020. LNCS, vol. 12214, pp. 176–186. Springer, Cham (2020). https://doi.org/10.1007/978-3-030-50788-6_13
36. Sottilare, R.A.: A comprehensive review of design goals and emerging solutions for adaptive instructional systems. Technol. Instr. Cogn. Learn. **11**(1), 5–38 (2018)
37. Sottilare, R.A., Sinatra, A.M., DeFalco, J.A.: Considerations in modeling adaptive instructions as a complex self-improving system. In: Sinatra, A.M., Graesser, A.C., Hu, X., Brawner, K., Rus, V. (eds.) Design Recommendations for Intelligent Tutoring Systems, Volume 7 Self-Improving Systems, pp. 29–40. US Army Research Laboratory, Orlando (2019)
38. Spain, R., Rowe, J., Smith, A., Goldberg, B., Pokorny, R., Mott, B., Lester, J.: A reinforcement learning approach to adaptive remediation in online training. J. Defense Model. Simul. Appl. Methodol. Technol. (2021). https://doi.org/10.1177/15485129211028317
39. Thistle, R., Veitch, B.: An evidence-based method of training to targeted levels of performance. SNAME Marit. Convention 2019, SMC 2019 (2019)
40. Veitch, E., Molyneux, D., Smith, J., Veitch, B.: Investigating the influence of bridge officer experience on ice management effectiveness using a marine simulator experiment. J. Offshore Mech. Arct. Eng. **141**(4) (2019). https://doi.org/10.1115/1.4041761, https://asmedigitalcollection.asme.org/offshoremechanics/article/doi/10.1115/1.4041761/475585/Investigating-the-Influence-of-Bridge-Officer
41. Walsh, M.M., et al.: Mechanisms underlying the spacing effect in learning: a comparison of three computational models. J. Exp. Psychol. Gen. **147**(9), 1325–1348 (2018). https://doi.org/10.1037/xge0000416

42. Weitekamp, D., Harpstead, E., Koedinger, K.R.: An interaction design for machine teaching to develop AI tutors, pp. 1–11. Association for Computing Machinery, New York (2020). https://doi.org/10.1145/3313831.3376226

43. Weitekamp, D., Ye, Z., Rachatasumrit, N., Harpstead, E., Koedinger, K.: Investigating differential error types between human and simulated learners. In: Bittencourt, I.I., Cukurova, M., Muldner, K., Luckin, R., Millán, E. (eds.) AIED 2020. LNCS (LNAI), vol. 12163, pp. 586–597. Springer, Cham (2020). https://doi.org/10.1007/978-3-030-52237-7_47

44. Wray, R., Stowers, K.: Interactions between learner assessment and content requirement: a verification approach. Adv. Intell. Syst. Comput. **596**, 36–45 (2018). https://doi.org/10.1007/978-3-319-60018-5_4

45. Wray, R.E.: Enhancing simulated students with models of self-regulated learning. In: Schmorrow, D.D., Fidopiastis, C.M. (eds.) HCII 2019. LNCS (LNAI), vol. 11580, pp. 644–654. Springer, Cham (2019). https://doi.org/10.1007/978-3-030-22419-6_46

Promoting Equity and Achievement in Real-Time Learning (PEARL): Towards a Framework for the Formation, Creation, and Validation of Stackable Knowledge Units

Patrick Guilbaud[1]([✉]) and Michael J. Hirsch[2]

[1] Winthrop University, Rock Hill, SC, USA
guilbaudp@winthrop.edu
[2] ISEA TEK LLC, Maitland, FL, USA
mhirsch@iseatek.com

Abstract. This paper presents and delineates Promoting Equity and Achievement in Real-time Learning (PEARL), a framework developed to help strengthen academic and career-related skills of students and professionals, and most specifically those from underserved and underrepresented backgrounds. PEARL uses problem-based learning pedagogy, competency-based education and artificial intelligence to break learning and assessment activities into manageable learning chunks. This allows for the formation, creation, and validation of stackable knowledge units. The paper also highlights how artificial intelligence along with learner-centered pedagogy can be used to improve knowledge gain and skill mastery.

Keywords: Problem-based learning · Artificial Intelligence · Digital badges stackable knowledge unit · Diversity and equity · BIPOC

1 Introduction

Cognitive psychology posits that learning and decision making involve a combination of memory use, motivation, and thinking [1–3]. It is likewise understood that better educational outcomes can be attained if learning content is partitioned into manageable chunks, particularly when focus is placed on solutions to real-world problems [4, 5]. In this context, Stackable Knowledge Units (SKU), which are validated by Mini/Micro-Certificates (MC) such as Digital Badges (DB) can serve as educational achievement markers and motivation vehicles. This is because they place focus on gradual development and then mastery of specific career-related skills, knowledge, and achievements. Further, as MC can serve as distinct proof of knowledge or skill mastery, SKU can strengthen learner's self-confidence and self-efficacy as they progress through the learning process [6, 7].

SKUs, which trace their roots within the orientation and focus on Competency-Based Education (CPE), offers the chance to expand educational opportunities to diverse community of learners including those who are historically underrepresented, underserved, and/or at-risk [5, 8–11]. In concert with the use of CPE and MC, SKU allows learning to be divided into sub-modules or chunks that are easier to remember and master. This new instructional paradigm and particularly the use of badges thus provides a new and more manageable pathway for learning and content mastery [11, 12]. However, there are a few challenges with the implementation and use of SKUs in both formal and informal education settings [13–15]. These include: 1) the facilitation of knowledge development and skills mastery in real time, 2) the external acceptance of a new academic credential, and 3) the desire of businesses and other organizations to employ personnel with the new academic credential. Moreover, while standards for traditional educational achievements such as a degree, certificate, credit hour, or continuing education unit are well established, there are currently no agreed-upon measures for what constitutes a SKU.

This paper presents Promoting Equity and Achievement in Real-time Learning (PEARL), which is a CPE-based micro-learning framework. PEARL breaks course or instructional content and assessment activities into manageable segments to facilitate mastery of educational contents and support personalization of problem-based learning (PBL) experiences. PEARL thus supports the issuance of SKUs (MC and DB), when and as skills are learned, adopted, and then mastered. The paper notes how learning technology tools and applications can help instructors: 1) adjust course content and regulate the pace of learning delivery; 2) manipulate assessment-related activities; and 3) promote optimal skill building and knowledge mastery. In addition, we discuss how the use of PEARL will stand to facilitate the acquisition of cognitive, affective, and domain skills and abilities. Finally, the paper highlights how PEARL can be used both as a formative and summative assessment method that will allow students and other learners the chance to demonstrate their newly acquired knowledge and skills. Thus, PEARL will give instructors and trainers the opportunity to adjust their teaching methods based upon students' progress and real-time feedback received.

2   Literature Review

One of the major challenges faced today by academic institutions across the country, including K-12 schools, colleges and universities, is the dearth of involvement and participation of underrepresented minority students in the Sciences, Technology, Engineering and Math (STEM) related disciplines [16, 17]. It is also noted that students from lower socio economic and other disadvantaged backgrounds – whether of minority or majority persuasion – are not sufficiently represented in high demand STEM-related careers [17–19].

Theorists and researchers in the fields of education, instruction, and human development have long offered that a commitment to knowledge acquisition, development and mastery along with a focus of self-efficacy will help learners achieve their lifelong dreams and goals [6, 20, 21]. As a result, educational approaches, tools, and strategies that offer opportunities for self-assessment, skills building and personal intrinsic motivation can go a long way towards eliciting intrinsic motivation of learners, irrespective of their ethnic or social-economic background. However, many students and most particularly Black, Indigenous, People of Color (BIPOC) have gaps in knowledge, skills, and abilities needed to be competent, effective, and ethical STEM knowledge workers and practitioners in the 21st century economy. [16–18] Current research shows that traditional education approaches and techniques are not always well-suited for all users and learners, particularly those from BIPOC backgrounds.

With the emergence of a STEM-focused future employment environment, it is clear that K-12, college, and other learners who have not mastered the set of skills and competencies required by employers stand a good chance to fall behind intellectually and economically compared to their peers [22, 23]. Moreover, a key lesson of the present knowledge-centered global economy is that innovations, creativity, and cross-cultural collaborations are important levers for career advancement, mobility, and renewal [24–26]. Thus, academic institutions, business organizations, and government-oriented entities that have not adequately prepared their youths, professionals, and other citizens for this STEM-driven future are likely to face a more uncertain future.

As noted by Aoun [22] and Bialik and Fadel [27], success in the global interconnected workplace requires familiarity with a wide range of skills and abilities. While cognitive ability, technical know-how and industry-related skills (e.g., marketing research, operations management, computing expertise, finance modeling, and patient care) might be adequate to land a job, they are most often insufficient to help someone maintain a solid career in the modern digital workplace [28]. Further, it is now clear that there is a need for all professionals to have both strong non-cognitive skills and technical know-how to "robot-proof" their careers [22, 29].

At present, instructors and educators of all types seek to find ways to utilize the digital technology tools and applications that are available at their institutions as learning vehicles to help students gain career knowledge and skills, particularly in high demand STEM fields [30–32]. However, many courses lack the "hands-on" and "minds-on" Active and Practical Learning (APL) focus that's specifically designed to promote or support progressive mastery of work-oriented knowledge and skills [33–35]. While the complete list of required skills for the modern workplace will vary depending upon the business organization or field of focus, they can nevertheless be categorized into three major groups: cognitive ability, affective skills, and domain or industry knowledge.

2.1 Cognitive Ability

Task performance is heavily dependent on cognitive ability in many technical fields, e.g., healthcare, engineering, computer science, and data analysis [36, 37]. For example, mechanical and structural engineers need to know the properties of steel in order to design and build a safe bridge, production managers must be familiar with safety guidelines for successful plant operations, and chemists need to know the conditions necessary for yeast to convert sugar into alcohol to make good beer. Research, however, reveals that successful task completion also requires the ability to deal with unstructured and ill-defined problems that very often occur in work situations [28, 38, 39].

2.2 Affective Skills

Affective or "people/soft" skills are in high demand by today's employers [27, 40, 41]. In fact, many organizations now indicate that soft skills are more important than hard skills [42]. For example, Google found that the top seven skills related to success in their firm were soft skills. Technical skills, on the other hand, came eighth on the list [43]. Key soft skills that are best-suited for career-related opportunities can vary as they are based upon the specifics of the field or industry. However, communication, collaboration, teamwork, diversity, and cultural sensitivity, discussed below, are typically included in most lists that delineate soft skills important in the workplace [44].

2.3 Domain/Industry Knowledge

Domain or industry knowledge is defined as expertise in a particular problem area or field that requires both a significant amount time of study and solid hands-on experience on the topic or field [45–47]. Therefore, one needs more than a good grasp of the technical language in a given field to acquire such knowledge. Wallace [48] and Hildreth and Kimble [49] emphasize that domain knowledge is tied to increased speed and accuracy of tasks in a given field. They also note that experts know more, make fewer errors, and work faster than novices. Thus, as experts continue to learn and gain more knowledge, they are able to use quicker and even unorthodox methods to accomplish assigned tasks successfully. Lave and Wenger [39], on the other hand, offers that those novices or newcomers to a given field need to be well-informed and comfortable with its methods, principles, tenets, skills, and core beliefs.

3   Problem Based Learning, Competency-Based Education, and Stackable Knowledge Units

PBL approaches create opportunities for students to wrestle with and solve a real-world issue or dilemma in small groups using repeated practice [38]. Consequently, PBL helps in fostering cognitive development, promoting social skills, and enhancing analytical and soft (e.g., critical thinking) skills of learners through the use of APL. Recent research studies and analytical reports have focused on factors that affect the efficacy of PBL [50, 51]. For example, Hmelo-Silver [38] notes that PBL works best when groups work well together and members have interactions with each other. Winarno, Muthu, and Ling [52], on the other hand, argue that blending group-based learning with PBL and traditional direct instruction presents some challenges. It is important, therefore, to carefully consider the implementation of a PBL approach within an existing curriculum or educational program. Instructors need to analyze whether PBL will provide measurable educational benefits to students and can support the learning outcomes of the existing course or program of study.

Recently, CBE has gotten a lot of attention in many education and training sectors [5, 53, 54]. At its core, CBE is focused on helping learners master a skill or ability at their own pace [55, 56]. Further, much like PBL, CBE can be used to help the learner cultivate work-related competencies such as communication, analytical, collaboration, teamwork, and intercultural skills and capabilities. Therefore, CBE when properly implemented can be used in conjunction with PBL to facilitate the acquisition of cognitive ability, affective skills, and domain or industry knowledge [57, 58].

Focus is also being placed on the issuance stackable credentials [9–13] in the form of DBs and MCs to serve as proof of competencies, skills, and abilities acquired. Unlike formal credentials such as associate, bachelors and master degrees, which require at least two years of academic study, badges can be earned in much shorter time periods such as months, weeks, and even days [13–15]. It is thus argued that badges help increase learner's engagement and motivation in their educational endeavors. Likewise, SKU is a learning approach and educational process that facilitates skills development and improvement both quickly and gradually [59]. Like CBE, SKU is focused primarily on learning mastery and not directly on formative assessment. In so doing, use of the SKU approach both formally and informally in the classroom or other learning setting along with the issuance of DBs can help learners with differing abilities obtain the scaffolding support and direct feedback that they require as they work on mastering new skills and competencies. Thus, SKUs provide the opportunity to enhance both academic performance and career-readiness of BIPOC and other underserved learners as they master new work-related skills and competencies.

4 The Artificial Intelligence Learning Approach

At its most basic construct, Artificial Intelligence (AI) deals with the simulation of intelligent behaviors within computers [60]. A major part of this process is the ability for an AI software application to learn from its experience with its environment (in an educational setting that's the interaction it offers with users such as students, faculty, and staff). AI encompasses many different approaches and algorithms. Some AI technologies are fairly simple e.g., learning and continually updating the estimates on a regression equation as new data is input. Other AI technologies can be exceedingly complex e.g., mimic the problem-solving and decision-making capabilities of a skilled researcher, scientist, or engineer [61]. However, some technologies labeled as AI are simply systems for automating mundane tasks. Therefore, education technology applications that do not have an adaptive learning component cannot be presented as an AI-centered solution.

Machine Learning (ML) is the fundamental way to make a computer 'intelligent' [61]. Modern ML techniques have moved from attempting to mimic the way a human learns are now more concerned with allowing a computer to increase the knowledge it has stored internally [62]. This shift in ML is thus less concerned with categorizing and processing data information as humans do. The new ML approaches can be further partitioned into the sub-category of deep learning, which does not need a pre-defined model (e.g., in contrast to linear regression) neural networks, which allow computers to find patterns that are very complex. For example, Ferlitsch [62] offers that deep learning is usually employed using an approach with multiple layers within which patterns, structure, or insight about the data is discovered. At present there are many different ML algorithms and approaches that have been used to solve specific problems [61, 63]. A key aspect of AI and ML is the use of feedback. As a concept, feedback is used in a system so that the outputs of it are fed back as inputs [64]. This feature thus helps the system with correcting assumptions and modifying the weights used on individual pieces of data, allowing it to influence not only future decisions but how, when, and why those decisions are made. Therefore, in a learner-centered education setting use feedback such as AI and ML will allow information presented to students to be adapted and corrected based upon their interactions with the system.

4.1 Adaptive Learning in Educational Settings

In the field of human development it is noted that as children grow, they are constantly interacting with their internal and external environments [65, 66]. External environments that those children experience include the world that is external to themselves, which they perceive through their normal senses [67]. In contrast, their internal environment includes thoughts and emotions, as well as memories. Thus the child grows and learns through their experiences and interactions with both their external and internal environment. Moreover, their internal knowledge base (i.e., stuff accumulated in the brain) stores these experiences and learned knowledge for future uses.

In addition to storing learned information, the brain updates incoming information as new data to be learned, if found to be relevant [68–70]. As an example, when children first learn multiplication, they start operating with one-digit numbers [71]. As children begin multiplying many numbers together, they eventually realize that the procedure is the exact same as one digit number multiplication. With the process repeated a few times, they start seeing the generalized process/algorithm. Eventually, children might learn of a short-cut approach to multiplying a number by 11. As they use this "trick" their approach to multiplication gets updated with this component/fact/approach. In this way, children (and adults) are constantly taking in information, combining it with what they already have stored in their brains, in order to update the information stored in their brains. In effect, the brain is a natural adaptive learning system.

As the example presented in the previous section shows, an adaptive learning system in an educational setting must (a) work similar to the processes that a human performs in learning; and (b) must understand how the human performs learning in order to best interact with the human and help the human learn correctly [72, 73]. Consequently, the learning system will take the learner down the path of material mastery while correcting any missteps in both the decisions made by the student and the underlying understanding of the material presented. Further the adaptive learning system must help students minimize the number of mistakes made. This can be done, by adjusting the teaching and/or testing mechanisms and/or tailoring it to the student's individual capability and background. Some learning systems are just reactive (e.g., when students make a mistake they are prompted review certain material again) as is the case of Intelligent Tutoring Systems [74, 75]. In contrast, an adaptive learning system should be anticipatory (e.g., use data to determine the likelihood that the student will make a mistake on a given question and adapt the content provided to the student appropriately). In this way, the adaptive learning system help students gain new knowledge and master new materials and concepts based upon their background, skills, and prior knowledge.

4.2 Opportunities

Two of the biggest uses of AI within an educational learning environment are: (a) Longitudinal improvement or enhancement of course content over multiple semesters based upon a design of experiments (e.g., considering material covered, time spent on individual concepts, student demographics and abilities, etc.) and (b) Personalized improvement or tailoring course content to each individual student, taking into account specific learning strategies and methodologies that will most resonate with the student [22, 62]. These tailored strategies can be learned by the AI system over time by interacting with the individual student. Moreover, it can do so by capturing features across multiple students and learning the strategies that resonate best with different subsets over the multi-dimensional student demographics to include those from BIPOC and underserved backgrounds.

4.3 Challenges

The main barrier to realizing these two opportunities deals with data [60, 61]. For each of these two opportunities, there is a large number of data that must be captured and analyzed, in order for an AI system to learn appropriate methodologies to present course content. For longitudinal improvement, many semesters worth of data must be captured, with the experimental design independent variables changed in each, in order for the AI system to determine the best way course content should be presented. For Personalized improvement, numerous students will be needed, from different ethnicities, religions, income levels, parenting situation, home life, and academic abilities [18]. All of this demographic data must be stored for all of these individual students, as well as their interaction with the AI system teaching them different material. The decisions that the AI system made with each of the students also needs to be captured, along with student end-state for the material. As more and more of this data is captured, the AI system will be able to begin tailoring its presentation approach to a student based upon how closely that new student matches with those that have already utilized the system [61, 62] as well as the approaches that worked the best for these students. As time progresses, the knowledge base upon which the AI system will use to continually adapt itself to the students' it is encountering will allow for each student to get a tailored course content presentation of material, best suited to help them learn the material.

5 The PEARL Approach

The PEARL approach leverages PBL, AI, and data analytics to provide alternative learning paths and tailored scaffolding support to students [60, 73–78]. Through this unique approach, PEARL allows the creation and use of SKUs to help students, irrespective of their personal, social, and academic backgrounds gain the knowledge, skills, and abilities required for success in STEM academic programs and careers, while keeping them engaged in the learning process [18, 79–83].

PEARL involves 5 learning steps, which are: 1) Profile and Skill Assessment (PSA), 2) Case Scenario Interaction (CSI), 3) Formative Self-Evaluation (FSE), 4) Scaffolding and Intervention Activities (SIA), and 5) Knowledge Mastery and Outcome (KMO). These 5 steps work together to help underserved students gain basic, advanced, and comprehensive knowledge about the topic being learned. In this way, the PEARL approach allows students to build, develop and obtain SKUs as they advance from the introductory to mastery level of a specific STEM-related skill and ability.

As illustrated in Fig. 1, during the PSA step, PEARL makes use of a learners' background along with AI to determine their appropriate initial learning preferences, cognitive styles, and STEM ability profile (e.g., low, medium, or high) [84–86]. This also involves gauging the current knowledge, capacity, and interest of a given learner in a subject. In the CSI step, students are presented with an "ill structured" problem scenario that they are tasked to solve [81–83]. During the FSE step, students are asked to formulate questions, hypotheses, and potential answers to the given problem or situation, which is based on real life cases. This relevance to real-life scenarios makes the learner more invested in learning the material, as they see first-hand the applicability of the material learned.

In the SIA step, PEARL utilizes data from multiple sources, including time on task, goals attained, levels attempted, etc., to help students make preliminary recommendations about the scenario of focus [87–89]. Finally, in the KMO step students are asked to offer solutions including prevention measures that eliminate the problem or reduce its impact. PEARL will also gather data to power its feedback, adaptive learning, mechanism and coaching approach that take place during the SIA and KMO steps.

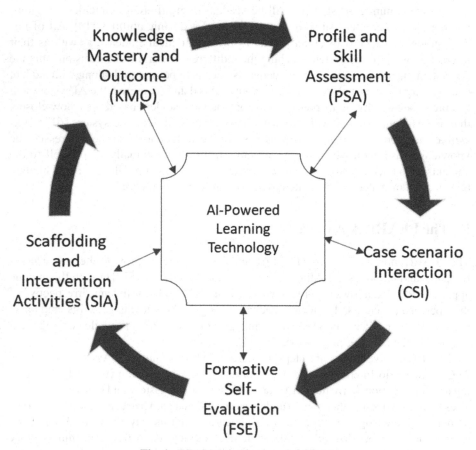

Fig. 1. PEARL adaptive learning model.

6   Earl in Practice

Attaining expert-level performance in practically any domain takes time. Nevertheless, novices or students who have the opportunity to engage in deliberate learning and practice can develop sufficient detailed knowledge to address practical business problems in the workplace [45–47]. It is also well noted that interventions, efforts, and activities that are part of a PBL and CPE-oriented educational endeavor or task will stand to facilitate sufficient gains in performance or knowledge acquisition for novices to work or engage comfortably in many fields and domains [33, 38, 39, 53, 54, 56]. Therefore, the use of SKUs as a learning progression tool stands to offer learners and most specifically those from underserved and underprepared backgrounds, the opportunity to develop and acquire new skills and competencies as they pursue a credential. This is because SKUs place the focus on gradual skill building, cognitive development, and knowledge mastery. In this way, the SKU approach ensure learner persistence, resilience, and success irrespective of their prior skill level or ability with a given topic.

The PEARL approach is presented in the next section of the paper via an illustrative PBL scenario. Information and scaffolding support will be provided to the learners on an as-needed basis, to help them reflect on the scenario presented and then take steps to resolve it.

6.1   Illustrative Healthcare-Related PBL Scenario

First, PEARL will secure the student's initial skill level during the PSA step. Then, the following illustrative learning scenario (CSI step) involving Evelyn will be presented to the student along with FSE and SIA to elicit KMO on the part of students, such as reflective inquiry, data analysis and reasoned judgment.

"Evelyn is a 6th grader who was found lying on the grass during recess. At the time, she was having difficulty breathing normally and sitting up. When questioned about what happened, Evelyn said she was playing by herself in the courtyard and started having pains in her leg and chest. Evelyn also reported that her arms felt very heavy. Evelyn then shared that she was playing around with her friend's new skateboard the day before and had to stop after unsuccessfully attempting a few jumps and tricks because she felt very tired."

PEARL prompts the student to help Evelyn understand why she is having the problems that she is experiencing (CSI step). Additional information such as Evelyn's age, medical history, fitness, regular diet, exercise routine, sleeping patterns, etc., will be available to the student. Information related to cramping and other symptoms being experienced will be subtly interlaid into the case. Additionally, the student will have access to narrated health texts that will explain the structures and functions of the human body.

As the students advance through the learning scenario, they will be required to take quizzes (FSE step) to show mastery of the information reviewed before they can proceed to address Evelyn's symptoms. Students will receive confirmation, feedback, and rewards for each positive and negative step they take in determining some of the underlying causes of Evelyn's physical and health situation. Students will be able to ask additional questions about the case's health condition. If students are having difficulty solving Evelyn's problem, they can ask a Virtual Assistant (VA) that's based on AI, such as an emergency medical technician, a nurse, or a physician for help (SIA). All of the analysis in the FSE and SIA steps take place within the AI system. As the student is reviewing the material, taking the quizzes, and asking questions to the VA, the AI system will be making use of the knowledge gained through many other previous students' interacting with the system to determine the best way to help this current student down the path toward understanding and mastery of the material.

All help from the VAs will be in the form of probing questions focused on helping the student come up with hypotheses about Evelyn's situation on their own. The selected VA will question the student to ensure that they know which aspect(s) of the human system is relevant to the problem faced, based on knowledge and skills level. This exchange will allow PEARL to "learn" through ML about typical plans of action by students at specific competency levels and then guide them toward new learning activities that can improve their knowledge and skills of the topic at hand (KMO).

6.2 Acceptance and Implementation Perspectives Regarding PEARL

PBL in general, and inquiry-based learning specifically, are not typically used in the average school or classroom environment [77–79]. This is often due to the time, organizational support, culture, and the costs that are required as part of introducing a new pedagogy like PBL to teachers and a school [83, 85, 86]. In the development of PEARL, focus is placed on soliciting input, suggestions, and recommendations from teachers and curriculum designers on all aspects of introducing the PEARL approach in an educational setting such as the workplace or classroom environment. In addition, the plan is to work with the leaders of the school to train the teachers in PBL pedagogy, as well as inquiry-based learning approaches, ensuring the teachers receive appropriate professional development credit e.g., certificate or continuing education unit (CEU) credit, appropriate to their knowledge and mastery of the subject matter.

6.3 Learner's Preparation and Readiness for PBL

Gagné [90] argues that an education foundation through prior learning must be in place before learners are ready to receive new information or learning stimulus. To ensure that students are ready for the PBL approach used in PEARL, an overview of the relevant or grade appropriate aspects of the pedagogy will be provided as part of the PSA diagnostic assessment step when they log in to the PEARL system. The student will be made aware that the use of PBL is intended to help strengthen their decision-making and collaborative working skills [76, 83, 84, 90]. Further, the student will be able to access scaffolding materials and VAs as the student uses PEARL to interact and solve simulated real-world problems or issues that are presented to them as learning activities. Moreover, the inquiry-based approach used in PEARL will support the development of higher order thinking that is needed to work on the PBL assignment.

6.4 Learning, Scaffolding, and Technology Integration

Advances in adaptive learning pedagogies, Edtech technology, gamification, AI, and ML offer the opportunity to create new models of educational and training interactions to improve skills development [91, 92]. The newer technologies can also thus serve as enablers and catalysts in the push to allow students and instructors to move beyond artificial experimental settings that have limited ecological validity. Thus, while modern technologies can be used to create highly personalized and tailored learning activities, there is also the need to place focus on appropriate pedagogies to achieve the goal of offering an understanding of real-world problems and issues to learners through multi-faceted educational interactions [16, 93, 94]. PEARL is anticipated to be used independently and hosted on learning management systems such as Blackboard, Moodle, and Canvas. Performance and running ability on external systems will need to be verified, validated, and refined prior to production-level roll-out of PEARL. Using open-source software and well tested engines, such as OPEN-PBL APP, Competency and Skills System (CaSS) and Google AI will significantly reduce implementation and deployment problems.

With a SKU focus, PEARL can be used to help facilitate increased motivation and engagement of all learners and most specifically those that are from underrepresented and underserved backgrounds. This unique focus will allow them to obtain the knowledge and help that they need to master key STEM concepts. Moreover, with the use of AI and adaptive learning, PEARL offers an optimum individualized learning path for each student, thereby facilitating their progression, performance, and success of the student in STEM-related disciplines and careers.

7 Conclusions and Future Research

Present-day learners have different academic backgrounds, socio-cultural contexts and personal characteristics [18, 95]. As a result, the opportunity exists to leverage AI-centered adaptive learning algorithms along with appropriate pedagogical approaches to foster both personalization of learning contents and gradual knowledge–building and mastery of STEM and inquiry skills. This paper argues that the use SKUs offer the possibility of enhancing learners' motivation and sense of self-fulfillment. At the same time, it is important to consider implementing the instructional activities and elements needed to create an effective learning environment when taking on the challenges of integrating a new instructional model or process in a course or training activity [23, 68]. Many of those items can be derived from established practices such as those advocated by many learning technology and design theorists and researchers [3, 20, 21]. Building upon these models and studies, the PEARL framework offers a pedagogically-grounded approach that uses advanced AI adaptive learning technology to help strengthen knowledge acquisition processes and practices of learners. This is so the students may successfully strengthen skills such as problem-solving, critical thinking, collaboration and teamwork, which are needed for the 21st century workplace.

References

1. Anderson, L., Krathwohl, D.: A Taxonomy for Learning, Teaching, and Assessing: A Revision of Bloom's Taxonomy of Educational Objectives, Abridged Longman, New York (2001)

2. Greeno, J.: The situativity of knowing, learning, and research. Am. Psychol. **53**(1), 5–26 (1998)
3. Salomon, G.: Technology's promises and dangers in a psychological and educational context. In: De Vaney, A. (Ed.), Theory to Practice: Technology and the Culture of Classrooms, pp. 4–10 (1998)
4. Guilbaud, P., Camp, J., Bruegge, A.V.: Digital badges as micro-credentials: an opportunity to improve learning or just another education technology fad?. In: Winthrop Conference on Teaching and Learning, vol. 25 (2016)
5. Levine, E., Patrick, S.: What is competency-based education? An Updated Definition. Vienna, VA: Aurora Institute (2019)
6. Bandura, A.: Self-efficacy: toward a unifying theory of behavioral change. Psychol. Rev. **84**(2), 191–215 (1977)
7. Ackerman, P., Heggestad, E.: Intelligence, personality, and interests: evidence for overlapping traits. Psychol. Bull. **121**(2), 219–245 (1997)
8. Nodine, T.: How did we get here? a brief history of competency-based higher education in the United States. J. Competency-Based Educ. **1**, 5–11 (2016)
9. Derryberry, A., Everhart, D., Knight, E.: Badges and competencies: new currency for professional credentials. In: Muilenberg, L., Berge, Z. (eds.) Digital Badges in Education: Trends, Issues, and Cases, pp. 12–20 (2016)
10. Pitt, C., Bell, A., Strickman, R., Davis, K.: Supporting learners' STEM-oriented career pathways with digital badges. Inf. Learn. Sci. (2018)
11. Aberdour, M.: Transforming workplace learning culture with digital badges. In: Ifenthaler, D., Bellin-Mularski, N., Mah, D.-K. (eds.) Foundation of Digital Badges and Micro-Credentials, pp. 203–219. Springer, Cham (2016). https://doi.org/10.1007/978-3-319-15425-1_11
12. Davis, K., Singh, S.: Digital badges in afterschool learning: documenting the perspectives and experiences of students and educators. Comput. Educ. **88**, 72–83 (2015)
13. Abramovich, S., Wardrip, P.: Impact of badges in motivation to learn. In: Muilenberg, L., Berge, Z. (eds.) Digital Badges in Education: Trends, Issues, and Cases, pp. 53–61 (2016)
14. Ifenthaler, D., Bellin-Mularski, N., Mah, D.-K. (eds.): Foundation of Digital Badges and Micro-Credentials: Demonstrating and Recognizing Knowledge and Competencies. Springer International Publishing, Cham (2016). https://doi.org/10.1007/978-3-319-15425-1
15. West-Puckett, S.: Making classroom writing assessment more visible, equitable, and portable through digital badging. Coll. Engl. **79**(2), 127–151 (2016)
16. Chemers, M., Zurbriggen, E., Syed, M., Goza, B., Bearman, S.: The role of efficacy and identity in science career commitment among underrepresented minority students. J. Soc. Issues **67**, 469–491 (2011)
17. Carpi, A., Ronan, D., Falconer, H., Lents, N.: Cultivating Minority scientists: undergraduate research increases self-efficacy and career ambitions for underrepresented students in STEM. J. Res. Sci. Teach. **54**, 169–194 (2017)
18. Guilbaud, P., Bubar, E., Langran, E.: STEM excellence and equity in K-12 settings: use of augmented reality-based educational experiences to promote academic achievement and learner success. In: Stephanidis, C., Antona, M., Ntoa, S. (eds.) HCII 2021. CCIS, vol. 1421, pp. 45–50. Springer, Cham (2021). https://doi.org/10.1007/978-3-030-78645-8_6
19. Beede, D., et al.: Education Supports Racial and Ethnic Equality in STEM. U.S. Department of Commerce, Economics and Statistics Administration, Washington (2011)
20. Chickering, A., Gamson, Z.: Seven principles for good practice in undergraduate education. In: Chickering , A., Gamson, Z. (eds.) Applying the Seven Principles for Good Practice in Undergraduate Education, pp. 63–69 (1991)
21. Grasha, A.: Teaching with Style: A Practical Guide to Enhancing Learning by Understanding Teaching and Learning Styles. Alliance Publishers, Pittsburgh (1996)

22. Aoun, J.: Robot-Proof: Higher Education in the Age of Artificial Intelligence. U.S (2017)
23. Wang, Y.: Education in a Changing World: Flexibility, Skills and Employability. World Bank, Washington DC (2002)
24. DiRenzo, M., Greenhaus, J., Weer, C.: Relationship between protean career orientation and work life balance: a resource perspective. J. Organ. Behav. **36**, 538–560 (2015)
25. Carr, P., Walton, M.: Cues of working together fuel intrinsic motivation. J. Exp. Soc. Psychol. **53**, 169–184 (2014)
26. Kuron, L., Schweitzer, L., Lyons, S., Ng, E.: Career profiles in the "new career": evidence of their prevalence and correlates. Career Dev. Int. **21**, 355–377 (2016)
27. Bialik, M., Fadel, C.: Skills for the 21st Century: What Should Students Learn? Center for Curriculum Redesign. Boston, Massachusetts (2015). http://www.curriculumredesign
28. Jessop, B.: The knowledge economy as a state project. In: The Nation-State in Transformation, pp. 110–129. Aarhus University Press, Århus (2010)
29. Strada Institute and Emsi: Robot-Ready: Human Skills for the Future of Work (2018)
30. Connell, G., Donovan, D., Chambers, T.: Increasing the use of student-centered pedagogies from moderate to high improves student learning and attitudes about biology. CBE Life Sci. Educ. **15**(1) (2016)
31. Miri, B., David, B., Uri, Z.: Purposely teaching for the promotion of higher-order thinking skills: a case of critical thinking. Res. Sci. Educ. **37**(4), 353–369 (2007)
32. Montealegre, R., Cascio, W.: Technology-driven changes in work and employment. Commun. ACM **60**, 60–67 (2017)
33. Pusca, D., Bowers, R., Northwood, D.: Hands-on experiences in engineering classes: the need, the implementation and the results. **15**, 12–18 (2017)
34. Nilson, L.: Creating Self-Regulated Learners: Strategies to Strengthen Students' Self-Awareness and Learning. Sterling, VA: Stylus (2013)
35. Rotgans, J., Schmidt, H.: Situational interest and academic achievement in the active-learning classroom. Learn. Instr. **21**(1), 58–67 (2011)
36. Cattel, R.: Intelligence: Its Structure, Growth, and Action. Elsevier Science Publishers, BV, Amsterdam, The Netherlands (1987)
37. Evans, J.: In two minds: dual-process accounts of reasoning. Trends Cogn. Sci. **7**(10), 454–459 (2003)
38. Hmelo-Silver, C.: Problem-based learning: what and how do students learn? Educ. Psychol. Rev. **16**(3), 235–266 (2004)
39. Lave, J., Wenger, E.: Situated Learning: Legitimate Peripheral Participation. Cambridge University Press, Cambridge (1991)
40. Guskey, T.: Mastery Learning in 21st Century Education: A reference Handbook, vol. 1 (2009)
41. Robles, M.: Executive perceptions of the top 10 soft skills needed in today's workplace. Bus. Commun. Q. **75**(4), 453–465 (2012)
42. Davidson, K.: Employers find 'soft skills' like critical thinking in short supply. Wall Street J. **3** (2016)
43. Strauss, V.: The Surprising Thing Google Learned About its Employees—and What it Means for Today's Students. The Washington Post (2017)
44. Karimi, H., Pina, A.: Strategically addressing the soft skills gap among STEM undergraduates. J. Res. STEM Educ. **7**(1), 21–46 (2021)
45. Fadde, P., Klein, G.: Accelerating expertise using action learning activities. Cogn. Technol. **17**(1), 11–18 (2012)
46. Ericsson, K.: The influence of experience and deliberate practice on the development of superior expert performance. In: Ericsson, K., Charness, N., Hoffman, R., Feltovich, P. (eds.) The Cambridge Handbook of Expertise and Expert Performance, pp. 683–703 (2006)
47. Ericsson, K., Charness, N.: Expertise: Its structure and acquisition. Am. Psychol. **49**, 725–747 (1994)

48. Wallace, D.: Knowledge Management: Historical and Cross-Disciplinary Themes (2007)
49. Hildreth, P., Kimble, C.: Knowledge networks: innovation through communities of practice, vol. 9 (2004)
50. Onyon, C.: Problem-based learning: a review of the educational and psychological theory. Clin. Teach. **9**(1), 22–26 (2012)
51. Loyens, S., Kirschner, P., Paas, F.: Problem-based learning. In: Harris, K., Graham, S. Urdan, T. (eds.) APA Educational Psychology Handbook, vol. 2 (2011)
52. Winarno, S., Muthu, K., Ling, L.: Direct PBL (DPBL): a framework for Integrating direct Instruction and PBL approach. Int. Educ. Stud. **11**(1), 119 (2017)
53. Walton, J., Ryerse, M.: Competency-based education: definitions and difference makers. Getting Smart (2017)
54. Bliven, A., Jungbauer, M.: The impact of student recognition of excellence to student outcome in a competency-based educational model. J. Competency-Based Educ. (JCBE). **6**(4), 195–205 (2021)
55. Henri, M., Johnson, M., Nepal, B.: A review of competency-based learning: Tools, assessments, and recommendations. J. Eng. Educ. **106**(4), 607–638 (2017)
56. Burke, J.: Competence-based education and training (1989)
57. Levine, E., Patrick, S.: What is competency-based education? an updated definition (2019)
58. U.S. Department of Education Office of Career, Technical, and Adult Education, Advancing Career and Technical Education in State and Local Career Pathways Systems (2015)
59. Zoogah, B.: Historicizing management and organization in Africa. Acad. Manage. Learn. Educ. **20**(3), 382–406 (2021)
60. Kotsiantis, S.: Use of machine learning techniques for educational proposes: a decision support system for forecasting students' grades. Artif. Intell. Rev. **37**(4), 331–344 (2012). https://doi.org/10.1007/s10462-011-9234-x
61. Hua, W., Cuiqin, M., Lijuan, Z.: A brief review of machine learning and its application. In: IEEE International Conference on Information Engineering and Computer Science (2009)
62. Ferlitsch, A.: Deep Learning Patterns and Practices (2021)
63. Hirsch, M.J., Crowder, J.A.: Machine learning to augment the fusion process for data classification. In: Rayz, J., Raskin, V., Dick, S., Kreinovich, V. (eds.) NAFIPS 2021. LNNS, vol. 258, pp. 154–165. Springer, Cham (2022). https://doi.org/10.1007/978-3-030-82099-2_14
64. Doyle, J., Francis, B., Tannenbaum, A.: Feedback Control Theory (1992)
65. Piaget, J.: Cognitive development in children: development and learning. J. Res. Sci. Teach. **2**, 176–186 (1964)
66. Lewis, M., Michalson, L.: Children's Emotions and Moods: Developmental Theory and Measurement. Plenum Press, New York (1983)
67. Kremenitzer, J.P., Miller, R.: Are you a highly qualified emotionally intelligent early childhood educator? Young Children **63**, 106–112 (2008)
68. David, S.Y.: Social and emotional learning programs for adolescents. Future Child. **27**(1), 73–94 (2017)
69. Bronfenbrenner, U.: The Ecology of Human Development Experiments by Nature and Design. Harvard University Press, Cambridge (1996)
70. Welsh, J.A., Nix, R.L., Blair, C., Bierman, K.L., Nelson, K.E.: The development of cognitive skills and gains in academic school readiness for children from low-income families. J. Educ. Psychol. **102**(1), 43–53 (2010)
71. Pantsar, M. The enculturated move from proto-arithmetic to arithmetic. Front. Psychol. **10** (2019)
72. Colchester, K., Hagras, H., Alghazzawi, D., Aldabbagh, G.: A survey of artificial intelligence techniques employed for adaptive educational systems within e-learning platforms. J. Artif. Intell. Soft Comput. Res. **7**(1), 47–64 (2016)

73. Brusilovsky, P.: Adaptive and intelligent web-based educational systems. Int. J. Artif. Intell. Educ. **13**(2–4), 159–172 (2003)
74. Diziol, D., Walker, E., Rummel, N., Koedinger, K.: Using intelligent tutor technology to implement adaptive support for student collaboration. Educ. Psychol. Rev. **22**(1), 89–102 (2010). https://doi.org/10.1007/s10648-009-9116-9
75. Du Boulay, B.: Artificial intelligence as an effective classroom assistant. IEEE Intell. Syst. **31**(6), 76–81 (2016)
76. Kolb, D.: Experiential Learning: Experience as the Source of Learning and Development, Second Edition (2014)
77. Barrows, H.: Problem-based learning in medicine and beyond: a brief overview. In: Wilkerson, L., Gijselaers, W. (eds.) New Directions for Teaching and Learning, vol. 68, pp. 3–11 (1996)
78. Bayat, S., Tarmizi, A.: Effects of problem-based learning approach on cognitive variables of university students. Procedia Soc. Behav. Sci. **46**, 3146–3151 (2012)
79. Ericsson, K., Chamess, N.: Expertise: its structure and acquisition. Am. Psychol. **23** (1994)
80. Starr, L.: Integrating Technology in the Classroom: it Takes more than just having Computers (2011)
81. Park, S., Ertmer, P.: Examining barriers in technology-enhanced problem-based learning: Using a performance support systems approach. Br. J. Edu. Technol. **39**, 631–643 (2008)
82. Rackley, C.: Mapping knowledge units using a Learning Management System (LMS) course framework. KSU Proc. Cybersecurity Educ. Res. Pract. **7** (2018)
83. Centers of Academic Excellence in Cyber Defense: Knowledge Units (2019)
84. Iwata, J., Clayton, J., Saravani, S.: Learner autonomy, microcredentials and self-reflection: a review of a Moodle-based medical English review course. Int. J. Inf. Commun. Technol. **10**, 42–50 (2017)
85. Fink, D.: Creating significant learning experiences: an integrated approach to designing college courses (2003)
86. Sweet, C., Blythe, H., Phillips, B., Carpenter, R.: Transforming Your Students into Deep Learners: A Guide for Instructors (2016)
87. Potvin, P., Hasni, A.: Interest, motivation and attitude towards science and technology at K-12 levels: a systematic review of 12 years of educational research. Stud. Sci. Educ. **50**(1), 85–129 (2014)
88. Kolb, D.: Learning Style Inventory (1976)
89. Henderson, M., Selwyn, N., Aston, R.: What works and why? student perceptions of 'useful' digital technology in university teaching and learning. Stud. High. Educ. **42**(8), 1567–1579 (2017)
90. Gagne, R.: The Conditions of Learning, 4th (edn.) (1985)
91. Bybee, R., Powell, J., Trowbridge, L.: Teaching Secondary School Science: Strategies for Developing Scientific Literacy (2008)
92. Magana, A.: Learning strategies and multimedia techniques for scaffolding size and scale cognition. Comput. Educ. **72**, 367–377 (2014)
93. Koedinger, K., Corbett, A., Perfetti, C.: The Knowledge-Learning-Instruction (KLI) framework: Bridging the science-practice chasm to enhance robust student learning. Cogn. Sci. **36**(5), 757–798 (2012)
94. Keengwe, J., Onchwari, G., Wachira, P.: The use of computer tools to support meaningful learning. AACE J. **16**(1), 77–92 (2008)
95. Torp, L., Sage, S.: Problems as Possibilities: Problem-Based Learning for K-12 Education, 2nd (edn.) (2002)

SatisfAI: A Serious Tabletop Game to Reveal Human-AI Interaction Dynamics

Moon K. Kim[✉] ⓘD and Ethan Trewhitt ⓘD

Georgia Tech Research Institute, Atlanta, GA 30318, USA
moon.kim@gtri.gatech.edu

Abstract. Interactions between humans and AI-enabled systems are occurring more frequently, often at times unbeknownst to the human end users. As AI-enabled systems become more pervasive and advanced in helping humans make decisions, there is a need to ensure the systems are designed around human needs and for the end users to have a general understanding of the benefits and limitations of AI. Humans who understand both the benefits and limitations of AI systems may then make more informed decisions that maximize the satisfaction of their own needs. Toward this end, we have developed a game based on Max-Neef's Fundamental Human Needs Scale to inform human-AI interaction design. We have designed and prototyped the game so that it may serve as a data collection and user research tool. The tool presents various human-AI interaction scenarios to players in an effort to reveal the level of agency humans are comfortable ceding to an AI-enabled system, based on their personal needs that are satisfied by that system. Through the design mechanics and gameplay that result from real people playing the game, insights may be gathered regarding human-AI interaction dynamics, which may then be used to inform the design of future AI systems.

Keywords: Human-AI interaction · Serious game · User-centered design

1 Introduction

AI-assisted and AI-enabled technologies are being employed increasingly in all aspects of life. Interactions between humans and AI systems are thereby occurring more frequently, often unbeknownst to the human end users. As AI-enabled systems become more pervasive and advanced in helping humans make decisions, there is a need to ensure the systems are designed around human needs and for the end users to have a general understanding of the advantages and tradeoffs to engage with AI-enabled systems. This is at the crux of human-AI interaction design.

Designing human-AI interactions is complicated. This is especially true for human-AI interactions involving multi-agent decision support, where decision making authority is shared by person(s) and machine(s). For which applications are human end users comfortable deferring to AI and when do they want to override? How should AI systems be designed to satisfy human needs and values to ensure trust while respecting human end users' preferred level of autonomy? Current methods for designing technology

R. A. Sottilare and J. Schwarz (Eds.): HCII 2022, LNCS 13332, pp. 174–189, 2022.
https://doi.org/10.1007/978-3-031-05887-5_13

(i.e., HCI methods, user experience design) were created for "non-intelligent solutions" and need to evolve, as they fall short on future-focused, speculative research around human-AI interactions (Xu 2019). In recent years, several frameworks and guidelines have been proposed to help researchers and developers take a human-centered approach to AI designs, such as the HAI framework (Xu 2019) and Microsoft's guidelines for Human-AI Interaction (Ribera and Lapedriza 2019). However, they are high-level, and, in some cases, are only applicable for evaluation of AI systems with specific attributes like graphical user interfaces (Ribera and Lapedriza 2019).

In this paper, we explore the value of game-based simulation to train users on the various aspects of AI-based systems that should be considered in decision making, specifically as it pertains to levels of automation. We have developed a game, called *SatisfAI*, based on human values and needs model (Max-Neef et al. 1991) to inform human-AI interaction design. We have designed and prototyped the game so that it may serve as a data collection and user research tool that aims to understand the level of agency humans are willing to cede based on the needs that the AI-enabled systems can satisfy.

For example, human end users may be comfortable utilizing a highly automated system and deferring decision-making/control to an AI-based system if the need to satisfy has to do with leisure or creativity (as oppose to needs related to existence), but be more conservative about ceding agency to an AI-based system that satisfied their subsistence need. The game elicits this decision-making in players throughout a variety of human-AI interaction scenarios. Through the design mechanics and gameplay that results from real people playing the game, insights may be gathered regarding human-AI interaction dynamics, which may then be used to inform the design of future AI systems.

2 Background

Human-AI interaction can be defined as communication, actions, and/or decision making that happen between human and an AI-enabled system. Similar terminologies include human-AI teaming, human-AI collaboration, human-in-the-loop, and human-on-the-loop. Types of human-AI interactions can vary based on the AI application and scenario in question. Additionally, human-AI interactions can entail varying degrees of human participation, ranging from those in which the human is fully in control, to those in which control is fully assumed by the AI-enabled system.

As more services and products rely on AI and machine learning algorithms to facilitate transactions and make decisions on behalf of humans, there are growing concerns over the risks of replacing humans with AI-enabled systems which could lead to biases, inequities, inaccuracies, and even bodily harm (Xu 2019). In response, a human-centered approach to AI is gaining attention in research universities and others in HCI community. For example, Stanford University's human-centered AI initiative calls for AI that focuses on enhancing humanity and not replacing it, among other goals (Li and Etchemendy 2018). Xu (2019) proposed a human-centered AI framework comprised of principles focused on 1) ethically aligned design that enhances (not replaces) humans; 2) technology that reflects human intelligence; and 3) human factors design that ensures "AI solutions are explainable, comprehensible, useful, and usable" by humans. The

HCI community agrees on the value of human-centered design approach to AI systems. However, HCI professionals must grapple with the challenges of designing AI systems.

Traditional human-centered/user-centered design methods typically cannot be readily applied to AI, due to challenges and uncertainties that characterize AI's capabilities and output complexities (Yang et al. 2020). Additionally, HCI practitioners may have difficulty prototyping and testing human-AI interactions without yet having access to the fully functioning AI system. For example, simple, inexpensive paper-based prototypes and sketches have limitations when it comes to designing complicated human-AI interactions that can often be difficult to predict, due to their varied and speculative nature. Hence, the often-utilized approaches have been a) to use "Wizard of Oz" rules-based simulators, which can only simulate known interactions and cannot recreate as-yet-unanticipated user experience scenarios and issues; or b) to conduct user studies after the AI system has been deployed, which deviates from accepted agile processes and can be costly/risky (Yang et al. 2020).

An alternative approach is to combine some aspects of prototyping and simulation to understand the speculative nature of human-AI interactions. First appearing in the 1990s, provocative prototypes (or provotypes) have been used to identify issues and requirements with current state of technology while also exploring the desired future end state by intentionally probing or provoking users on current realities as well as innovative possibilities (Boer and Donovan 2012). Provotyping activities can take various forms, including interviews, observations, and design workshops, wherein the process focuses on discovery, rather than conceptualization, similar to the use of simulations. The design and implementation of the *SatisfAI* game is inspired by the provotyping approach.

2.1 Human Needs Models

Human behaviors are determined by their motivation to satisfy a need. In the field of motivation psychology, many studies have been conducted and theories proposed to understand how human behaviors and actions can be explained by the pursuit of satisfying needs (Maslow 1943; Alderfer 1969; Max-Neef et al. 1991; Ryan and Deci 2000). In summary, a need presents a motive, which is a reason or a driver for humans to act or behave a certain way to satisfy their need. For the purposes of this paper, we examined prominent theories that could encompass a wide range of needs since AI-enabled systems and applications typically exist to satisfy a similar range of human needs and services. As such, we looked at two of the most widely known theories: Maslow's Hierarchy of Needs (Maslow 1943) and Max-Neef's Fundamental Human Needs Scale (Max-Neef et al. 1991).

In Maslow's Hierarchy of Needs (summarized in Fig. 1), human needs are hierarchical and assumes order, starting with psychological needs, safety needs, belonging and love needs, esteem needs, and finally self-actualization. While Maslow's Hierarchy of Needs is widely known and cited, the theory has been criticized for its ambiguity with self-actualization need, absence of recreation and entertainment needs, assumption of sequence of satisfaction, and not being able to test empirically (Yu 2018).

Proposed decades later, Max-Neef's Fundamental Human Needs Scale (1991) is comprised of nine needs that are interrelated and interactive. These needs include subsistence, protection, affection, understanding, participation, leisure, creation, identity,

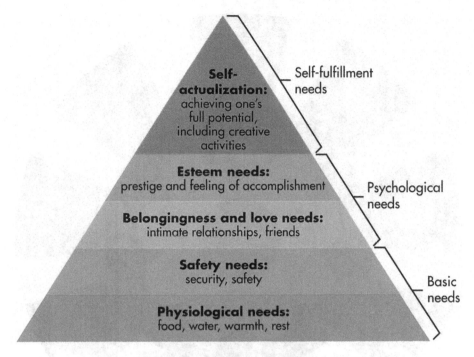

Fig. 1. Maslow's hierarchy of needs (McLeod 2020)

and freedom. Except for subsistence, there are no hierarchies and sequence to the order of human needs; the needs are viewed as a system. The nine needs are either existential (needs related to existence) or axiological (needs related to human values). Max-Neef postulates that these fundamental human needs are finite, few, and classifiable. Although needs rarely change, what changes are the ways in which humans satisfy their needs. For example, in-person instruction, internet, books, and Google search engine, though widely different, can be satisfiers of the need for understanding. As such, satisfiers (or ways in which humans satisfy needs) vary among individuals, groups, and cultures, as needs are satisfied within three contexts: oneself, social group, and environment. Figure 2 shows an overview of the Max-Neef model.

In comparing the Maslow and Max-Neef models, we chose the latter as a basis for the game design due to its inclusion of a broader range of needs, e.g. leisure, that are often satisfied by modern AI-based systems.

2.2   Dimensions of AI Systems

Our effort investigates the complexities and pitfalls of human-AI interactive systems and uses those as elements in a serious game. Not only will this be a learning tool for users, it will also inform researchers of the concerns vs. comfort level of human users who experience specific types of AI behavior, both good and bad. Additionally, the game is structured so that different scenarios and parameters can be introduced to the gameplay

Fig. 2. Max-Neef's fundamental human needs (Koren and Soni 2010)

to customize it to a domain. This way, the game is not limited to a single scenario but versatile across multiple problem spaces.

The term "artificial intelligence" describes a broad set of solutions with diverse design characteristics. Some AI approaches are designed to address problems across many domains, while some are more domain-specific. As AI science has progressed, more and more techniques are invented within the AI ecosystem, broadening the mechanisms and techniques used to solve these problems. As such, AI systems occupy a large variety of design and problem space dimensions. We use some of these dimensions to categorize the systems within the game with the goal of identifying which aspects of AI system design affect human interaction and to what extent. The dimensions of interest in this project are listed here.

Levels of Automation. Parasuraman et al. (2000) provides a framework for describing the extent to which an AI system may perform its tasks automatically, with or without varying degrees of human interaction. The paper defines ten (10) levels of automation.

At the top level, the system performs its task in a fully automated fashion, with zero opportunity for human interaction. At the bottom level, no automation occurs at all. Nearly all real-world AI systems exist at some point between these two extremes, with "human in the loop" and "human on the loop" existing within. We propose a small extension to this framework that groups several levels together, as described later.

Explainability. AI systems are often described in terms of how "explainable" their approach and solutions are (Arya et al. 2019), which is the extent to which human users can understand a) the mechanisms within the system, and b) the relationship between input problems and output solutions. Explainability is not solely an issue with AI algorithms, but the complexity of many AI algorithms can lead to limited explainability. AI systems with poor explainability act as black boxes, with limited ability for users, and sometimes even system designers, to understand how problems are solved. This often means that it is unclear which future problems will be solvable by the system, or why one class of problem yields good solutions while other problems yield poor solutions, leading to surprises. This can create issues with trust, as described next.

Trust. Whether or not users and designers of an AI system can trust the solutions of that system will greatly impact the utility of that system in real-world applications (Siau and Wang 2018). Improved trust in AI systems may lead people to remain comfortable in increasing levels of AI automation.

3 Game Design

The primary goal of *SatisfAI* is to determine the amount of agency a player is willing to yield to an AI-based system to satisfy each of their nine Max-Neef needs dimensions. Players choose a level of automation of the AI-enabled system that best satisfy their needs for the application of that AI. Based on the (Parasuraman et al. 2000) design for AI levels of automation, we have proposed a grouping of these ten automation levels into five AI "roles" as shown in Table 1.

The purpose of the grouping is to simplify these levels into a more manageable set of choices from among which players will choose, as well as to provide a convenient label to represent a more complex concept for level of automation. A redefinition of the human-AI relationship is defined for each of these roles in Table 2.

Scenario Cards. Players are provided with Scenario Cards that contain four pieces of information:

- **Objective**: a description of a real-world problem in which an AI-based system might be used to help produce a solution.
- **Need**: the highest human need, from among the nine in the Max-Neef model, that is satisfied by the solution of the objective problem. This represents the key mapping needed to relate the player's choice of level-of-automation and the need satisfied by the achievement of that scenario objective.

Table 1. Levels of automation with additional "AI Role" groupings

Level	The AI ...	AI Role
10	...decides everything and acts autonomously, ignoring the human.	Surrogate
9	...informs the human only if it, the computer, decides to.	
8	...informs the human only if asked.	Collaborator
7	...executes automatically, then necessarily informs the human.	
6	...allows the human a restricted time to veto before automatic execution.	
5	...executes that suggestion if the human approves.	Assistant
4	...suggests one alternative.	Informer
3	...narrows the selection down to a few.	
2	...offers a complete set of decision/action alternatives.	
1	...offers no assistance: human must take all decisions and actions.	Non-AI

Table 2. Simplified definitions of the AI roles

AI Role	The AI ...
Surrogate	...acts autonomously and decides for itself whether it wants to involve a human in the process.
Collaborator	...takes action by default, though a human can decide whether they want to participate.
Assistant	...executes an action if (and only if) the human approves.
Informer	...provides information to the human but does not take direct action.
No AI	...offers no assistance: human must take all decisions and actions.

- **Time**: the time scale (order) of the problem's solution. Examples: months, minutes, milliseconds.
- **Criticality**: the severity of consequences (bodily injury, potential damage, financial impact) if the objective is not met. Examples: insignificant, minor, moderate, major, catastrophic (Markovic 2019).

Gameplay in *SatisfAI* is centered around the "player judge" mechanic. During each round, one player acts as Judge, selecting an AI Role that matches their comfort level within a single AI scenario. The other players attempt to predict the AI Role selected by the Judge. All players secretly choose a role with a face-down card, and once all players have chosen, all cards are revealed. The Judge receives a single point for each matching card from among the non-Judge players, and the corresponding non-Judge players each receive a single point if their card matched the Judge's selection. This mechanic enables players to compete to maximize their individual score, yet enables them to achieve higher scores by finding agreement with other players during individual rounds. Likewise, this means that the primary success metric is mutual understanding, rather than a "correct" value for any given problem-solution set that might otherwise be provided externally by the game itself. A sample of the details shown on these cards is included in Table 3.

Table 3. Sample objectives included on game scenario cards

Objectives	
• File your taxes • Create a piece of visual art • Obtain a loan in order to purchase a house • Determine whether someone is lying or telling the truth • Check for signs of cancer cells or other anomalies in medical imaging of patients	• Translate text between languages • Decode the content of over-the-air radio communications signals • Adapt radar settings to avoid being jammed • Predict the formation or path of a tornado and warn at-risk residents • Pilot a drone on an uninhabited planet

Scorecards and Tokens. To monitor each player's score, they are given a single score-card that contains an area divided into nine segments. Also available is a pool of five different tokens, each representing one of the five AI Roles. As players succeed in matching between Judge and non-Judge, they indicate the corresponding earned point by placing one of the "correct" AI Role tokens into the scorecard region that corresponds with the Need indicated on the Scenario Card. While each player's score is simply the total number of tokens on their scoreboard, tracking scores by Need and AI Role provides valuable research information regarding human preferences that will be used during the data collection efforts of this project.

Player Decision-Making Process. Table 4 shows various example scenarios for a given Scenario card's objective. All five AI Roles are shown for the objective "File tax returns," ranging from human-only to computer-only. While the Scenario card and its information are provided by the game, the table represents the situations a player might imagine to represent a human-AI interaction in which the objective is hypothetically solved by each AI Role. Thus, when the player thinks "What would a Collaborator-style AI look like solving the objective of filing tax returns?", they will be imagining a scenario in which the AI is performing its tasks nearly autonomously, with limited ability for a human to intervene. Once the player evaluates these options, they will then try to determine (if they are the Judge) the highest AI Role level that would not make them uncomfortable, or (if they are the other non-judge players) the same preference that the Judge is likely to have.

The scenario card shown includes the objective "File tax returns" with a mapping to the need "Protection" (see Fig. 2 in the Background section for context). This mapping, provided by the game itself, is based upon the logic that filing one's tax returns will typ-ically lead to a refund (which is a form of planning), and contributes to the government-provided social security (which will provide a social safety net to the taxpayer later in life).

From a purely game-theoretical standpoint, since the Judge receives a point for each matching player, they are incentivized to pick a realistic (predictable) answer rather

Table 4. Specific scenario examples, across all five AI Roles, for a given objective

Objective:
File tax returns

Need: Protection

Time: Months

Criticality: High

AI Role	Scenario Example
No AI	The human fills out a 1040 and uses e-File or mails it in.
Informer	AI-based software evaluates the human's finances and provides advice for how to minimize taxes for the current and future year.
Assistant	AI-based software determines the optimal tax management strategy and offers to submit updated W-4s digitally.
Collaborator	Once configured by the human, the computer automatically evaluates taxes each year and files them unless instructed to wait or cancel.
Surrogate	A fully automated system evaluates options for investments, charitable donations, and costs, making adjustments to each to minimize tax exposure, all without human intervention.

than gaming the system. Similarly, other non-Judge players are incentivized to make a realistic selection for the Judge.

3.1 Prototype Gameplay

Game Components. The game consists of the following physical components, many of which are shown in Fig. 3:

- One scoreboard per player
- 60 Scenario cards
- 5 AI Role cards per player (one of each AI Role)

Fig. 3. Playable prototype of the *SatisfAI* serious game

- A large set of AI Role tokens representing the 5 different AI Roles

Setup. These are the steps to be performed at the start of the game:

1. Each player receives a game board for scoring, set of tokens, and each of the five AI Role cards
2. Shuffle the Scenario card deck and place it face down in the middle of the table
3. Select a player to start as Judge

Round Play. Each round consists of the following steps:

1. The player who is the Judge for the round draws a Scenario card from the Scenario deck and reads it out loud. Each player, including the Judge, chooses the AI Role that they think would best satisfy the Judge's need for that scenario. Each player, including the Judge, puts their selected AI Role card face down in front of themselves.

   Players are encouraged to review the quick reference card that shows the definitions of the five AI Roles, and to imagine what an AI-based solution to the objective would be for each role.
2. Once everyone has selected an AI Role card, the Judge reveals the AI Role card they selected that satisfies the given Scenario Card. Then, each of the other players overturns their selected AI Role card. Each non-Judge whose card matches the Judge's card places one of their tokens on their Scoreboard at the appropriate need + role position. Then, for each non-Judge whose role card matches the judge's role card, the Judge places one of their tokens on the respective need + role position on their Scoreboard. Thus, the Judge may score 0, 1, or more points for the round, depending on the number of matches.

   Players are encouraged to discuss their choices and why theirs might be preferred over the choices made by others. This may be of particular value to researchers watching the gameplay, to better understand the motivations behind players' choices.
3. The next round continues in counterclockwise order, with the person to the left of the current Judge taking on the role of Judge for the next round.

Game End. When the scenario deck has run out, the game ends. Each player's final score is determined by the total number of tokens on their individual scoreboard. The winner is the player with the most tokens on their scoreboard. If there is a tie, the first tiebreaker is the number of different needs a player was able to satisfy. If the top players are still tied, they share the victory.

The figures that follow show a visual representation of gameplay (Figs. 4, 5, 6, 7 and 8).

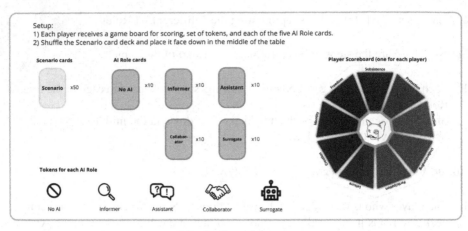

Fig. 4. Gameplay setup

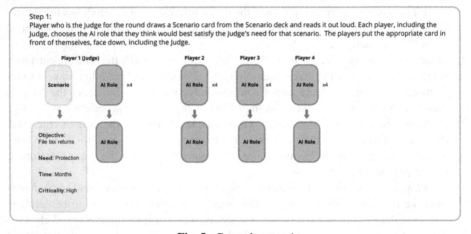

Fig. 5. Gameplay step 1

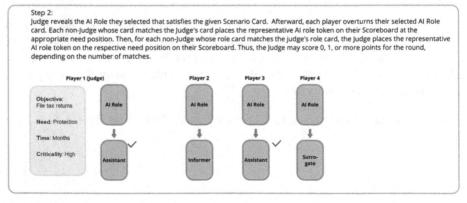

Fig. 6. Gameplay step 2

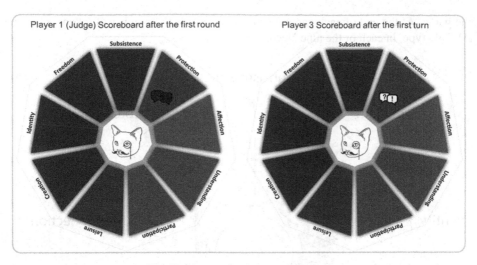

Fig. 7. Player scoreboards after step 2

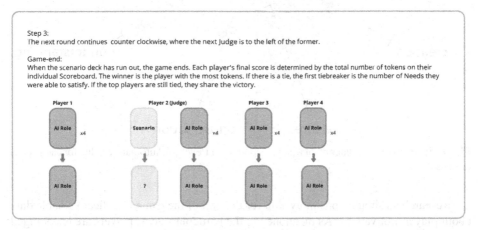

Fig. 8. Gameplay step 3

4 Data Collection Design and Approach

4.1 Data Collection

The function of the "player judge" mechanic in *SatisfAI* is to provide a metric for success within each player interaction in which the "correct" AI Role for a given objective (and thus its mapping to one of the Max-Neef human needs) is not predetermined by the game, but instead by other players. This is the fundamental mechanic that makes it possible for researchers to evaluate the relative comfort of players who imagine themselves interacting in a human-AI system to solve a real-world problem they may face. The scoreboard and AI Role-based tokens track players' scores, but more importantly, they provide us (as researchers) the ability to record players' preferences as they play the game. At the end of each game, our project records the number of tokens, of each AI Role type, in each of the nine Needs categories.

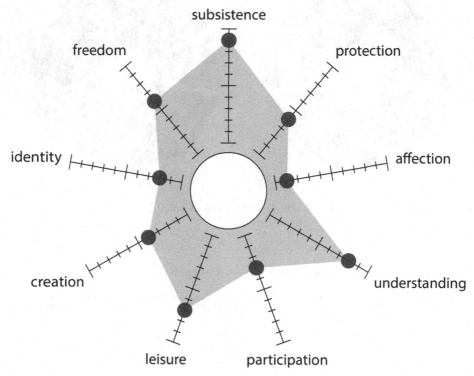

Fig. 9. Example profile diagram of a player's preferred Level-of-Automation value for each needs dimension

Researchers should not simply wait for the end of the game to collect valuable data about player motivations. As mentioned in the gameplay steps, players are encouraged to discuss their choices during each round, and researchers should take note of any particular insights gained during this process.

Once a game is complete, each player's board is recorded, providing the information needed to produce the diagram shown in Fig. 9. As more games are played and these human-AI interaction "comfort zones" are identified, our aim is to evaluate the patterns of automation levels, specific to each human need dimension, selected by large numbers of players.

4.2 Assumptions and Risks

Assumption: Game Metrics Will Win Over Natural Human Motivations.
Our learning objectives are best achieved with players who make decisions that are based on their personal beliefs regarding the interactions of AI systems and human users. However, when players participate in a game with defined metrics of success, it

is possible that they will naturally switch to a role-playing mode in which they make decisions that will maximize their achievement in terms of those metrics. Thus, it is vital that the gameplay be designed so that the in-game definition of success most directly mirrors the learning objectives of this research project. **Risk**: Our gameplay is not sufficiently designed to maximize natural decision making and the results collected are driven by players' game-theory approach rather than true preferences in the domain. A key challenge in the design of *SatisfAI* is enabling players to maximize their success metrics in the game while still allowing their real-world motivations regarding AI and their individual human needs to shine through. There is a risk that the artificial metrics of success designed by the game will unintentionally encourage players to "role play" artificial decisions that yield a higher in-game score but do not represent their true opinions regarding the interactions between humans and AI systems. Our objective as game designers is that players will play the game as themselves rather than altering their behavior to meet a game-provided success metric.

Assumption: The Mappings of Objectives to Needs is Correct. It is important that the objectives described on the scenario cards match the needs shown. While the game designers have made best-effort decisions regarding the need listed on each card, each individual mapping may be subject to debate. **Risk**: It may be difficult to provide a robust justification for these mappings. The mappings chosen may fail to represent the need internalized by each player as they imagine the problem-solution scenarios for each AI Role. Thus, their final choice of AI Role would be misattributed to the wrong need, which would reduce the accuracy of data collection and analysis. During play testing we will discuss these mappings with players, modifying the mappings as necessary to match players' expectations while maintaining the integrity of the data-collection/research efforts of the game.

5 Play Testing and Evaluation Approach

To ensure human-AI design/research utility and game playability (i.e. entertainment value), we aim to test and evaluate the game prototype with internal peers who have varying levels of experience in the AI domain.

Our notional design of the play test and evaluation are based on the user studies and Heuristics Evaluation for Playability developed by Desurvire et al. (2004).

1. Identify 10–12 individuals (participants) from within our organization to participate in two-hour play test and evaluation sessions of the prototype game. Each session will have a varied number of player participants ranging from three to four players.
2. Each play test and evaluation session will consist of the following:
a. Introduction to the instructions and rules of the prototype game to the participants. The participants are asked to share their thoughts and questions during the game play as they normally would when learning and playing a new game in a group setting.
b. One hour of game play where the research team observes and documents pertinent information about the participants, game play dynamics, results and scores, and other information (e.g. participants' game play decisions, questions, comments, errors, misunderstandings).

c. Evaluation of the game play and the game itself consists of a brief survey (independently completed by each participant) and group discussion, conducted in that order. The survey and the group discussion will include questions to probe the participants' thoughts on specific game mechanics, game play experience, and any resulting thoughts/insights on AI or their personal interactions with AI-enabled systems, products, or services.
3. Collate and analyze resulting data from the play test and evaluation sessions. As part of the analysis, pain points or points of opportunity to improve the game playability will be identified, along with potential solutions to incorporate into the next version of the game.

5.1 Evaluating Data Collection

In addition to traditional game testing, this game also has the purpose of being a data collection tool. While we play test our prototype, we will also be evaluating the fitness of the game as a data collection tool. To evaluate the utility of the prototype game to inform the design process of human-AI interaction, we will examine if and how well the resulting insights from the game play and players' actions/decisions serve to address the human-AI design challenges identified by Yang et al. (2020). In conjunction, we will assess and compare the resulting data from the play test and evaluation sessions with existing studies/measures that examine human trust, values, and preferences in levels of automation for AI applications. These could include, but not limited to the following:

- Studies involving the application and measurement of Max-Neef's Fundamental Human Needs (1991).
- Studies involving the application and measurement of levels of automation (Parasuraman et al. 2000).
- Studies that examine the various aspects of human-AI interactions and the factors that could potentially increase or decrease human trust and needs satisfaction.

Upon completion of the play test and evaluation sessions, we will improve the prototype over one or two additional iterations, considering the lessons learned from play testing.

6 Conclusions

The *SatisfAI* game, described within this paper, aims to enable a greater understanding of the level of agency a human is willing to cede or share with AI-enabled systems in order to satisfy their human needs. Research based on the results of gameplay can use this understanding to help inform the design of AI applications and multi-agent decision support. The game also acts as an educational tool to help players understand how level-of-automation is a critical aspect of human-AI interaction, which could dictate their trust with the technology.

For future studies, we intend explore how the differences in personas of players affect decisions and outcomes of the game. As the prototype is designed and implemented, its value as a research tool will be a primary consideration. Follow-on research using the prototype game should investigate whether, given the same scenario and AI technology, people's preferred levels of automation differ based on needs satisfied. It should investigate factors contribute to these differences and how AI-based systems may be designed around these differences in order to maximize trust and minimize bias.

References

Alderfer, C.P.: An empirical test of a new theory of human needs. Organ. Behav. Hum. Perform. **4**(2), 142–175 (1969). https://doi.org/10.1016/0030-5073(69)90004-X

Arya, V., et al.: One explanation does not fit all: a toolkit and taxonomy of ai explainability techniques. https://arxiv.org/abs/1909.03012v2 (2019)

Boer, L., Donovan, J.: Provotypes for participatory innovation. In: Proceedings of the Designing Interactive Systems Conference, DIS 2012, pp. 388–397 (2012). https://doi.org/10.1145/231 7956.2318014

Desurvire, H., Caplan, M., Toth, J.A.: Using heuristics to evaluate the playability of games. In: Conference on Human Factors in Computing Systems - Proceedings, pp. 1509–1512 (2004). https://doi.org/10.1145/985921.986102

Koren, G., Soni, S.: Basic Human Needs. Mooze Design (2010). http://www.moozedesign.in/ima ges/portfolio/actual/bhn.jpg

Li, F.-F., Etchemendy, J.: Introducing Stanford's Human-Centered AI Initiative (2018). https:// hai.stanford.edu/news/introducing-stanfords-human-centered-ai-initiative

Markovic, I.: How to use the risk assessment matrix to organize your project better. TMS (2019). https://tms-outsource.com/blog/posts/risk-assessment-matrix/

Maslow, A.H.: A theory of human motivation. Psychol. Rev. **50**(4), 370–396 (1943). http://www. abika.com/

Max-Neef, M.A., Elizalde, A., Hopenhayn, M.: Human Scale Development: Conception, Application and Further Reflections. The Apex Press, Lexington (1991)

McLeod, S.: Maslow's Hierarchy of Needs. Simply Psychology (2020). https://www.simplypsy chology.org/maslow.html

Parasuraman, R., Sheridan, T.B., Wickens, C.D.: A model for types and levels of human interaction with automation. IEEE Trans. Syst. Man Cybern. Part A Syst. Hum. **30**(3), 286–297 (2000). https://doi.org/10.1109/3468.844354

Ribera, M., Lapedriza, A.: Can we do better explanations? A proposal of User-Centered Explainable AI (2019)

Ryan, R.M., Deci, E.L.: Self-determination theory and the facilitation of intrinsic motivation, social development, and well-being. Am. Psychol. **55**(1), 68–78 (2000). https://doi.org/10.1037/0003-066X.55.1.68

Siau, K., Wang, W.: Building trust in artificial intelligence, machine learning, and robotics. Cut. Bus. Technol. J. **31**, 47–53 (2018). https://www.researchgate.net/publication/324006061_Bui lding_Trust_in_Artificial_Intelligence_Machine_Learning_and_Robotics

Xu, W.: Toward human-centered AI: a perspective from human-computer interaction. ACM Interact. Mag. 42–46 (2019). https://interactions.acm.org/archive/view/july-august-2019/tow ard-human-centered-ai

Yang, Q., Steinfeld, A., Rosé, C., Zimmerman, J.: Re-examining whether, why, and how human-AI interaction is uniquely difficult to design. In: Conference on Human Factors in Computing Systems – Proceedings, 21 April 2020. https://doi.org/10.1145/3313831.3376301

Yu, O.: A new model of human needs as the foundation for innovation management. IEEE Eng. Manage. Rev. **46**(3), 40–45 (2018). https://doi.org/10.1109/EMR.2018.2870431

A Framework for the Design and Development of Adaptive Agent-Based Simulations to Explore Student Thinking and Performance in K-20 Science

Tyler Kinner[✉] and Elizabeth T. Whitaker

Georgia Tech Research Institute, Atlanta, GA, USA
tyler.kinner@gtri.gatech.edu

Abstract. In this paper, we describe the design and development of an agent-based modeling and simulation framework to support adaptive science learning based on modeling and simulation using NetLogo. This exploration is informed by the need for new curricular resources aligned with the modern vision of science teaching and learning, as put forth by the National Research Council's *Framework for K12 Science Education* [1]. This paper seeks to support the development of interactive, adaptive resources for science education that meet the 3-Dimensional expectations of science education described in the National Research Council's *Framework*. In addition to the primary focus on the student learning experience, the *Framework for Agent-Based Adaptive Learning in Science* (FABALS) considers both the developer and practitioner perspectives. As such, the modeling and simulation development framework and modular procedures for facilitating the adaptive training were developed with technological, pedagogical, and curricular perspectives in mind.

Keywords: Science learning · Modeling and simulation · Agent-based modeling and simulation · Curriculum · Intelligent tutoring · Representations · Science and engineering practices · NGSS · Develop and use models

1 Introduction and Motivation

This paper discusses the development and use of FABALS, an adaptive and interactive instructional simulation-based framework for developers and practitioners to use that meets the 3-Dimensional expectations of science education described in the National Research Council's Framework for Science Education [1]. The 3-dimensional nature of modern science learning focuses on the disciplinary core ideas of science, the broad cross-cutting concepts of the universe, and how scientists and engineers use these specific concepts and broad ideas through their everyday practices. FABALS has a strong focus on the student learning experience as well. The implementation of standards and curriculum built with this 3-dimensional view of science learning is an ongoing process [2]. In particular, there is a need for more instructional and assessment resources for practitioners to use that include all three dimensions of science learning, as well as resources that bridge the gap between research and practice in science education [2–9]. To support the development of adaptive, agent-based simulations for instruction and assessment, this paper reviews fundamental pedagogical and technical concepts, shares a development framework for those seeking to build useful tools for K-20 science education, and illustrates the use of the FABALS framework in an example centered on particulate-level understandings for students of physical science and chemistry.

2 Instruction and Assessment for 3D Science Learning

Since the publication of The Framework for K-12 Science Education, many states and educational authorities have adopted new expectations for the teaching and learning of science [2, 4, 6]. The broadest adoption comes in the form of the *Next Generation Science Standards* (NGSS) [4, 6]. These expectations continue a tradition of evolving thoughts in science education reform that students can, and should, engage in science authentically [10]. Whereas previous reforms, such as the AAAS benchmarks described scientific habits of mind that were informed and related to the practice of science, the Framework describes a vision of science learning where students are engaged in vertically aligned content area learning through the practices of scientists and engineers [2, 10] (Fig. 1).

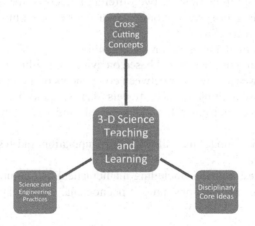

Fig. 1. The connected nature of 3-D science teaching and learning [1, 5]

These two concepts referred to the Disciplinary Core Ideas (DCIs) and the Science and Engineering Practices (SEPs) comprise two-thirds of the Framework's recommended expectations for science instruction. The third dimension, the Crosscutting Concepts (CCCs) are unifying ideas, or concepts, that occur throughout science in all its fields [2].

Adaptive instructional systems have a unique role to play in the curriculum landscape for this new vision of science education. Perhaps the most salient feature of 3D science teaching and learning put forth by the National Research Council's *Framework for K12 Science Education* is the shift to students actively practicing the skills of scientists and engineers as a component of the learning process.

Instruction and assessment tools for K-20 science education that meet the expectations of 3D science learning are still being developed [2–9]. There is a particular need for resources that can probe or assess the ability of students to incorporate the Science and Engineering Practices and the Cross-Cutting Concepts and offer teachers the opportunity to provide feedback and responsive learning experiences [7, 8]. This is perhaps expected, as these two dimensions are the most unlike previous science education standards and frameworks [10]. While some Science and Engineering Practices may be more easily practiced and assessed using strategies familiar to practitioners, others require the development of new resources and strategies [6, 8]. Given the nature of agent-based models and simulations, this works focuses on the practice of *Developing and Using Models*.

2.1 Developing and Using Models in K-20 Science Education

The practice of *Developing and Using Models* calls for students to engage in grade-level appropriate model making, analysis, evaluation, and revision. As students advance throughout their K-12 careers, their use of models is expected to advance in sophistication and purpose. Beyond 12th grade, similar use of the practice of *Developing and Using Models* has been incorporated into undergraduate science curriculum [12]. At the high school level, the following expectations are embedded into the NGSS [2]:

- Evaluate merits and limitations of two different models of the same proposed tool, process, mechanism, or system in order to select or revise a model that best fits the evidence or design criteria.
- Design a test of a model to ascertain its reliability.
- Develop, revise, and/or use a model based on evidence to illustrate and/or predict the relationships between systems or between components of a system.
- Develop and/or use multiple types of models to provide mechanistic accounts and/or predict phenomena, and move flexibly between model types based on merits and limitations.
- Develop a complex model that allows for manipulation and testing of a proposed process or system.
- Develop and/or use a model (including mathematical and computational) to generate data to support explanations, predict phenomena, analyze systems, and/or solve problems.

2.2 The Importance of Making Models for Science Learning

Within the NGSS, the practice of Developing and Using Models is often used to address Disciplinary Core Ideas that are microscopic or otherwise invisible. Topics such as the relationship between the needs of organisms and their environments, the changes in particle motion with changes in system temperature, lunar phases, and the force of gravity are addressed in the NGSS through students engaging in the practice of Developing and Using Models [2]. No matter the type of model or underlying subject matter for which it was developed, models are always representative; therefore, how students interpret and use representations are an important consideration for resources that seek to engage students in practicing their ability to develop and use models [11, 13–19].

Representations such as cell diagrams, models of atoms of a gas, and wave diagrams allow practitioners and learners to communicate ideas about concepts invisible to the human eye. Symbolic representations, such as vectors, Punnett Squares, and chemical equations create the ability to communicate information in a way that is succinct and useful [18]. The ability to translate between representations and other forms of information in science is critical to the learning process [13–19].

Johnstone [18] illustrates the relationship among different types of information and representations in science learning. These representational domains are interconnected in science and science learning; however, ubiquity does not mean universal understanding. Using a sorting task comprised of concept-alike videos, graphs, animations, and equations, Kozma and Russell [14] found that students sorted the representations based upon the visual features in the representations more than the experts, who sorted based upon underlying principles and concepts in chemistry. Similarly, the novices were much less adept at producing transformations of the representations. Even when representations are of the same medium, students have difficulty in relating representations and translating among them [19] (Fig. 2).

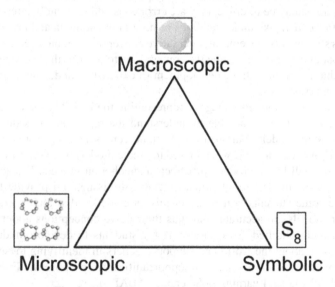

Fig. 2. Illustration of Johnstone's domains [18]

Student difficulties in understanding representations are particularly troubling when these representations are used to communicate information about topics for which students often have naïve conceptions. Take for example the phenomenon of a phase change, which is often represented in diagrams, such as that shown in Fig. 3.

When interviewing eighth-grade students who had previously undertaken a year of physical science instruction, Novick and Nussbaum [20] determined that they tended to use static or continuous visualizations of gases. When tasked with visualizing the arrangement of agas before and after the partial evacuation of a flask, students more

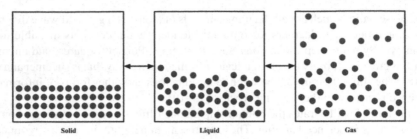

Fig. 3. Example phase change diagram

frequently selected representations depicting the gases as a continuous fluid or as non-continuous clusters of particles. When examining the Law of Conservation of Mass and gases, Stavy [21] found that younger students were the most likely to report that a closed tube containing a substance lost mass as the substance evaporated. As Novick and Nussbaum [20] note, "the aspects of the particle model least assimilated by pupils in this study are those most in dissonance with their sensory perception of matter". Thus, while these naïve conceptions may be intuitive, they are erroneous and ultimately interfere with the development of expert-like understandings of natural phenomena and processes. Often referred to as student misconceptions, alternative conceptions, or naïve conceptions [20–31], in this paper, we use the term naïve conceptions to refer to those conceptions held by students that interfere with the development of expert-like understandings of natural phenomena and processes.

As many scientific models engage in representing the invisible forces and elements of the world, students must be able to understand the representations they are incorporating into their models. Failure to understand representations commonly used in instruction means that students, when tasked to create models that will draw upon these representations, will likely produce incorrect or incoherent models. If teachers do not call these incorrect models to attention, diagnose the issue, and provide feedback, it is surmisable some students will interpret this as evidence that their conceptions are correct, their models are accurate, and thus their naïve conceptions about the natural world are further solidified. To create tools for students to develop models that use their conceptions about the world and support teachers in identifying these interpretations and providing responsive learning opportunities, we propose the Framework for Agent-Based Adaptative Learning in Science (FABALS).

3   Framework for Agent-Based Adaptive Learning in Science

3.1   Agent-Based Models and Simulations

Our exploration in design and development utilizes NetLogo. NetLogo is a multi-agent programmable modeling and simulation environment from Northwestern. NetLogo is based on the programming language Logo, developed in the 1960s by Wally Feurzeig, Seymour Papert, and Cynthia Solomon [33]. The Logo language was designed for teaching children; it is particularly useful to teach programming concepts related to agent-based modeling and simulation. Due to its ease of use, it provides access to agent-based model and simulation (ABMS) in a variety of domains. NetLogo is opensource and easily downloadable for teachers, students, and researchers.

ABMS is an approach that is used to explore the behaviors and interactions of individual agents in particular situations or environments. This could include individual people, individual animals, or even smaller agents, such as particles, molecules, or cells. The use of agents in modeling and simulation provides the unique opportunity for students to create and use models that simulate many concepts within science, and beyond, that are impossible to witness within any classroom. It also creates a window into real-life applications of computer programming, as ABMS is used to simulate a wide variety of scenarios [33–46].

Each agent in an agent-based model represents a discrete individual or organization with a set of characteristics and a reasoning system that governs its behaviors, decision-making, and interactions with other agents in the model. The agents exist in an environment defined by the model. The agents interact according to a set of protocols defined in the simulation to represent how the individuals might interact in the real world. Each agent can recognize and reason about the traits of other agents that it comes into contact within the simulation. The agents are autonomous, that is they have their own reasoning and decision-making approaches. Each agent has a memory or state and learns and adapts from experience. It changes its behaviors in response to new knowledge, experiences, and communications.

3.2 Intelligent Tutoring Systems and Simulation

An intelligent tutoring system (ITS) is a computerized system for instruction that is based on artificial intelligence.

- To provide tailored instruction and immediate feedback to learners
- Is guided by theories of teaching and learning
- Contains expert domain knowledge
- Can include the opportunity to explore domains through simulation
- Can include a game-based approach
- Can include simulations and models for student experimentation

Theory and applications of intelligent tutoring systems began to be developed in the 1960s and have advanced a great deal since then, based on new technologies in both hardware and software [33, 44]. Classical intelligent tutoring systems are built around four modules containing the information and reasoning modeled after human teaching: a domain expert which contains knowledge of the subject being taught, a student model which tracks the student history through the activities the student has experienced in the tutoring sessions along with any inferences about what concepts the student understands and weakness in those understandings, a dialogue manager that manages the communications among the modules and the communication with the student, and a tutor module which has knowledge about how to teach. The simulation framework that we are developing does not claim to be a full-fledged intelligent tutoring system with all these components, but rather focuses on the simulation aspects with human aspects which can fill in for those components. One significant advantage of a simulation-based learning system is the ability for the student to experiment with different situations in a safe and inexpensive environment.

The primary goal of ITS or simulations developed using the FABALS framework is to facilitate student development of expert-like model creation and interpretation through experimentation, visualization, and system and/or teacher delivered feedback. This supports the development of resources that meet the new science learning expectations for students to develop and use models capable of visualizing behavior, predicting changes to a system, or generating data [1, 5]. In addition, the lightweight nature of this framework and accessibility of NetLogo empowers practitioners to create tools aligned to the curriculum and needs of their own classrooms. Lastly, the need for the teacher to integrate feedback and instruction into any ITS or simulations designed with FABALS gives the opportunity for teachers to recognize their own naïve conceptions and develop more expert-like thinking themselves.

3.3 FABALS Architecture

FABALS applies intelligent tutoring features [45] to support new science learning curriculum objectives. The term "intelligent tutoring features" is used to convey the concept that FABALS is not a full-fledged intelligent tutoring system, but it includes light student modeling or s5tudent tracking, the ability to use knowledge about the student activities to select and adapt teaching content, an interface to communicate with the student, and domain knowledge encoded into the scenarios. The approach integrates the teaching of modeling and simulation with related programming concurrently with providing scenarios that enable students to explore a given set of science concepts. The framework is also being designed in a way that will enables instructors to set up sequences of scenarios that allow students to explore complex learning examples.

Although we envision that future version might include the use of further extensions to NetLogo and external module development to build a stronger intelligent tutoring system, the FABALS Architecture is completely hosted in NetLogo. The Architecture consists of the following modules encoded as procedures or list representations in the NetLogo Code (Fig. 4):

o   A Scenario Set
o   A Content Selector
o   A Student Model
o   An Interface (built in the Interface Tab of NetLogo)

Central to the simulations learning sessions enabled by FABALS is the concept of scenarios. We are describing the content driving a learning simulation as a scenario. Each learning session is intended to consist of a set of scenarios that the student experiences to learn a set of concepts contained in the science lesson being taught. Each of these scenarios includes a set of parameters that can be varied to give the student multiple views and examples of each concept. It is intended that a learning session will include a set of scenarios that get progressively more complex in the visualization of the concepts being taught and that the student will be directed to new scenarios will include new concepts (Table 1).

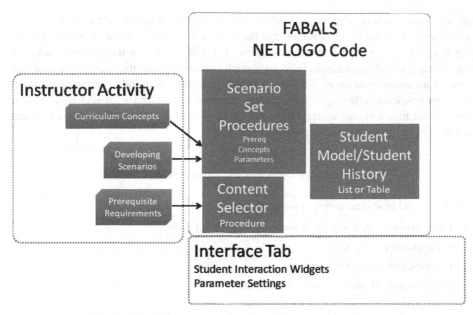

Fig. 4. FABALS architecture with intelligent tutoring features

Table 1. Scenario features

Scenario features
Scenario ID
Name
Primary concept
Secondary concept
Parameter list
Prerequisite scenario list
Prerequisite concept list

The Content Selector organizes the student's simulation lessons. It reasons about and selects the next scenario to be experienced by the student, guided by the instructor's reasoning and prerequisite setup and the current state of the student's experience.

The purpose of a student model is to provide a snapshot of the student's experience history and mastery levels at any time during or after the student's learning session. It is used to tailor the student's learning session, and it is used by the Content Selector to select and order content/scenarios to drive the simulation learning session. The Student Model will be built using lists or tables in NetLogo. Currently, this is more representative of a Student History; however, future iterations may grow in sophistication. It will contain the list of scenarios experienced by the student, the parameters associated with each scenario and the sets of parameter values that the student has run with that scenario. It will also include the set of concepts that the student has learned about. NetLogo has the capability to receive input from the student. We will include in the implementation a small sequence of inputs associated with each scenario where the scenario design can include an assessment to provide information to the instructor and to the student model for content selection decisions. At this stage the FABALS design is for very light student assessment mechanisms, with their primary goal being to provide information to select the next scenario in the learning session (Table 2).

Table 2. Student model/student history configuration

Student model or student history configuration
List of scenarios played with list of parameter sets for scenario
List of concepts experienced
List of questions asked of the student
Triples (question ID, answer, answer evaluation)

The NetLogo Interface Tab provides the capability to design interactions with the student through buttons that will set up and execute scenarios and the ability to input question answers related to a scenario as well as parameter settings. Some of this will also be enabled by the user-input command in the code. The ease of setting up multiple plots in NetLogo will enable the instructor to associate multiple data plots with each scenario that the student is to experience.

Adaptations in the simulation consisting of moving to a new scenario appropriate to the student's stage of learning will mean that the instructional designer can make a current simulation more complex, such as through adding more turtles, more parameters, more obstacles, more variables associated with the patches that turtles are navigating through) or resetting the environment entirely and providing a simulation focused on a new set of concepts.

Some parts of this will be automated, as one could do via branching chains in NetLogo, to move the student through different scenarios of advancing difficulty and more expert-like conceptual understandings. One approach to moving through multiple scenarios can also be done with multiple scenario buttons that the student can choose and move through, with perhaps some prerequisite scenarios. The instructor may wish to define a set of parameters that the student can use to rerun the scenarios multiple times to answer a set of what-if questions about the phenomenon or scenario. The what-if questions will consist of a set of parameter settings that the student can use to answer questions about the behaviors of the system; the design of these what-if scenarios is perhaps best informed by existing inquiry into student thinking in science [20–32]. Instructional designers and developers will need to define a set of triggers to be used by the Content Selector to determine when the student moves from one scenario to another.

3.4 ABMS Design and Development with FABALS

This project and associated framework operates on the foundation that experimentation, visualization, and feedback are key components in building expert-like student scientific knowledge and skills [12–22]. To shrink the gap between development and practice, FABALS is an intentionally light-weight, teacher-involved framework informed by educational standards and research into student thinking in science.

The agent-based model will select simulation scenarios based on a state variable in the learning system which will be used to reset the simulation and provide a next step in learning for the student based on the student's responses or activities in interacting with the system.

The framework is being designed such that scenarios can be coded into NetLogo procedures or built from parameters or read from an input file. The scenarios will be designed to take the student in gradually increasing complexity in learning and understanding the disciplinary core ideas associated with the modelled phenomena, as well as the science and engineering practices used in the student's engagement with the simulation. This includes algorithmic design and agent-based modeling and simulation as they relate to the practice of Developing and Using Models as described in the National Research Council's Framework for K12 Science Education.

3.5 Framework Application

The application of the FABALS framework into teaching and learning is intended to:

- Create tools for students to develop models using their conceptions about the natural world, even naïve conceptions
- Support teachers in identifying student naïve conceptions and provide feedback through the practice of developing and using models (Fig. 5)

This lightweight, conception-centered framework is also intended to minimize the need of the instructor to teach additional skills beyond those embedded within the NGSS, such as specific programming languages [46], animation tools [47], or simulation software [11]. This is intentional, as the amount of available time for instruction is a mediating factor in teacher use of NGSS-aligned activities, such as those that engage students in inquiry via model-making [48]. Therefore, FABALS is designed to facilitate the development of time-effective activities that support students in model creation.

To achieve this, the development of learning stages is informed by previous research into students' potential naïve conceptions. Relevant visualizations are pre-programmed and provided to students as various selections, with the general directions to "*Configure the model to best represent the phenomena*".

Example Application. To illustrate the utilization of the framework, we present a discussion of the design of an agent-based adaptative simulation based upon particulate-level misconceptions in chemistry is discussed below. The following NGSS was identified as the target standards for which this adaptive ABMS is designed: *MS-PS1-4 Develop a model that predicts and describes changes in particle motion, temperature, and state of a pure substance when thermal energy is added or removed* [5].

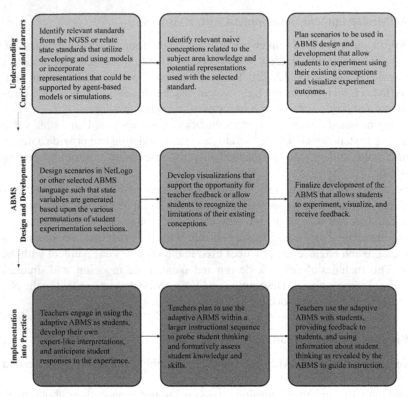

Fig. 5. Concept of operations for FABALS -informed ABMS development and implementation

Several naïve conceptions students may have that interfere with the development of expert-like understandings and models have been uncovered by researchers:

- Gases have less mass than liquid or solid counterparts [23]
- Gases have no mass [24]
- The liquid and/or gas phases are continuous, rather than particulate [25]
- Particles expand when heated [23]
- Phase changes involve a chemical change in the substance [21]
- The weight of a gas depends on the volume of the container [27]

Guided by the described Framework, an adaptive ABMS to support the development of expert-like conceptions was designed using a three-stage process informed by students' potential naïve conceptions. Students' progress through Stages A through C by configuring the ABMS as they believe best represents the phenomenon. Configurations built using naïve conceptions are then addressed by various interventions that visualize a more expert-like model. The base model shows ideal particles in a fixed volume container. Students first select whether they want to heat or cool the system, as well as the system's initial temperature in Kelvin. Students then run the ABMS to observe what happens within the base simulation. Afterward, students are instructed to move through adaptive Stages A through C by configuring the ABMS to *best* represent the phenomenon according to their conceptions.

In Stage A, the adaptive ABMS probes students into their thinking about the size of particles and if the size of a particle is impacted by changes to their speed. As this naïve conception may be reinforced by the phenomenon of gases expanding when increasing temperature due to changes in the volume of the system, the fixed-volume container is replaced by a flexible container that expands with increasing particle speed. Here, the goal is to provide students with an opportunity to view an expert-like model; in the classroom, the visual cue provided by the flexible volume container allows the teacher to observe students who have reported a naïve conception in Stage A and provide feedback and additional guidance to the student. This immediate opportunity to provide feedback is valuable in facilitating the student development of expert-like interpretations and the ability to develop and use models. Stages B and C are similarly configured, with students first reporting their understandings about particle mass in Stage B and then system mass in Stage C. Naïve conceptions reported in these stages result in the display of two plots: one plot represents the student's reported conception, and the other represents the expert-like conception (Fig. 6).

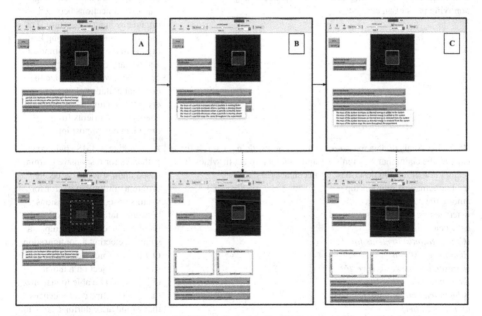

Fig. 6. Snapshots of an adaptive ABMS designed to facilitate expert-like model development for particle behavior; built in NetLogo [34].

3.6 Areas for Future Development

To support the development of adaptative agent-based models and simulations for science teaching and learning, we present a selection of NGSS-aligned student naïve conceptions and potential ways to design adaptative ABMS (Table 3).

Table 3. Potential adaptive ABMSs for science teaching and learning.

NextGen standard [5]	Student naïve conception	Opportunities for adaptive ABMS
4-PS4-1 Develop a model of waves to describe patterns in terms of amplitude and wavelength and that waves can cause objects to move MS-PS4-2 Develop and use a model to describe that wave are reflected, absorb, or transmitted through various materials	Sound is an object or substance that requires a clear pathway from transmitter to receiver [28] Air is not required for sound to travel [29]	An adaptive ABMS could create pathways for responsive learning based upon students selecting different configurations that would allow a sound wave to travel from transmitter to receiver. For example, a student selecting an empty space as the required medium for sound wave travel would be able to visualize that the lack of particles within the medium results in the end of the wave's propagation
MS-LS2-3 Develop a model to describe the cycling of matter and flow of energy among living and nonliving parts of an ecosystem	Relationships among organisms in an ecosystem are linear and not interdependent [27]	An adaptive ABMS could create pathways for responsive learning based upon students selecting various connections between organisms in an ecosystem. The lack of connections among key organisms, such as decomposers and tertiary or quaternary consumers would visualize the eventual collapse of an ecosystem without the return of energy and nutrients from organism decomposition
HS-PS2-1 Analyze data to support the claim that Newton's second law of motion describes the mathematical relationship among the net force on a macroscopic object, its mass, and its acceleration (*Clarification statements for this NGSS suggest that representations, such as vector diagrams, fall within its scope. It is included here as students need to use and interpret representations, despite it not being the specific practice of Develop and Use Models.*)	Objects at rest do not exert a force upon another object in which it is in contact [27]	An adaptive ABMS could create pathways for responsive learning based upon students selecting different configurations for force vectors based upon collisions between stationary objects and moving objects. For example, a student selecting a configuration that has no force exerted by a stationary object on a moving object would be able to visualize the non-intuitive consequences that would arise during a collision

(*continued*)

Table 3. (*continued*)

NextGen standard [5]	Student naïve conception	Opportunities for adaptive ABMS
HS-LS2-5 Develop a model to illustrate the role of photosynthesis and cellular respiration in the cycling of carbon among the biosphere, atmosphere, hydrosphere, and geosphere	The Sun is one of multiple sources of energy for plants to carry out photosynthesis [30] Plants get energy directly from photosynthesis and do not carry out cellular respiration [27]	An adaptive ABMS could create pathways for responsive learning based upon students selecting different experimental setups for plants during their growth. This would allow students to visualize that in the absence of light, a plant may grow for a short period of time based upon their stored energy, but eventually will die without a light source for photosynthesis
HS-PS3-5 Develop and use a model of two objects interacting through electric or magnetic fields to illustrate the forces between objects and the changes in energy of the objects due to the interaction	Magnets require a medium, such as air, to act upon another object [31] A magnet only acts upon magnetic objects in direct contact with it [32]	An adaptive ABMS could create pathways for responsive learning based upon students selecting various configurations for a two-magnet system, or a system with one magnet and one magnetized material. This would allow students to visualize the ability for magnets to act at a distance without the need for a medium, such as air

4 Conclusions

The ultimate goal of the Framework for Agent-Based Adaptive Learning in Science is to increase access to resources for students and teachers that allow for visualization, experimentation, and feedback for model development and usage in a way that meets curricular expectations and supports the development of expert-like conceptions in students. To meet this goal, FABALS is being developed using both pedagogical and technical considerations; this is intended to increase access to new information and ways of thinking for both developers and practitioners. We believe that for practitioners, the ability to develop curriculum-aligned adaptive learning tools will improve teaching and learning in classrooms. For developers, we believe a framework beginning with understanding curriculum and learners will improve the reception of educational interventions developed outside the classroom. We believe that this will result in effective learning resources to introduce students to science concepts as well as modeling and simulation and computer programming to help explore those concepts.

Although NetLogo does not have the flexibility and richer representations of some general-purpose programming languages, it was chosen because it allows an easy implementation of agent-based simulations, the programming language is easily available to instructors and students, and the system has a short learning curve. This is particularly useful for resources related to the practice of Developing and Using Models in K12

Science, and specifically for those that are of systems or many-agent phenomena [5]. In, later versions we use NetLogo extensions and develop external modules to extend the intelligent tutoring features.

Next steps in the FABALS development include the implementation of a detailed demo that explores and tests the intelligent tutoring features in the FABALS architecture. We also anticipate leveraging NetLogo extensions and the development of external modules to extend the intelligent tutoring features. This will include a more detailed evaluation of the NetLogo framework features that enable adaptivity in FABALS. Future versions will be evaluated for instructor scenario development capabilities and student learning results.

References

1. Council, N.R.: A Framework for K-12 Science Education: Practices, Crosscutting Concepts, and Core Ideas. The National Academies Press, Washington, DC (2012)
2. NGSS Lead States. Next Generation Science Standards: For States, By States. The National Academies Press, Washington, DC (2013)
3. Underwood, S.M., Posey, L.A., Herrington, D.G., Carmel, J.H., Cooper, M.M.: Adapting assessment tasks to support three-dimensional learning. J. Chem. Educ. **95**, 207–217 (2017). https://doi.org/10.1021/acs.jchemed.7b00645
4. Penuel, W.R., Reiser, B.J.: Designing NGSS-aligned curriculum materials. Committee to Revise America's Lab Report, pp. 1–51 (2018)
5. Lowell, B.R., Cherbow, K., McNeill, K.L.: Redesign or relabel? How a commercial curriculum and its implementation oversimplify key features of the NGSS. Sci. Educ. **105**, 5–32 (2021). https://doi.org/10.1002/sce.21604
6. Harris, K., Sithole, A., Kibirige, J.: A needs assessment for the adoption of next generation science standards (NGSS) in K-12 education in the United States. J. Educ. Train. Stud. **5**, 54–62 (2017). https://doi.org/10.11114/jets.v5i9.2576
7. Yoon, S.A., Goh, S.E., Park, M.: Teaching and learning about complex systems in K–12 science education: a review of empirical studies 1995–2015. Rev. Educ. Res. **88**(2), 285–325 (2018)
8. Underwood, S.M., Posey, L.A., Herrington, D.G., Carmel, J.H., Cooper, M.M.: Adapting assessment tasks to support three-dimensional learning. J. Chem. Educ. **95**(2), 207–217 (2018). https://doi.org/10.1021/acs.jchemed.7b00645
9. Gale, J., Koval, J., Ryan, M., Usselman, M., Wind, S.: Implementing NGSS engineering disciplinary core ideas in middle school science classrooms: results from the field. J. Pre-Coll. Eng. Educ. Res. (J-PEER) **9** (2018). https://doi.org/10.7771/2157-9288.1185
10. Barrow, L.H.: A brief history of inquiry: from dewey to standards. J. Sci. Teach. Educ. **17**, 265–278 (2017). https://doi.org/10.1007/s10972-006-9008-5
11. Krajcik, J., Merritt, J.: Engaging students in scientific practices: what does constructing and revising models look like in the science classroom? Sci. Teach. **79**, 38 (2012)
12. Cooper, M., Klymkowsky, M.: Chemistry, life, the universe, and everything: a new approach to general chemistry, and a model for curriculum reform. J. Chem. Educ. **90**, 1116–1122 (2013). https://doi.org/10.1021/ed300456y
13. Rau, M.A.: Making connections among multiple visual representations: how do sense-making skills and perceptual fluency relate to learning of chemistry knowledge? Instr. Sci. **46**, 209–243 (2017). https://doi.org/10.1007/s11251-017-9431-3

14. Kozma, R.B., Russell, J.: Multimedia and understanding: expert and novice responses to different representations of chemical phenomena. J. Res. Sci. Teach. **34**, 949–968 (1997). https://doi.org/10.1002/(SICI)1098-2736(199711)34:9%3C949::AID-TEA7%3E3.0.CO;2-U
15. McClean, P., et al.: Molecular and cellular biology animations: development and impact on student learning. Cell Biol. Educ. **4**, 169–179 (2005). https://doi.org/10.1187/cbe.04-07-0047
16. Jones, L.L., Kelly, R.M.: Visualization: the key to understanding chemistry concepts. In: Sputkin to Smartphones a Half-Century of Chemistry Education, pp. 121–140. American Chemical Society, Washington, DC (2015). https://doi.org/10.1021/bk-2015-1208.ch008
17. Tasker, R., Dalton, R.: Research into practice: visualisation of the molecular world using animations. Chem. Educ. Res. Pract. **7**, 141–159 (2006). https://doi.org/10.1039/b5rp90020d
18. Johnstone, A.H.: Why is science difficult to learn? Things are seldom what they seem. J. Comput. Assist. Learn. **7**, 75–83 (1991). https://doi.org/10.1111/j.1365-2729.1991.tb00230.x
19. Dori, Y.J., Kaberman, Z.: Assessing high school chemistry students' modeling sub-skills in a computerized molecular modeling learning environment. Instr. Sci. **40**, 69–91 (2011). https://doi.org/10.1007/s11251-011-9172-7
20. Novick, S., Nussbaum, J.: Junior high school pupils' understanding of the particulate nature of matter: an interview study. Sci. Ed. **62**, 273–281 (1978). https://doi.org/10.1002/sce.3730620303
21. Stavy, R.: Children's conception of changes in the state of matter: from liquid (or solid) to gas. J. Res. Sci. Teach. **27**, 247–266 (1990). https://doi.org/10.1002/tea.3660270308
22. Bielik, T., Stephens, L., McIntyre, C., Damelin, D., Krajcik, J.S.: Supporting student system modelling practice through curriculum and technology design. J. Sci. Educ. Technol. 1–15 (2021). https://doi.org/10.1007/s10956-021-09943-y
23. Griffiths, A.K., Preston, K.R.: Grade-12 students' misconceptions relating to fundamental characteristics of atoms and molecules. J. Res. Sci. Teach. **29**, 611–628 (1992). https://doi.org/10.1002/tea.3660290609
24. Mas, C.J.F., Perez, J.H., Harris, H.H.: Parallels between adolescents' conception of gases and the history of chemistry. J. Chem. Educ. **64**, 616 (1987). https://doi.org/10.1021/ed064p616
25. Nakhleh, M.B.: Why some students don't learn chemistry: chemical misconceptions. J. Chem. Educ. **69**, 191 (1992). https://doi.org/10.1021/ed069p191
26. Novick, S., Nussbaum, J.: Pupils' understanding of the particulate nature of matter: a cross-age study. Sci. Ed. **65**, 187–196 (1981). https://doi.org/10.1002/sce.3730650209
27. Driver, R., Squires, A., Rushworth, P., Wood-Robinson, V.: Making Sense of Secondary Science: Research into Children's Ideas. Routledge, London (1994)
28. Asoko, H., Leach, J., Scott, P.: Classroom research as a basis for professional development of teachers: a study of students' understanding of sound. New prospects for teacher education in Europe II. In: Proceedings of the 16th Annual Conference of ATEE (1992)
29. Watt, D., Russel, T.: Sound. Primary SPACE Project Research Report. University Press, Liverpool, UK (1990)
30. Barker, M.A., Carr, M.D.: Photosynthesis—can our pupils see the wood for the trees? J. Biol. Educ. **23**, 41–44 (1989). https://doi.org/10.1080/00219266.1989.9655022
31. Bar, V., Zinn, B., Rubin, E.: Children's ideas about action at a distance. Int. J. Sci. Educ. **19**, 1137–1157 (1997). https://doi.org/10.1080/0950069970191003
32. Barrow, L.H.: Do elementary science methods textbooks facilitate the understanding of magnet concepts? J. Sci. Educ. Technol. **9**, 199–205 (2000). https://doi.org/10.1023/a:1009487432316
33. Solomon, C.J.: Computer environments for children as reflections of theories of learning and education (logo, seymour papert, patrick supps, plato system, basic). Harvard University (1985)
34. Tisue, S., Wilensky, U.: NetLogo: a simple environment for modeling complexity. In: International Conference on Complex Systems, vol. 21, pp. 16–21, May 2004

35. Zhang, W., et al.: Agent-based modeling of a stadium evacuation in a smart city. In: 2018 Winter Simulation Conference (WSC), 9 December 2018, pp. 2803–2814. IEEE (2018)
36. Loper, M., Whitaker, E.T., Register, A.: Introduction to modeling and simulation. In: Raiszadeh, B., Batterson, J. (eds.) CK-12 Modeling and Simulation for High School Teachers: Principles, Problems, and Lesson Plans. NASA CK-12 Flexbook (2012)
37. Whitaker, E., Gonzalez, M., Trewhitt, E., MacTavish, R., Loper, M.: Cultural considerations in a simulated emergency. In: Proceedings of DHSS 2011, Rome, Italy (2011)
38. Intelligent agents in electronic commerce NCR Technol. J. (1997)
39. Henninger, A.E., Whitaker, E.T.: Modeling behavior. In: Loper, M. (ed.) Modeling and Simulation in the Systems Engineering Life Cycle, pp. 75–87. Springer, London (2015). https://doi.org/10.1007/978-1-4471-5634-5_8
40. Whitaker, E.T., Simpson, R.: Active Learning Framework, US Patent US6905341. http://www.google.co.uk/patents/US6905341
41. Whitaker, E.T., Lee-Urban, S.: Intelligent agent representations of malware: analysis to prepare for future cyber threats. In: Nicholson, D. (ed.) Advances in Human Factors in Cybersecurity, vol. 501, pp. 391–400. Springer, Cham (2016). https://doi.org/10.1007/978-3-319-41932-9_32
42. Whitaker, E.T.: System dynamics. In: Loper, M. (ed.) Modeling and Simulation in the Systems Engineering Life Cycle, pp. 157–165. Springer, London (2015). https://doi.org/10.1007/978-1-4471-5634-5_13
43. Whitaker, E.T., Trewhitt, E.B., Hale, C.R., Veinott, E.S., Argenta, C., Catrambone, R.: The effectiveness of intelligent tutoring on training in a video game. In: Proceedings of IGIC 2013, The IEEE International Games Innovation Conference, pp. 267–274 (2013)
44. Vie, J.J., Popineau, F., Bruillard, E., Bourda, Y.: A review of recent advances in adaptive assessment. In: Peña-Ayala, A. (ed.) Learning Analytics: Fundaments, Applications, and Trends, vol. 94, pp. 113–142. Springer, Cham (2017). https://doi.org/10.1007/978-3-319-52977-6_4
45. Whitaker, E., Trewhitt, E., Veinott, E.S.: Intelligent tutoring design alternatives in a serious game. In: Sottilare, R., Schwarz, J. (eds.) Adaptive Instructional Systems, vol. 11597, pp. 151–165. Springer, Cham (2019). https://doi.org/10.1007/978-3-030-22341-0_13, ISBN: 978-3-030-22341-0
46. Xiang, L., Passmore, C.: A framework for model-based inquiry through agent-based programming. J. Sci. Educ. Technol. **24**, 311–329 (2015). https://doi.org/10.1007/s10956-014-9534-4
47. Wu, H.K., Krajcik, J.S., Soloway, E.: Promoting understanding of chemical representations: students' use of a visualization tool in the classroom. J. Res. Sci. Teach. **38**, 821–842 (2001). https://doi.org/10.1002/tea.1033
48. Kolbe, T., Steele, C., White, B.: Time to teach: instructional time and science teachers' use of inquiry-oriented instructional practices. Teach. Coll. Rec. **122**, 1–54 (2020). https://doi.org/10.1177/016146812012201211

Adaptive Instructional System for Complex Equipment Trainings in the Post-covid Era: Breaking the Ice of Time-Consuming Tasks

Elena Nazarova[1]($\boxtimes$), Alexander Butyaev[1], Mohamed Youssef Bouaouina[1], Dominic Filion[2], and Jerome Waldispuhl[1] (ID)

[1] School of Computer Science, McGill University, Montreal, Canada
`elena.nazarova@mcgill.ca`
[2] Montreal Clinical Research Institute (IRCM), Montreal, Canada

Abstract. Complex equipment trainings frequently rely on in-person trainings to describe instrument parts and personalize explanations based on training objectives, prior knowledge, and cognitive abilities of trainees. In this study, we assess the main challenges of adapting intelligent instructional system by training centers with limited human and technological resources. We found that the preparation of the training material is the most time-consuming task. During the COVID-19 pandemic, many training centers were forced to remotely conduct their trainings, thus generating a massive amount of digital training content. Here, we explore this unique opportunity to recycle this training material to design an adaptive instructional system (AIS) for bioimaging training. In this paper, we discuss the functional features of AIS that facilitate autonomous training for trainees and instructors: progress bar, notification system, built-in teleconferencing, and chatting tools. To address a high level of customization of in-person trainings, we designed AIS trainings as module-based instructions that can be easily tailored to accommodate the objectives and needs of the trainees. We also demonstrate that modular design of the training material database accelerates allocation and preparation of training content for similar types of equipment. We set up a framework for implementing a recommendation system that would accommodate the training material to the trainee's experience. Our study shows that over the short or medium term, the potential of AIS solution for equipment trainings significantly outweighs the most time-consuming tasks like preparation of the training material.

Keywords: Adaptive instructional system · Equipment training · Personalized learning · Intelligent training

1 Introduction

Adaptive instructional systems (AIS) provide a great potential for increasing effectiveness and efficiency of trainings by better accommodating the needs and abilities of different learners (1). AIS development for equipment trainings is a separate area among intelligent training systems and is extensively studied as part of flight, driving simulators and trainings of power plant operators [2–7]. The initiative to adapt classical verbal

© The Author(s), under exclusive license to Springer Nature Switzerland AG 2022
R. A. Sottilare and J. Schwarz (Eds.): HCII 2022, LNCS 13332, pp. 207–225, 2022.
https://doi.org/10.1007/978-3-031-05887-5_15

in-person trainings towards more automated regimes in these areas comes from the strict requirements to high level of standardization of obtained knowledge and skills.

A very different dynamics of adaptation AIS can be observed in training environment driven by self-motivation of the trainees such as academic core facilities. Core facilities (CFs) represent a strategic part of the research infrastructure in universities, institutes and health centers that concentrate different types of scientific equipment (e.g., microscopes) shared between different research teams [8, 9]. Due to the high cost and complexity, all the equipment of the same type is centralized in dedicated entities like CFs, which provide routine maintenance and on-demand training services for these different equipment units. The limited labor force in CFs and high self-motivation of the trainees in on-demand trainings resulted in the lack of prerequisites for transitioning towards AIS-based trainings in CFs.

However, the COVID-19 pandemic forced training centers like institute/university CFs to conduct trainings remotely. While previous studies showed the great potential of introducing VR-based equipment trainings as a remote visual alternative [10, 11], it is not always possible to quickly adapt a VR solution for equipment trainings with limited human and technological resources. Due to the complexity and diversity of the equipment, CF provide 1-on-1 training sessions that are highly customized to the objectives, experience, and engagement of the trainees. Therefore, creating digital material for such sessions is a time-consuming task that is largely restrained by limited labor-force. During the pandemic, training centers like CFs completely stopped training new users during lockdowns and switched to teleconferencing type of trainings for existing users (a.k.a. verbal training) [12, 13]. This provided a unique opportunity to re-purpose the digital material created for visual support for remote trainings (photos of equipment parts) and videos of the controlling software recorded during the training sessions. The goal of this study is to design an AIS solution for 1-on-1 equipment trainings and to demonstrate its potential as a cost-efficient solution for next generation CF trainings.

To address this goal, we examine two research questions: (RQ1) What are the factors that can increase the cost-efficiency of adapting an AIS training solution (MicroAssistant) for equipment trainings without compromising its quality? (RQ2) How to personalize MicroAssistant AIS for different training objectives and levels of explanation to achieve better assimilation of training material (to prevent trainee from being overwhelmed and keep the trainee's attention)?

In this study, we explore the difficulties of adapting an autonomous training solution for equipment trainings in CFs. The first research question examines several features of the instruction system and training material designed to increase the reliability and performance of AIS equipment training solution. The second research question discusses an application of modular design to customize trainings based on the trainee's objectives and without exposing the trainee to unnecessary parts of the training. As a part of the training customization, we also implement a recommendation system that tracks the trainee's activity and suggests adjustments to the training material according to the experience level.

2   Material and Methods

2.1   Participants and Equipment

This study was conducted with students as trainees who intended to use CF equipment for their research projects. The instructor who participated in the trainings can be considered as an expert instructor. All participants signed an informed consent form describing the purpose and benefits of the study, stating that participation was voluntary. Their input was confidential, and they could withdraw from the study at any time. As an example of equipment units, we used Leica confocal microscope TCS SP8 equipped with 3 PMTs (Photo-Multiplier Tubes) and one hybrid detector (HyD) and Leica widefield microscope DM6 equipped with a motorized XYZ stage and two cameras for different acquisition regimes.

2.2   Training Quality Evaluation

The training quality was evaluated by the instructor using mixed methods after each training session, by assessing the quality of the microscopy images of the samples taken during the microscopy trainings and as a part of a post-training interview with a trainee.

2.3   Pre-training Form

A pre-training form asked the trainees to (i) explain planned imaging experiments, (ii) respond to questions assessing previous experience with other microscopes, and (iii) point to equipment functionalities they are interested in.

2.4   Instructor and Trainee Feedback Forms

We use the trainee feedback form to obtain Kirkpatrick Level 1 feedback on the different parts of training instructions, different functionalities of MicroAssistant, overall experience with MicroAssistant, and the will to continue using MicroAssistant for trainings on other imaging instruments (14). In addition, we use the instructor feedback form to obtain Kirkpatrick Level 1 feedback on different functionalities of MicroAssistant as well as the satisfaction with the single and double-training experience with MicroAssistant solution.

2.5   Software Architecture of MicroAssistant Platform

MicroAssistant is designed as a web application that can be accessed remotely through a browser by any authenticated account holder. The frontend client consists of a React application that uses Bootstrap for component design and styling, along with Redux and React Hooks for state management. It is serviced by an HTTP API created using Flask web-development framework. It is responsible for analysing and fetching all application data from the PostgreSQL database deployed on the server. Efficient data retrieval is implemented using standard Postgres indexing techniques (B-Tree indexes). The PostgreSQL database is used to store all metadata for a particular block/module/instruction. The storage and distribution of all visual resources is outsourced to a cloud service (Amazon Web Services).

2.6 MicroAssistant UI Design

MicroAssistant has 3 main components. A training interface also referred as the User Cabinet (Fig. 1), an instructor interface allowing to monitor the progresses of the trainees, and a training module manager allowing an administrator to edit and organize the training material.

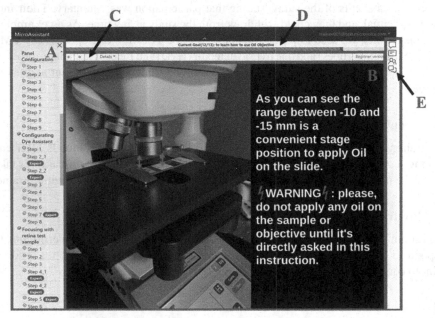

Fig. 1. Training interface of user cabinet. (A) The retractable sidebar with step-by-step training instruction. (B) The content placeholder displaying current training step. (C) The navigation bar with back/forward buttons and details section with supplemental training modules. (D) Progress bar with the current training goal and an overall training progress. (E) The tollbar with additional functionalities: notes, comments, external links, MicroTalk and MicroChat.

User Cabinet UI. The User Cabinet represents our training interface (Fig. 1). The retractable left sidebar (Fig. 1A) displays a tree-like view on the instruction in progress, where each step consists of a resource block with training content. The Content placeholder (Fig. 1B) is designed to occupy the central part of the screen and displays an instruction training step represented by an annotated image with (optional) audio support. The Navigation bar (Fig. 1C) includes Back and Forward buttons and a Details dropdown button including all the supplemental training modules that are referred to in the main instruction. The trainee can access information about the current goal and progression through the training material as a progress bar on the top panel (Fig. 1D). The toolbar on the right (Fig. 1E) provides trainee with additional functionalities of MicroAssistant: leaving notes (can be exported by the trainee at the end of the training), comments, external links, MicroTalk (a built-in videoconferencing tool to contact instructor directly) and MicroChat (a built-in chat with an instructor).

Progress (Instructor Monitoring) UI. A progress interface enables instructors to monitor in a real-time the progress of trainees. It highlights when a trainee approaches a training milestone, which requires additional instructor attention. It also displays comments left by the trainee and provides access to a built-in videoconferencing tool (MicroTalk) and chatting feature (MicroChat) (Fig. 2). The real-time functionalities mentioned above are enabled through the Socket.IO library which allows low latency, bi-directional communications between the multiple React frontend clients (instructor and trainees) and the Flask backend.

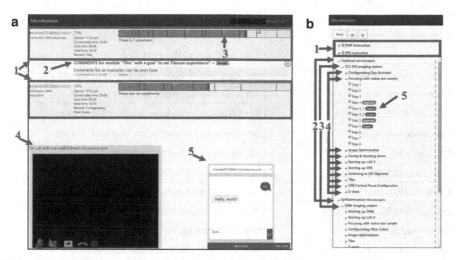

Fig. 2. Progress interface design (A) includes progress of different trainees (1), comments left by a trainee (2), marking critical steps (3), built-in teleconferencing tool (4) and chatting option (5). (B) training modules database includes instruction-type of blocks (1) and training material blocks sorted by type of the imaging equipment (2), microscope name (3), training subjects (4) and assigned with levels used by recommendation system (5).

Training Modules UI. The training material is composed of training modules, which are themselves consisting of series of several training steps. The database of modules was organized by type of the microscope, name of the imaging instrument, training subject explained inside of the module, and difficulty level (used by the recommendation system) (Fig. 3). To create a new training module, an instructor (or database administrator) needs to create new steps (or copy steps from pre-existing modules). Each step contains an annotated image (.pdf or.png file) and an optional audio file that is played to the trainee during the training session. By using fuzzy search tool, the instructor can rapidly navigate through the training database, select the necessary modules to form an instruction-type of block (i.e., instruction), and assign it to multiple trainees. Any module can be linked to multiple instructions or can be copied to be edited as a separate module. The modularity of the content creation facilitates assembly and reuse of modules across the training material database.

2.7   MicroAssistant Recommendation System

The MicroAssistant application tracks the trainee's actions during the course training session. The full list of actions tracked by MicroAssistant are shown in Table 1. User activity is recorded using the react-tracking NPM package and all tracking logs are persisted to the application's PostgreSQL database to analyze the triggering parameters for notification and recommendation systems.

Table 1. User actions that are identified by the tracking system

Action name	Description
Trainee-step-previous	User navigates to the previous step
Trainee-step-next	User navigates to the next step
Trainee-step-sidebar	User switches to another step in the sidebar
Trainee-mouse-idle	User's mouse stays idle for over a minute
Trainee-visibility-change	User switches to another browser tab/window
Trainee-toggle-sidebar	User opens or closes the instructions sidebar
Trainee-note	User opens or closes the notes section
Trainee-comment	User sends a comment to the instructor
Trainee-finish-instruction	User completes an instruction
Trainee-details-select-block	User selects supplemental module
Trainee-details-step-{…}	User navigates inside supplemental material
Trainee-details-return	User returns to the main instruction

The notification system activates when the trainee 1) is inactive on the MicroAssistant training platform for a significantly unusual amount of time (baseline established by the control group) or 2) expresses abnormal behavior by repeated switching between different training modules. Once triggered, the notification system displays messages to the instructor inside the Progress interface to alert about an unusual behavior.

The recommendation system uses the time spent by the trainee at each step (step-time) as the triggering parameter to adapt the instruction material. The baseline data for training levels (Beginner and Expert) were established for each instruction step based on 2 control Expert trainings. When Eq. (1) or (2) is satisfied respectively for n consecutive steps, the recommendation system suggests to the trainee leveled up (Eq. 1) or leveled down (Eq. 2) the training material (Fig. 3). After being upgraded to the Expert level, the trainee can at any time revert manually to the Beginner version in the Navigation bar (Fig. 1C). The number of consecutive steps were selected as the average number of steps in the instruction required to explain 2 different equipment functionalities (n = 4 for TCS SP8 instruction used in this study):

$$time_i < avg_i + std_i \tag{1}$$

$$time_i \geq avg_i + std_i \tag{2}$$

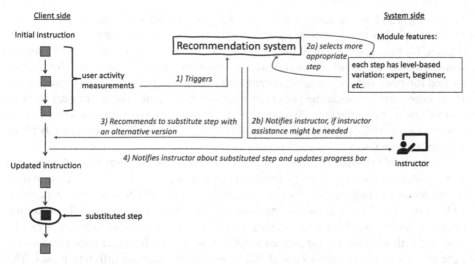

Fig. 3. Schematic design of the recommendation system and dynamic adjustments to the training instruction during the training session.

3   Results

3.1   Assessing the Difficulty of AIS Adaptation for Core Facilities

The main difficulty for training centers like CFs to automate personal trainings is to absorb the time-consuming task of preparing the training material. This study was conducted in collaboration with an institute bioimaging core facility that consists of 20 imaging instruments. The role of CF is to perform on-demand trainings (with an average of 200 trainings/per year) for all of these imaging instruments and ensure the operability of the equipment. All the trainings are performed in-person in 1-on-1 sessions to customize explanations to the trainee's objectives and level of expertise in bioimaging. Each training includes a theoretical part, hardware explanations, and a description of software functionalities and controls. However, each instrument has multiple functionalities. Each training is customized to only explain the necessary equipment features required to reach the objectives of the trainee. For example, using the same equipment unit, one trainee needs to be trained how to do 3D reconstruction of their specimen, while another trainee may need to do a 2D scan of the specimen. In this example, the instructor customizes the explanations to describe different functionalities of the equipment. Based on an interview with the instructor, digitizing the training material for one equipment unit would mean generating at least 3–4 versions of the instruction to address different objectives. This would require the instructor to generate 20–50 photos of equipment parts and up to 100 screenshots of the states of the software controlling the equipment unit, which would cost a prohibitive amount of time. In the interview, the manager of the core facility underlined that the cost-efficiency of preparation of training content for the facilities with limited labor-force would be too high to adapt autonomous trainings with an instructional system.

During pre-pandemic times, it was considered as a highly time-consuming task (i.e., to create digital instructions with a lack of hardware and software images), the necessity to assemble this material during the pandemic turned out to be a unique opportunity to re-use this digital content. When research laboratories were partially open, they implemented rotations and measures to prevent the accumulation of people in shared spaces. Then, core facilities made the photos of equipment locations and instrument parts to conduct the trainings remotely. We recycled this digital material as well as videos of the controlling software that were recorded during the training sessions to design a database of training modules.

Each module consists of a series of images (in PDF or PNG format) explaining one training topic and is designed to be followed step-by-step (Fig. 1). The preparation time for producing training modules for one imaging instrument (TCS SP8 microscope) took 20 h followed by 15 min for uploading them in MicroAssistant (Fig. 6A). This training content consisted of 12 training modules with 108 steps in the main instruction and 10 modules with 33 steps for supplemental information. Supplemental training modules were created to provide trainees with additional information on different topics. The whole training content for this instruction consisted of 85.1% the images derived from digital content created from pandemic remote trainings; 7.1% of images of the general theory and 7.8% of images for explaining navigation in MicroAssistant. We could not evaluate the cost-efficiency of creating training content as an isolated phase of AIS adaptation in CFs due to the lack of data of content preparation during pre-pandemic time (all the trainings were conducted in person at that time). Instead, we proceed in the evaluation of cost-efficiency of training content preparation as a part of the process of implementing of an AIS training solution for CFs.

3.2   RQ1: Different Tools to Overview and Intervene Training Sessions Reduces the Active Presence of the Instructor During Trainings with MicroAssistant

To evaluate the cost-efficiency of AIS adaptation in CFs, we first studied the factors that will increase the performance of MicroAssistant as a digital solution for equipment trainings. Based on the interviews with an instructor, we found that an important factor for the adaptation of an AIS in CFs is to increase the trackability of the trainee's activity for the instructor. Therefore, we designed tools that allows an instructor to monitor the training as needed. It includes 1) a progress bar allowing an instructor to access the trainee's progress and comments in real-time (Fig. 2A, 1); 2) a built-in chat and teleconferencing tools to communicate with the trainee during the training session (Fig. 2A, 4–5); 3) a warning system notifying the instructor and trainee of critical steps that must be followed to prevent possible damage to the equipment (Fig. 2A, 3); 4) a notification system tracking abnormal behaviour of the trainee (prolonged silence or jumping between steps) (Fig. 3). These features of the system made the training session more "transparent" and manageable, thus encouraging the instructor to proceed with autonomous trainings with the MicroAssistant AIS solution. Although, an instructor preferred to keep the remote desktop connection with the trainees, we observed a decrease in the time of active remote presence of the instructor during training sessions. Active remote presence was estimated as the active time the instructor spent on the Progress interface or the remote desktop connection to follow the trainee's actions. The trainings without progress bar showed that the instructor track the trainee's actions through the remote desktop connection 12.5% of the training session time (std = 3.5%, n = 2 trainings) (Fig. 4). After introducing in MicroAssistant the progress tracking system, teleconferencing tools, notifications and comments, the time of active remote presence dropped to 5.5% for the next 4 training sessions (std = 1%, n = 4 trainings). We conclude that the development of features to remotely track the progress and communicate with the trainee, as well as the implementation of a notification system facilitated the adoption of AIS-based trainings in CF, as it decreases the need for active presence of the instructor and therefore, increases the cost-efficiency of the MicroAssistant solution.

3.3   RQ1: Modular Design of Database of Training Content Decreases the Time for Preparation of Autonomous Trainings for Other Equipment

Another feature that increased the performance of MicroAssistant is the application of modular design principles to organize the training material. There are limited number of companies (Leica Microsystems, Nikon, Zeiss, Olympus) that produce bioimaging instruments with high resolution level (<180 nm) [15]. Thus, imaging stations in CFs often have similar parts and are controlled with similar software. As an example, we used the DM6 and TCS SP8 microscopes which both represent products of Leica Microsystems and are operated by Leica imaging software (LAS X). Although, the imaging properties and principles of image acquisition are very different between these two microscopes (DM6 is a widefield microscope with cameras for light detection; SP8 is a confocal microscope with special photon detectors), some parts of the microscope stand (touchpad, focus knobs) and some functionalities of the controlling software (tilescan, z-stack and project saving options of LAS X) are similar.

Structuring training material based on the training topics and storing them as modules with several steps allows to quickly search the necessary modules (using keyword search) and share training modules between the TCS SP8 and DM6 instructions. Training content for the DM6 microscope consisted of 9 modules with 83 steps in the main instruction and 8 modules with 16 steps in supplemental modules Two modules in the main instruction and 4 modules in the supplemental information (19.22% of training material) for the DM6 microscope were identical to the TCS SP8 training content (Fig. 5B, dark blue segment). Additionally, 34.03% of the steps of the DM6 main training content were adapted from similar steps in the TCS SP8 training material (Fig. 5B, light blue segment). DM6 training content preparation took 8 h followed by 15 min of uploading to MicroAssistant (Fig. 6A). We attribute such acceleration of content creation for DM6 to 1) the interchangeable properties of training content for equipment units in bioimaging facility, 2) the availability of templates from previous training material. The modular design of the MicroAssistant database of training content allowed us to significantly accelerate the time-consuming steps of training content preparation. Therefore, further increasing the cost-efficiency of the MicroAssistant solution.

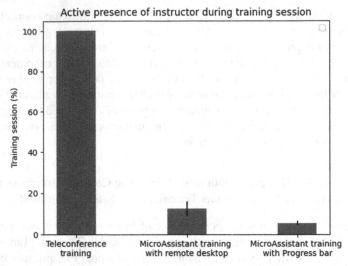

Fig. 4. Amount of time instructor actively interacts during teleconference trainings (100%), MicroAssistant trainings with remote desktop (12.5%, std = 3.5%, n = 2 trainings), with progress interface (5.5%, std = 1%, n = 4 trainings).

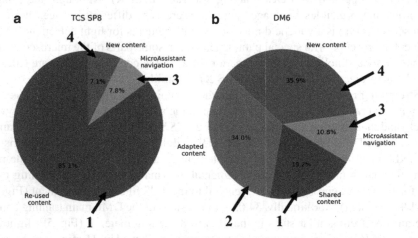

Fig. 5. Training content composition for TCS SP8 (A) and DM6 (B) microscopes. Re-used training material and shared modules indicated by dark blue (1), adapted training modules – by light blue (2), microAssistant program navigation modules – by orange (3) and newly created training modules – by green colors (4). (Color figure online)

3.4 RQ1: Introducing Simultaneous Trainings as a New Way to Increase Cost-Efficiency of Adaptation of Automatic Training Solution for Equipment Trainings in CFs

Simultaneous equipment trainings of users in CFs are only organized in the form of a separate training course. It is based on explaining and demonstrating to multiple trainees how to operate one type of imaging instrument. Such courses usually involve multiple copies of the same equipment units assembled in one room and arranged together with the institute or university and the equipment manufacturer. These trainings are conducted by one or a few instructors and are always organized as in-person trainings. The usual practice of equipment trainings in CF does not involve any aspect of simultaneous trainings, because imaging instruments in CF are never identical and often located in different rooms or different floors of the building.

In this study, we used the AIS training solution for training two equipment units simultaneously. The DM6 and TCS SP8 microscopes used in this study were physically located on different floors of the building and could not be used for simultaneous trainings during pre-pandemic in-personal trainings. Such double-training was also impossible to perform through verbal remote trainings conducted during the pandemic lockdown. By using the MicroAssistant step-by-step instructions, we set up 3 double-training sessions and the instructor was able to follow their progress in real-time (Fig. 2A). Trainees were unaware of the simultaneous aspect of their training session and rated the training experience as 5 out of 5 (n = 6 trainees). Based on the after-training interviews and instructor survey-form, the instructor did not notice a significant increase in active participation time per training (5%). However, there was a concern raised of a potential risk to accommodate 2 trainees in case of simultaneous emergencies with the equipment. The usual routine of remote desktop connection to the equipment computers does not support multiple connections at the same time. We suggest that additional double-trainings are required to test MicroTalk together with screen sharing options as a potential way to provide a trainee an immediate support. Therefore, we propose MicroAssistant AIS as a solution for potential simultaneous equipment trainings if there is a well-established connection routine for troubleshooting issues of multiple trainees.

3.5 Cost-Efficiency of AIS Training Solution for Equipment Trainings in CF

To evaluate the cost-efficiency of adapting the MicroAssistant solution for autonomous equipment trainings in CFs, we used time as a proxy of the cost. Our assumption is that either an instructor is also a designer of the training modules, or they are two staff members of CF with the same hourly wage. Using the TCS SP8 and DM6 microscope trainings, we calculated the cost-efficiency of using the MicroAssistant autonomous training solution instead of in-person verbal trainings. Based on the booking calendar of CF in 2018 and 2019, the standard in-person training session for TCS SP8 lasted 65.2 min (std = 33.37, n = 29 trainings) and for DM6, it lasted 50.7 min (std = 29.3, n = 29 trainings). We estimate that the trainings with MicroAssistant involved 5.5 min of the instructor's active involvement per training for both microscopes (std = 0.55, n = 6 trainings) based on time spent on following the trainee's progress. None of the trainings required an intervention of the instructor or direct communication with the trainees through teleconferencing tools. During double-trainings, we estimated that the instructor spent twice as much time to follow both trainees.

We define the efficiency of adaptation of MicroAssistant for equipment trainings as the time required for CF to liberate the necessary amount of instructor time during MicroAssistant-based trainings to be able to prepare training content for 1 equipment unit:

(Eq. 3).

$$CE = \frac{TF \times (DT - AP)}{(CP + CU)} \tag{3}$$

CE - Cost-efficiency of MicroAssistant solution per equipment unit (months $^{-1}$), TF-training frequency with MicroAssistant (trainings/year), CP- preparation time of training content, CU- time of uploading in MicroAssistant (15 min), DT – training duration for in-person training on that equipment unit, AP- time of active presence of the instructor.

We estimated the cost-efficiency of the MicroAssistant AIS training solution in 3 scenarios (Fig. 6B):

1. preparation of training content by re-using digital material from teleconferencing remote trainings (TCS SP8 example);
2. creation of training content using modular design of training content database created in the first scenario (DM6 example);
3. application of simultaneous training strategy for autonomous trainings with MicroAssistant (double-training example).

There was a dramatic increase in the cost-efficiency of MicroAssistant-based trainings for instructions prepared using a database of training modules from another equipment unit. After 11 in-person trainings on DM6 microscope, every next training with MicroAssistant will save $(DT-AP)$ of instructor time. In case of double-trainings, it may double the cost-efficiency of the MicroAssistant solution per equipment unit (Fig. 6B) as it will allow to conduct twice as more trainings per calendar year. Based on the frequency of trainings for DM6 and TCS SP8 microscopes in 2018–2019 years (14.5 training/year), we estimate that the adaptation of the MicroAssistant instructional platform can be considered as a cost-efficient solution after 8.5 months. However, considering the advantage of conducting simultaneous autonomous trainings the cost-efficiency with double trainings using the MicroAssistant solution could be reached within 4.25 months.

3.6 RQ2: Modular Design Allows Rapid Customization of Training Session

One of the difficulties to adapt autonomous solution for equipment trainings in bioimaging CF was the nature of the multifunctional characteristic of the imaging equipment. This requires multiple versions of the trainings and therefore increases the time to prepare training material for autonomous trainings.

The modular design of the training content used in MicroAssistant allows the instructor to select only necessary training modules relevant to the objectives of the trainee. In this study, the trainee filled a Pre-training form with required information about imaging objectives and experiment needs. The instructor was selecting the required training material out of the database of training modules (through keyword search) and either assembling them in the new instruction or selecting pre-assembled versions of the instruction (Fig. 2B). Therefore, during the training session, the assigned trainee could access only the necessary training content. In this study, these steps were manually done by the instructor because selecting and assembling instruction takes less than 5 min and

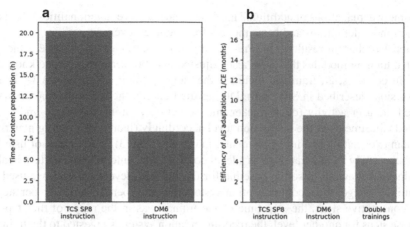

Fig. 6. Time required for training material preparation by creating training modules *de novo* (for TCS SP8 training) and by using training modules database (for DM6 trainings) (A). Time required for CF to reach cost-efficiency with MicroAssistant trainings in 3 scenarios: 1) creating training material by re-using digital material (TCS SP8 trainings), 2) creating training material by using database of training modules (DM6 trainings); 3) MicroAssistant application for double-trainings.

does not significantly change the cost-efficiency of MicroAssistant adaptation. However, upon further collection of pre-training form responses, we plan to automatize this step to reduce instructor involvement.

We used the TCS SP8 and DM6 microscopes as an example of the equipment units to build the training content. While 9 out of 12 modules were mandatory for operating the TCS SP8 microscope, the other 3 modules (A, B, C) and any of their combinations are included in the training according to the trainee's needs/objectives. Therefore, creating a modular database of the training content allowed the instructor to compose at once 7 different instructions that could include A, B, C, AB, AC, BC, or ABC modules. 7 out of 9 modules were mandatory for any training on the DM6 microscope, while including any combination of the 2 other modules could vary based on the trainee's needs/objectives. Therefore, the modular design allows instructors to easily customize trainings according to the different objectives of the trainee.

3.7 RQ2: Recommendation System Increases Engagement of Trainees During Autonomous Trainings

In-person trainings of equipment units in CFs are usually based on the objectives and experience level of the trainee with imaging instruments. To address experience level in AIS, we designed a recommendation system inside MicroAssistant. During the training, the recommendation system of MicroAssistant monitors the performance and engagement of the trainee by tracking the user activity (Table 1) and recommends the trainee to adjust the instruction according to their experience level.

In this study, we used TCS SP8 microscope for trainings with the recommendation system. We design the training content for 2 levels (Beginner and Expert) to include either more explanations of additional hardware and software functionalities (for Expert level) or more definitions and basic theory (for Beginner level).

First, based on the results of the Pre-training form, the trainee was assigned objective-oriented training modules that were also adjusted to appropriate experience/knowledge level. In our tests, all trainings included the same training objectives (ABC instruction version, described in Sect. 3.6). During the training, the recommendation system tracked the user activity (click tracker, mouse activity, timing the steps) and recommended adjustments to the instruction level accordingly (including training material in the main explanation and in supplemental materials) (Fig. 3). Due to lack of user base during this study (n = 2 trainings as a level control), we could not do a pattern analysis of trainees' behaviour to set up baseline for the different levels. Instead, we used as a threshold to trigger the recommendation system the time spent by a trainee per step. If, for 4 consecutive steps, the step time of the trainee was in the range of the step-time for those steps for another level, the recommendation system suggested to the trainee to adjust the instruction by leveling up (Expert) or leveling down (Beginner) the training material. If the trainee followed the recommendation, MicroAssistant modified the current instruction with training modules and supplemental information that corresponded to the new level.

We conducted 2 trainings with the recommendation system. Both trainees were previously trained in-person on the TCS SP8 microscope and were going through a re-training process. Despite expert level assignment by the Pre-training form, to minimize the risk of equipment damage during the testing of the recommendation system, we set all users to start the trainings as Beginners. Trainee 1 accepted the initial suggestion of the recommendation system to switch to Expert level after 16.7% completion of the training (Table 2). However, after multiple instruction restarts due to wi-fi network issues, the trainee denied afterwards to switch to the Expert level and finished the training as Beginner. Trainee 2 accepted levelling up to Expert level after 13.9% from the beginning of the instruction (Table 2); however, trainee 2 reverted multiple times to the Beginner level manually. Based on the comments of trainee 2, we attribute such behaviour to the trainee curiosity to check both versions of some modules. As indicated by the post-training feedback form, both re-trained users found some of the level-adjusted modules to be useful/new for them. Trainee 2 also found useful the Expert parts of supplemental topics. Both trainees appreciated the initiative of an autonomous training and rated the overall experience as 4 out of 5. Therefore, we conclude that adjustments to the training material through the recommendation system increased the engagement of trainees by providing useful training material for re-trained and new users. However, to improve the experience of the re-trained users, we suggest further customization of the training material that will reduce the mandatory parts of the training. We also propose further testing of the recommendation system with inexperienced users to establish the baseline behaviour of the control groups for each level.

3.8 Feedback-Based Evaluation of the System

Based n the results of the Feedback forms and after-training interviews, the overall training experience with MicroAssistant was evaluated as 4.7 out of 5 (n = 9 users). 75% of trainees expressed a wish to be trained with MicroAssistant for other imaging

Table 2. Summary of MicroAssistant trainings with the recommendation system

Performance of the recommendation system	Trainee 1	Trainee 2
Training duration (min)	89	72
Recommendation system first triggered at (% of instruction)	16.67%	13.89%
Recommendation system accepted (number of times)	1	4
Recommendation system denied (number of times)	7	1
Manual switching between levels (number of times)	0	3
Restarting instruction	4	0
Number of useful/new topics recommended by MicroAssistant	1	2
Overall experience evaluation (out of 5)	4	4

instruments. The main advantage of the system reported during interviews were: 1) an opportunity to follow the training in a self-paced manner; 2) an opportunity to learn equipment functionalities autonomously rather than observing instructor's actions during in-person trainings; 3) it was easy to follow organisation of training material. The main concerns raised by re-trained users were related to the big volume of the training material. We attribute such responses to the misfitting of Expert level training material for re-training sessions and propose in the future to prepare for re-training users a separate experience level. The instructor appreciated the elevated level of standardization in AIS trainings versus in-personal trainings but raised concerns regarding MicroAssistant trainings with trainees that are completely novice in microscopy. The instructor feedback included an appreciation of the reduced active time during MicroAssistant trainings, although pointing to the challenges of remote trainings that may rise in case of equipment failure.

4 Discussion

In this study we assessed the difficulty of introducing an AIS for equipment trainings conducted by training centers like CFs. Due to the complexity and diversity of the equipment, CFs usually provide 1-on-1 verbal training sessions that are highly customized to the objectives, experience, and engagement of the trainees. Therefore, creating digital material for such sessions is a time-consuming task largely restrained by the limited labor-force.

By re-using the digital material created for remote trainings during the pandemic lockdown (photos of equipment parts and videos of training sessions), we could create a database of training modules and design the AIS solution (MicroAssistant) for autonomous trainings. This study demonstrates that the material created in the aftermath of the pandemic offers a unique opportunity to assemble digital content created during COVID-lockdowns and to implement AIS in a field where its previous adaptation was not considered as cost-efficient. Based on our experience of developing AIS-based trainings for bioimaging CFs, we see a new post-pandemic potential for adaptation of

AIS training solutions in other different fields with complex equipment trainings (e.g., manufacturing operators, heavy equipment operators) [16, 17].

Despite time-consuming and labor-intensive tasks of training content preparation, we show that similarities between equipment units significantly facilitate the preparation of the training material (Fig. 6A), increasing the cost-efficiency of applying an AIS solution for CF trainings (Fig. 6B). Based on the limited number of manufacturers of imaging equipment, we anticipate that even greater similarities can be observed between different academic and medical bioimaging CFs as they operate with similar types of imaging instruments. This should even further facilitate the preparation of training material. Therefore, we suggest that AIS adaptation for medical and bioimaging training centers can be accelerated once the MicroAssistant training database reaches a higher level of diversification of training material and becomes able to accommodate the most common types of equipment and operating software.

To address the high level of personalization of in-person trainings, we designed MicroAssistant trainings as module-based instructions that can be easily altered to accommodate the objectives/needs of the trainees. We also developed a customization solution for different expertise levels of the trainees. The high self-motivation characteristic of trainees in CF allowed us to design an adaptive part of MicroAssistant based on recommendations proposed to the trainees by the tracking system. Once triggered, it provided an option to adjust the instruction with more relevant training content to the trainee's experience. Due to the small number of control training sessions for Expert and Beginner levels, we could not apply any probabilistic learner models (e.g., Bayesian Networks) for behavioral pattern analysis of the trainees [18, 19]. Instead, we used step-time as a parameter to trigger the recommendation system. Although MicroAssistant trainings with the recommendation system were part of a re-training initiative, all trainees appreciated and found useful the training material provided by the recommendation system (Table 1). Therefore, we conclude that the recommendation system is a useful tool to personalize equipment trainings. We expect that the continuous use of the MicroAssistant AIS will provide a more extensive user base and will allow further developments of the triggering conditions of the recommendation system.

Another advantage of using the MicroAssistant AIS for equipment trainings is the possibility of conducting simultaneous trainings by the same instructor. The progress bar and the built-in chat and teleconferencing tools provide an instructor with an overview of the progress of the trainee and facilitate remote communication if needed. Our study showed that the overall active presence of instructor per trainee did not increase during simultaneous trainings. Notably, all new trainees expressed interest in being trained with MicroAssistant for other imaging instruments. Based on our experience with double-trainings, we propose that the MicroAssistant AIS solution can introduce a new practice of simultaneous equipment trainings. However, we also conclude that simultaneous trainings require the development of a robust way to access multiple equipment computers remotely for troubleshooting if equipment issues arise at the same time.

The MicroAssistant AIS received positive evaluations from CF through an instructor feedback form, instructor interviews and trainee feedback forms (4.7 out of 5, n = 9 users). We attribute a successful user perception of AIS trainings to elevated self-motivation of the trainees in acquiring new skills.

5   Limitations

There are several limitations that must be considered in this study. First, the trainees in theses study already had some experience with other imaging equipment in CF. Therefore, they were not completely novice to the use of microscopes. Due to the potential risks of damaging the equipment, we could not test our autonomous training solution with first-time users. Second, this study used mixed methods to evaluate the quality of the training with MicroAssistant resulting in a lack of standardized evaluation of the results of the trainings. The on-demand trainings of self-motivated trainees prevents CFs from introducing routine after-training knowledge tests (i.e., it cannot be run as an exam). Third, the instructor concerns regarding simultaneous troubleshooting with MicroAssistant need further assessment with the potential implementation of simultaneous remote desktop connections. Additionally, this study does not reveal the full potential of the recommendation system for equipment trainings. Due to the lack of control trainings that would establish a baseline behaviour of trainees from distinct experience groups, we could not fully utilize the tracked user activity to set up a more advanced triggering mechanism for the recommendation system. Our focus was more concentrated on addressing the time-consuming tasks of training content creation as the main barrier for adaptation of AIS for equipment trainings in training centers with limited labor forces.

6   Conclusion and Future Directions

The results of this study show that, in the aftermath of the pandemic, there is a greater potential for designing AIS solutions for equipment trainings than before. By processing the digital content created during pandemic remote trainings, we designed the MicroAssistant AIS training platform for equipment trainings. We implemented MicroAssistant for bioimaging trainings. Based on our results, we conclude that MicroAssistant-driven trainings can be an efficient way to provide autonomous personalized equipment trainings if 1) the most time-consuming task (preparation of training material) is minimized by providing baseline training modules for equipment units, 2) the design of AIS enables the instructor to remotely monitor the training and getting notified in case of "abnormal" activity of a trainee, 3) the adaptive part of the AIS can provide trainee with only the necessary explanatory information.

In the future studies, we propose establishing an open-source training database for different medical and bioimaging CFs that could represent the source for training material for different imaging instruments. This should presumably further facilitate the adaptation of AIS for equipment trainings.

Another consideration for the future of the MicroAssistant solution is the development of its ability to adapt the training content to the expertise and motivation of the trainee. First, more reference trainings will need to be conducted with trainees deemed as Experts by the CF instructor as well as Beginner trainees. This further testing will help construct a better model of the differences between the distinct experience levels of CF trainees. Once the reference training data is gathered, we could apply a variety of learner models to characterize the behavior of a trainee in real-time. Probabilistic learner models (e.g., Bayesian Networks) and machine learning based models including

Deep Neural Networks [18, 19] could also be used by the recommendation system to analyze the trainee's activity and alter training material if needed. Additionally, we can also explore cognitive modelling methods like ACT-R to extract a more comprehensive and interpretable model of the trainee behavior in real-time [20].

References

1. Park, O.-C., Lee, J.: Adaptive instructional systems. In: Jonassen, D.H. (ed.) Handbook of Research for Education Communications and Technology, 2nd edn., pp. 651–684. Erlbaum, Mahwah (2004)
2. Yang, S., Yu, K., Lammers, T., Chen, F.: Artificial intelligence in pilot training and education-towards a machine learning aided instructor assistant for flight simulators. In: Stephanidis, C., Antona, M., Ntoa, S. (eds.) HCI International 2021 - Posters, vol. 1420, pp. 581–587. Springer, Cham (2021). https://doi.org/10.1007/978-3-030-78642-7_78
3. Bell, B., Kelsey, E., Nye, B., Bennett, W.: Adapting instruction by measuring engagement with machine learning in virtual reality training. In: Sottilare, R., Schwarz, J. (eds.) Adaptive Instructional Systems, vol. 12214, pp. 271–282. Springer, Cham (2020). https://doi.org/10.1007/978-3-030-50788-6_20
4. Zahabi, M., Park, J., Razak, A.M.A., McDonald, A.D.: Adaptive driving simulation-based training: framework, status, and needs. Theor. Issues Ergon. Sci. **21**(5), 537–561 (2020)
5. Ropelato, S., Zünd, F., Magnenat, S., Menozzi, M., Sumner, R.: Adaptive tutoring on a virtual reality driving simulator. In: 1st Workshop on Artificial Intelligence Meets Virtual and Augmented Worlds (AIVRAR) in Conjunction with SIGGRAPH Asia 2017, ETH Zurich (2017)
6. Faria, L., Silva, A., Ramos, C., Vale, Z., Marques, A.: Cyber-ambient intelligent training of operators in power systems control centres. In: Proceedings of IEEE 15th International Conference on Intelligent System Applications to Power Systems, Brazil, pp. 1–7 (2009)
7. Turati, P., Cammi, A., Lorenzi, S., Pedroni, N., Zio, E.: Adaptive simulation for failure identification in the advanced lead fast reactor European demonstrator. Prog. Nucl. Energy **103**, 176–190 (2018)
8. Meder, D., Morales, M., Pepperkok, R., Schlapbach, R., Tiran, A.: Institutional core facilities: prerequisite for breakthroughs in the life sciences. EMBO Rep. **17**, 1088–1093 (2016)
9. Haley, R.: A framework for managing core facilities within the research enterprise. J. Biomol. Tech. **20**, 226–230 (2009)
10. Hayrea, C.M., Kilgour, A.: Diagnostic radiography education amidst the COVID-19 pandemic: current and future use of virtual reality (VR). J. Med. Imaging Radiat. Sci. **52**(4), S20–S23 (2021)
11. Singh, R.P., Javaid, M., Kataria, R., Tyagi, M.T., Haleem, A., Suman, R.: Significant applications of virtual reality for COVID-19 pandemic. Diab. Metab. Syndr.: Clin. Res. Rev. **14**(4), 661–664 (2020)
12. Kigenyi, J., Mische, S., Porter, D.M., Rappoport, J.Z., Vinard, A.: Preparing for the unprecedented: the association of biomolecular resource facilities (ABRF) community coronavirus disease 2019 (COVID-19) pandemic response part 1: efforts to sustainably ramp down core facility activities. J Biomol Tech. **31**(4), 119–124 (2020)
13. Dietzel, S., et al.: A joint action in times of pandemic: the German bioimaging recommendations for operating imaging core facilities during the SARS-Cov-2 emergency. Cytometry Part A: J. Quant. Cell Sci. **97A**, 882–886 (2020)
14. Kirkpatrick, D., Kirkpatrick, J.: Evaluating Training Programs: The Four Levels. Berrett-Koehler Publishers, San Francisco (2006)

15. Chandler, D., Roberson, R.W.: Bioimaging: Current Techniques in Light & Electron Microscopy. Jones & Bartlett Publishers, Burlington (2008)
16. Roldán, J.J., Crespo, E., Martín-Barrio, A., Peña-Tapia, E., Barrientos, A.: A training system for Industry 4.0 operators in complex assemblies based on virtual reality and process mining. Robot. Comput. Integr. Manuf. **59**, 305–316 (2019)
17. Akyeampong, J., Udoka, S.J., Park, E.H.: A hydraulic excavator augmented reality simulator for operator training. In: Proceedings of the 2012 International Conference on Industrial Engineering and Operations Management, pp. 1511–1518. IEOM Society, Istanbul (2012)
18. Ganapathy, P., Rangaraju, L.P., Kunapuli, G., Yadegar, J.: Skill mastery measurement and prediction to adapt instruction strategies. In: Sottilare, R.A., Schwarz, J. (eds.) Adaptive Instructional Systems. Adaptation Strategies and Methods. LNCS, vol. 12793, pp. 45–61. Springer, Cham (2021). https://doi.org/10.1007/978-3-030-77873-6_4
19. Orji, F.A., Vassileva, J.: Modelling and quantifying learner motivation for adaptive systems: current insight and future perspectives. In: Sottilare, R.A., Schwarz, J. (eds.) Adaptive Instructional Systems. Adaptation Strategies and Methods. LNCS, vol. 12793, pp. 79–92. Springer, Cham (2021). https://doi.org/10.1007/978-3-030-77873-6_6
20. Zapata-Rivera, D., Arslan, B.: Enhancing personalization by integrating top-down and bottom-up approaches to learner modeling. In: Sottilare, R.A., Schwarz, J. (eds.) Adaptive Instructional Systems. Adaptation Strategies and Methods. LNCS, vol. 12793, pp. 234–246. Springer, Cham (2021). https://doi.org/10.1007/978-3-030-77873-6_17

Learning Support and Evaluation of Weight-Shifting Skills for Novice Skiers Using Virtual Reality

Shigeharu Ono[1]([⊠]) [iD], Hideaki Kanai[1] [iD], Ryosuke Atsumi[1,2], Hideki Koike[2] [iD], and Kazushi Nishimoto[1]

[1] Japan Advanced Institute of Science and Technology, Nomi, Japan
s2120010@jaist.ac.jp
[2] Tokyo Institute of Technology, Meguro, Japan

Abstract. In this study, we propose a virtual reality learning support system designed to help train novice skiers. In previous research, we extracted the differences between the weight shifting movements and skiing postures of experts and beginners using deep learning. The obtained results showed weight shifting to be a more important feature than posture. Accordingly, we focused on supporting the weight-shifting technique. The support system provides real-time feedback to a user on their current weight-shifting status. We conducted an experiment to verify the effectiveness of the proposed approach, in which we defined evaluation criteria for a user's level of skiing proficiency. The experimental results demonstrate that the system successfully facilitated participants' acquisition of the weight-shifting skill.

Keywords: Ski · Learning support · Virtual reality · Visual feedback · Deep learning

1 Introduction

Computer-assisted training systems for sports have been presented in several studies [5,13]. These systems analyze a user's performance in terms of measured data, and compare data of novice and expert users to determine the differences.

In this study, we focus on the development of a learning support system for skiing. Some ski-training systems for beginners do exist [8,18], although most are designed for intermediate and advanced users [3,6]. Moreover, existing support systems do not support communicate or display the differences between the movements of user and expert in real time.

Prior to the development of the learning support system, we extracted the differences in the weight shifting and posture of experts and beginners in a virtual reality training system to identify the most important skills required to learn to

This work was supported by JST SPRING (Grant number: JPMJSP2102) and JAIST Research Center for Cohabitative-AI×Design.

ski [2]. We classified 11 participants based on their posture and weight shifting using Darknet-53 [12] to extract features. Results obtained from Grad-CAM [15] indicated that weight shifting affected the classification more than posture. This result supports previous findings, such as balance test comparisons conducted before and after participants learned to ski [17].

Accordingly, we focused on weight-shifting technique in this work. Hence, we proposed a method designed to support learning the weight-shifting skills used in skiing with an awareness of the current status of the user's gravity center in real time. As part of the proposed approach, we implemented a visual feedback process to display the differences between the gravity center of a user and an expert. Moreover, we defined evaluation criteria to distinguish proficiency levels based on user performance to verify the effectiveness of the system.

2   Related Research

Several existing systems have adopted various methods to provide awareness of certain techniques to users. Heike and Ohgi [3] developed a ski-jump training system using deep learning. This system extracted a user's posture and the V-angle of their skis from inertial sensor data, and divided them according to each ski-jump phase. The system then extracted features from these data and compared them using a deep-learning model. Moreover, Fasel et al. [6] proposed an alpine skiing training system. In addition to multiple inertial sensors, this system utilized magnetic sensors placed at each gate to estimate the posture of a user.

Such methods have also been explored in other sports. For example, Sato and Tokuyasu [13] developed a pedaling-skill training system for cyclists using principal component analysis. The authors established evaluation criteria based on the muscle-activity data of skilled cyclists, and the system compared them with muscle-activity patterns extracted from a user via principal component analysis. Moreover, Chen et al. [5] developed a yoga training system using image recognition. This system compares feature points obtained from user data including skeletons and contours, topological skeletons, and main axis with those obtained from experts.

These systems analyzed a user's movement and compared them with those of an expert to help users improve their technique. In particular, existing skiing learning systems [3,6] do not provide feedback in real time because they are designed for advanced users who are able to improve their skills even without real-time feedback.

However, although these methods can inform users as to their technique and may increase users awareness of their form, they do not always help users to acquire the technique. In particular, it is difficult for beginners to acquire skills purely by receiving information on the differences between the techniques of beginners and experts. Therefore, in this study, we adopt real-time feedback to facilitate natural learning.

As a training system using real-time feedback, Hasegawa et al. [8] presented a skiing training system for beginners. The system enabled real-time correction of the positions of both feet using auditory feedback. This system measures the center of gravity of users' feet as they move forward. The system varied the pitch of the sound depending on these positions to inform a user as to the position of their feet.

However, this system was developed for actual skiing sites. Due to seasonal and geographical limitations, it is difficult to practice skiing at any arbitrary time and place. Consequently, learning-support environments for beginners are often insufficient. Hence, we adopted a virtual reality training environment in this work.

3   Development of a Ski Learning Support System

The proposed system is based on a previously developed approach [18]. The movements of an expert skier are displayed to the user via a virtual reality training system to guide them to adjust their movements to match those of the expert. We adopted the same virtual reality system (HTC Vive Pro[1]) and ski simulator (Pro Ski-Simulator: POWER SKI SIMULATOR[2]).

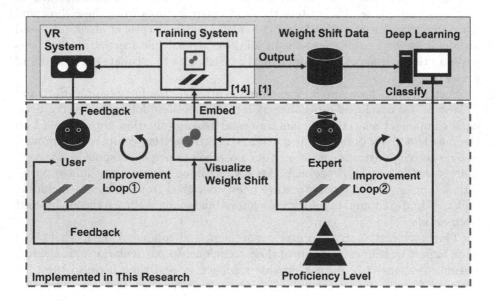

Fig. 1. Proposed system: this research implemented short and long-term improvement loop systems in the existing system.

---

[1] https://www.vive.com/eu/product/vive-pro/.
[2] https://www.ski-simulator.com/power-ski-simulator-en.

Figure 1 illustrates the proposed system. Overall, it comprises two loops to help beginners acquire the target skill, including a short-term loop (improvement-loop 1) and a long-term loop (improvement-loop 2). In the short-term loop, the system provides visual feedback to the user via a virtual reality training system in real time as they shift their weight. This enables beginners to become more aware of the difference in how they shift their weight compared to the technique demonstrated by the experts. In addition, in the long-term loop, the system provides the user's proficiency level in the current trial based on the classification results obtained by deep learning. This helps beginners consolidate their weight-shift skills.

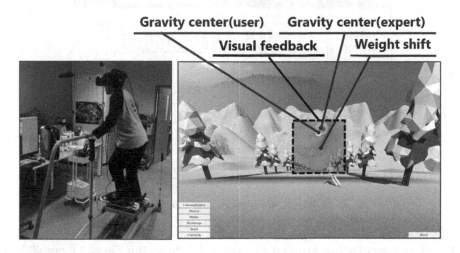

Fig. 2. Condition of the users while using the system

Figure 2 illustrates the conditions of users while using the system. In this study, we adopted visual feedback because it is effective in weight-shifting [11] and balance training [10,14]. The system includes a feedback layer displaying the user's current skiing status, which includes the user's weight shift and center of gravity, along with an expert's center of gravity. Accordingly, the user can instantaneously recognize all the relevant information simultaneously at a glance, without moving their eyes.

Figure 3 shows the pressure sensor module. In the proposed system, data on users' weight shifting are collected by 16 pressure sensors attached to the soles of ski shoes and transmitted via a wireless network. In a previous work [2], we used eight pressure sensors on each sole. These sensors were based on the system of Fukahori et al. [7]. However, their system focused on the approximate recognition of foot gestures, mainly toes and heels. Therefore, to measure the weight shift data more finely and accurately, we used double the number of sensors attached in previous research [2]. To implement this improvement, we used a compatible Arduino Mega 2560 microcontroller (What's Next Green) instead of

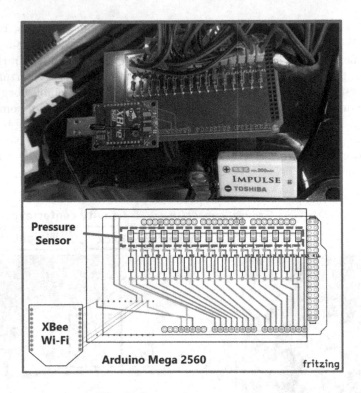

Fig. 3. Pressure sensor module

the older alternative (an Arduino Fio-compatible Sparkfun Fio v3ATmega32U4). Furthermore, in contrast to the previous module, the new module was able to drive with a battery by sending data via a wireless network using an XBee S6B Wi-Fi module.

4   Evaluation Experiment

We conducted an experiment to evaluate the efficacy of the proposed system.

4.1   Measurement Methods

As noted above in Sect. 3, the system allows users to observe the movements of experts via a virtual reality training system. In the experiment, participants used the system for one minute and were required to adjust their movements to match those of the expert. The participants used the system three times in each measurement, and we adopted the second and third attempts as data to perform classification.

To classify users using deep learning, the system outputs an image combining the user's weight shift and the expert's posture data, as presented in Fig. 4.

Fig. 4. Example of outputted image

In our prior research [2], we correctly classified 11 participants into four proficiency levels based on their skiing experience and turning method. In the present work, Darknet-53 [12] was adopted to perform feature extraction.

4.2 Participants

Initially, we measured the weight-shift data of 16 participants to configure the user experience level classifier, similar to approach adopted in the previous study [2]. Table 1 presents the participants' skiing experience and turning method.

At this point, we classified nine skiers able to perform snowplow turns and five beginners into four experience levels according to their number of days of experience: 0 days (F, J, K, L, M), 1–9 days (A, B, G, I, N, P), 10–49 days (E, O), and 50 days or more (D). Skiers who habitually performed parallel turns (C, H) were excluded to eliminate the influence of differences in turning methods.

4.3 Verification Method

We divided the beginners into two groups according to whether they used the system; participants J, L, and M used it while F and K did not. The participants did not have snowboarding experience, and exercised with a frequency of once a week or less. We measured the users' weight-shift data two to three times per week, for a total of eight sessions.

Eventually, we obtained 37 693 images of weight-shift data from both the skiers and beginners. A total of 23 991 images were used as training data, and the remaining images were used as testing data.

Table 1. Participants of the evaluation experiment

ID	Skiing experience	Turn method	Used system
A	4 days	Snowplow	Excluded
B	3 days	Snowplow	Excluded
C	Above 100 days	Parallel	Excluded
D	Above 50 days	Snowplow	Excluded
E	Above 10 days	Snowplow	Excluded
F	None	None	No
G	3 days	Snowplow	Excluded
H	Above 20 days	Parallel	Excluded
I	5 days	Snowplow	Excluded
J	None	None	Yes
K	None	None	No
L	None	None	Yes
M	None	None	Yes
N	1 day	Snowplow	Excluded
O	Above 10 days	Snowplow	Excluded
P	3 days	Snowplow	Excluded

4.4   Evaluation

We define "proficiency level" as the rate of image data misclassified as non-beginner in each session, as given below.

$$Proficiency\ level = 1 - \frac{Data\ classified\ as\ 0\ days}{All\ data}$$

For example, we obtained 100 images from a user in one measurement, of which 50 were classified as "0 days", such that their proficiency level was recorded as "0.5". We applied this equation to beginners. Changes in users' proficiency levels over time were used to ascertain whether the participant were able to acquire the weight-shifting skill. Higher proficiency levels indicate that the performance of the user is closer to the weight shift of experienced skiers. Hence, a proficiency level of "1" indicates that the user has progressed beyond the skill level of a beginner. In this research, users reaching a proficiency level of 1 indicates that they were able to acquire the weight-shifting skill successfully.

However, to help users improve their skills, expert examples must be selected carefully. If the expert's example was excessively close to the users' level, they the users were unable to sufficiently improve their skills. Conversely, if the expert example was excessively far from the users' level of skill, the users were unable to sufficiently realize the experts' technique. In this research, we used participant D's weight-shift as expert data because they were the most experienced in the basic snowplow turning method.

4.5 Result

Figures 5 and 6 present the recorded proficiency levels of users against each group's number of practice sessions. The dotted lines in these figures represent linear approximations of the change in each participant's proficiency level with increasing numbers of sessions.

The group that did not utilize the proposed system did not always exhibit an improvement, whereas all participants in the group that utilized the system showed an improvement. Therefore, the proposed system may be considered effective in facilitating the acquisition of weight-shifting skills by beginners with no skiing experience.

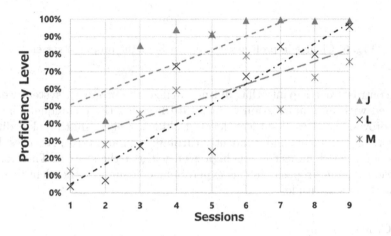

Fig. 5. Change in proficiency level (with the proposed system)

5 Discussion

5.1 Findings

The experimental results show that the feedback provided users with an awareness of how to improve their weight-shifting skill. However, in the group that utilized the proposed system, the proficiency levels of participants exhibited greater variably than those of other groups. The presented expert's data was only composed of gravity center collected from 32 pressure sensors. Therefore, even if a user's center of gravity was same with that of the expert, the actual pressure pattern may differ. This suggests that they improved their weight shifting skill via trial and error to determine how they should move to adjust more easily to the expert's center of gravity. Although all of the available data from the expert could be displayed, we considered that such feedback would be difficult to understand for beginners. Also, from the result, we observed that excessively simple feedback reduced the cognitive demands of the system while remaining understandable by beginners. Hence, these results suggest that trade-offs are involved between the fineness, accuracy, and comprehensibility of the feedback.

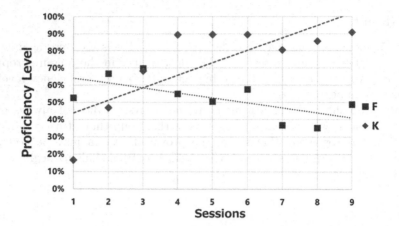

Fig. 6. Change in proficiency level (without the proposed system)

In contrast, participants that did not utilize the proposed system were not presented with any information to facilitate learning the technique. Therefore, the improvement of each subject' level of skill at weight-shifting skill differed. However, they did learn independently through trial and error. As a result, the proficiency level of subject K was high even without the system. Therefore, the evaluation system proposed in this experiment can support the identification of users with high aptitude.

5.2   Contribution

This type of visual feedback can be used to checking how we are moving. In this research, we focused on supporting users in learning weight-shifting skills. We consider this approach suitable for skills in which it may be difficult to directly recognize the correctness of one's technique. In ski, edging and posture are also important in correctly executing turns. Similarly to the proposed method, supporting edging technique and correct posture can help users to realize improvements in skill by using data from gyroscopic sensors and motion capture to classify their performance.

As another practical use, we can also support correcting strategy in contexts such as racing lines by using time score and sensor data. Although the behavior must be accurately defined, if the relationship between the sensor data and movement are known, unsupervised learning can be used to provide an expert's racing line, and the obtained strategy may differ with the conventional approach (as an example, see Praveen et al. [16]).

This feature can also provide a new approach to skill learning. For example, this feedback and virtual training system can be used to perform distance learning of various skills (Fig. 7). Existing distance learning environments remain inadequate [1]. In the case of physical education, it is difficult for instructors to check learners' movements in detail and provide them with direct guidance.

This may result in inadequate education. However, as mentioned above, users can check their movement using the proposed system. Then, if the system sends students' measured data to the teacher, the teacher can then evaluate students' movement more easily than with conventional distance learning methods. Similarly, if the system sends the teacher's measured data to the students, they can check their technique against the teacher's. Such direct interaction allows users to acquire skills more easily.

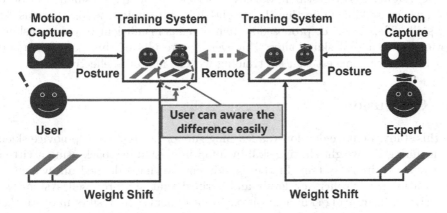

Fig. 7. Distance learning system for ski implemented this system

5.3   Limitation

This research involves the following limitations.

Participants. In this study, we focused on supporting beginners in acquiring weight-shifting skills. We used a supervised learning method for classification, and the criteria were only based on the number of days of experience, similar to previous research [2]. Therefore, as a limitation of this method, it was necessary to define the tentative borderline between beginner and non-beginner skiers. Although the previous research classified participants with 0–3 days of experience as beginners, in the present work, we considered participants without skiing experience to classify the difference between their performance and those of participants with some skiing experience. However, some non-beginner participants had not skied for years. It proved difficult for the supervised learning method to determine accurately whether such participants really deserved to be classified as non-beginners.

System. In this study, we displayed participant D's weight-shift data statically to simplify the system. However, if the system is applied to all beginners, the amount of support should be adjusted based on users' proficiency. Typically, in creative or physical activities, a user must develop and master skills. For example, children learn to ride bicycles through the following steps. 1) First, they ride with training wheels. 2) Then, they might ride without training wheels while someone holds the seat. 3) Finally, the helper must eventually release the seat when the child starts pedaling. Similarly, we consider that the amount of support should be adjusted according to users' levels of proficiency. In fact, in educational contexts, some research has explored gradual learning [4,19].

Experiment. In this research, we only experimented with a ski simulator. Moreover, we focused on supporting the weight-shifting technique among various skiing skills. Therefore, the proposed system remains insufficient to practice skiing optimally on actual skiing sites. However, the results of this experiment and other studies [9] suggest that ski training with simulator is effective.

6 Conclusion

In this study, we have developed a learning support system to help novice skiers in acquiring the weight-shifting skill by adopting visual feedback using a virtual reality ski training system. In the experiment, we have defined an evaluation criteria for user proficiency levels and verified that the proposed system was effective in facilitating skill acquisition by beginners. The results indicate that this system successfully helped participants achieve this objective.

In future research, further verification will be required with a larger number of participants and a longer duration to further verify the results. As the first step in this investigation, we focused on learning weight-shifting skill in the present work. However, other elements are also involved in learning skiing, such as posture and edging techniques. Hence, we plan to support these elements using the proposed method.

References

1. Adira, M.F., Ahmad, K.A.: The impact of e-learning on university students' learning of sport skill subjects during the Covid-19 pandemic. In: EjSBS - Issue 3, pp. 201–216. European Publisher, London (2021). https://doi.org/10.15405/ejsbs.2021.08.issue-3
2. Atsumi, R., Kanai, H., Wu, E., Koike, H.: A feature extraction method for classifying beginner and expert skier on a ski simulatour using deep learning. In: ACM CHI2021 Workshop: Human Augmentation for Skill Acquisition and Skill Transfer (HAA2021) Collocated with The ACM CHI Conference on Human Factors in Computing Systems (2021)
3. Brock, H., Ohgi, Y.: Assessing motion style errors in ski jumping using inertial sensor devices. IEEE Sens. J. **17**(12), 3794–3804 (2017)
4. Cazzola, W., Olivares, D.M.: Gradually learning programming supported by a growable programming language. IEEE Trans. Emerg. Top. Comput. **4**(3), 404–415 (2016). https://doi.org/10.1109/TETC.2015.2446192

5. Chen, H.-T., He, Y.-Z., Hsu, C.-C.: Computer-assisted yoga training system. Multimed. Tools Appl. **77**(18), 23969–23991 (2018). https://doi.org/10.1007/s11042-018-5721-2

6. Fasel, B., Gilgien, M., Spörri, J., Aminian, K.: A new training assessment method for alpine ski racing: estimating center of mass trajectory by fusing inertial sensors with periodically available position anchor points. Front. Physiol. **9**, 1203 (2018). https://doi.org/10.3389/fphys.2018.01203

7. Fukahori, K., Sakamoto, D., Igarashi, T.: Exploring subtle foot plantar-based gestures with sock-placed pressure sensors. In: Proceedings of the 33rd Annual ACM Conference on Human Factors in Computing Systems, pp. 3019–3028 (2015)

8. Hasegawa, S., Ishijima, S., Kato, F., Mitake, H., Sato, M.: Realtime sonification of the center of gravity for skiing. In: Proceedings of the 3rd Augmented Human International Conference, pp. 1–4 (2012)

9. Moon, J., Koo, D., Kim, K., Shin, I., Kim, H., Kim, J.: Effect of ski simulator training on kinematic and muscle activation of the lower extremities. J. Phys. Ther. Sci. **27**(8), 2629–2632 (2015). https://doi.org/10.1589/jpts.27.2629

10. Noamani, A., Lemay, J.F., Musselman, K.E., Rouhani, H.: Characterization of standing balance after incomplete spinal cord injury: alteration in integration of sensory information in ambulatory individuals. Gait Posture **83**, 152–159 (2021). https://doi.org/10.1016/j.gaitpost.2020.10.027, http://www.sciencedirect.com/science/article/pii/S096663622030607X

11. Pignolo, L., et al.: A body-weight-supported visual feedback system for gait recovering in stroke patients: a randomized controlled study. Gait Posture **82**, 287–293 (2020). https://doi.org/10.1016/j.gaitpost.2020.09.020, http://www.sciencedirect.com/science/article/pii/S096663622030566X

12. Redmon, J., Farhadi, A.: YOLOv3: an incremental improvement. arXiv preprint arXiv:1804.02767 (2018)

13. Sato, T., Tokuyasu, T.: Pedaling skill training system with visual feedback of muscle activity pattern. J. Biomech. Sci. Eng. **12**(4), 17–00234 (2017)

14. Sayenko, D.G., et al.: Positive effect of balance training with visual feedback on standing balance abilities in people with incomplete spinal cord injury. Spinal Cord **48**(12), 886–893 (2010)

15. Selvaraju, R.R., Cogswell, M., Das, A., Vedantam, R., Parikh, D., Batra, D.: Grad-CAM: visual explanations from deep networks via gradient-based localization. In: Proceedings of the IEEE International Conference on Computer Vision, pp. 618–626 (2017)

16. Venkatesh, P., Rana, R., Palanthandalam-Madapusi, H.: Fast and real-time end to end control in autonomous racing cars through representation learning. CoRR abs/2111.15343 (2021). https://arxiv.org/abs/2111.15343

17. Wojtyczek, B., Pasławska, M., Raschner, C.: Changes in the balance performance of polish recreational skiers after seven days of alpine skiing. J. Hum. Kinet. **44**, 29–40 (2014). https://doi.org/10.2478/hukin-2014-0108

18. Wu, E., Perteneder, F., Koike, H., Nozawa, T.: How to VizSki: visualizing captured skier motion in a VR ski training simulator. In: The 17th International Conference on Virtual-Reality Continuum and Its Applications in Industry, VRCAI 2019. Association for Computing Machinery, New York (2019). https://doi.org/10.1145/3359997.3365698

19. Yuan, J.X., Yang, K.Y., Ma, J., Wang, Z.Z., Guo, Q.Y., Liu, F.: Step-by-step teaching method: improving learning outcomes of undergraduate dental students in layering techniques for direct composite resin restorations. BMC Med. Educ. **20**(1), 1–6 (2020)

ILoveEye: Eyeliner Makeup Guidance System with Eye Shape Features

Hange Wang, Haoran Xie[✉], and Kazunori Miyata

Japan Advanced Institute of Science and Technology, Nomi, Ishikawa 923-1292, Japan
xie@jaist.ac.jp

Abstract. Drawing eyeliner is not an easy task for whom lacks experience in eye makeup. Everyone has a unique pair of eyes, so they need to draw eyeliner in a style that suits their eyes. We proposed ILoveEye, an interactive system that supports eye-makeup novices to draw natural and suitable eyeliner. The proposed system analyzes the shape of the user's eyes and classifies the eye types from camera frame. The system can recommend the eyeliner style to the user based on the designed recommendation rules. Then, the system can generate the original patterns corresponding to the eyeliner style, and the user can draw the eyeliner while observing the real-time makeup guidance. The user evaluation experiments are conducted to verify that the proposed ILoveEye system can help some users to draw reasonable eyeliner based on the specific eye shapes and improve their eye makeup skills.

Keywords: Eyeliner makeup · Eye features analysis · Makeup system · Interactive guidance

1 Introduction

Makeup is a daily skill like other creative activities such as painting and sculpting, which requires the users to master skillful techniques of makeup and possess a high-level understanding of makeup based on facial features. Eye makeup plays an important role in the daily facial makeup. However, it is still a challenging issue for common users with few makeup experiences to achieve satisfactory outcome. Usually, the users often seek makeup tutorials from video sites or social medias, such as Youtube and TikTok. These video tutorials commonly do not match personal facial features, which have significant relationship with the strategies of eye makeup. In this work, we focus on the guidance system for eyeliner makeup which is the basic facial makeup in our daily lives.

About makeup guidance systems, it is a common approach to provide the corresponding makeup suggestions based on the user's facial features and show the anticipated makeup effect on the face in real time. We found that it is beneficial to provide visual recommendation makeup styles and visual makeup tutorials for whom has few knowledge and experience of makeup. This work aims to provide intuitive and effective eyeliner makeup guidance in real time.

© The Author(s), under exclusive license to Springer Nature Switzerland AG 2022
R. A. Sottilare and J. Schwarz (Eds.): HCII 2022, LNCS 13332, pp. 238–254, 2022.
https://doi.org/10.1007/978-3-031-05887-5_17

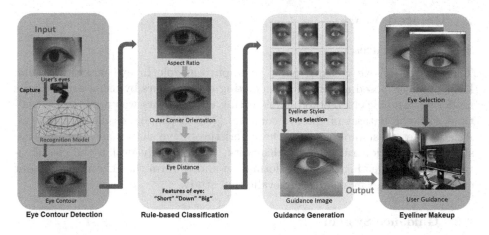

Fig. 1. The proposed makeup guidance framework of ILoveEye.

In this paper, we propose, ILoveEye, an interactive rule-based makeup system based on the analysis of user's eye shape features for less-experienced users as shown in Fig. 1. To obtain the feature points of eyes, we utilize state-of-the-art deep learning based facial recognition models in this work. To classify the features of eye shapes, we conduct eye shape classification and extract the features of the eye contour. After analyzed the typical eye shapes including almond, round, downturned, and close-set eyes, we propose a rule-based classification model to obtain the labels of the eye shape features with three feature values of the eyes: eye aspect ratio, outer corner orientation, and eye distance. For the makeup guidance, we investigate rules of typical eyeliner styles according to the relationship between eye shape and eyeliner styles. In terms of the eye feature points obtained from the proposed system, we adopt the classification rules to match current feature points of user's eye. The proposed system can reproduce the eyeliner styles and display in real-time as guidance to support the user to complete the eyeliner makeup.

We designed and conducted an evaluation experiment to verify the feasibility of the proposed ILoveEye system. In our user study, we invited participants to join the experiments, who were less-experienced in facial makeup with few skills of drawing eyeliner. To clarify the relationship between system evaluation and user's skill background, we designed a questionnaire which were conducted after experienced the proposed system. From the questionnaire results, we found that the proposed system could effectively classify the eye shape and support the user to draw eyeliner interactively. In addition, we analyzed the Pearson correlation coefficients between the user's makeup background and system evaluation. The analysis results indicated a significant and positive correlation between the frequency of wearing makeup and the skill improvement after using the system. We conclude that the users who normally wear makeup are acquired with experience and knowledge of makeup, thus they could draw better eyeliner after learning the exact contours of their eye shape features using the proposed system.

2   Related Work

2.1   Eye Features

The analysis of eye features plays an important role in eye makeup system. Bhat et al. proposed a method for detecting eye contours by using active shape model [4]. An efficient method for detecting eye contours was proposed in real-time with the dataset of the eye contours [7]. For shape classification of eyes, Alzahrani et al. summarized the characteristics of eye shapes and designed eye shape classification rules [3]. In this work, we originally design the classification rules by collecting relevant information and confirm the effectiveness of our classification rules by conducting evaluation study.

2.2   Guidance System

Along the development of augmented reality and image processing technologies, the guidance systems for supporting our daily activities have been explored extensively. The deep learning based human mesh modeling was used to compare the user's postures and target ones for core training [20]. To help take a good selfie, a voice guidance system was proposed with crowdsourcing evaluation of photo aesthetic [6]. The spatial augmented reality has been used for calligraphy practice [10], golf training [11] and dance support [9]. To meet the user's intentions in design activities, the user sketches were utilized for domino arrangement [16] and lunch box decoration [19]. In this work, we especially focus on the guidance system for eye makeup using augmented reality approach.

2.3   Makeup Guidance

With the development of facial recognition technologies, it becomes feasible to provide suitable makeup guidance based on facial features. Chiocchia et al. developed a mobile application to help users select cosmetics and showed users makeup tutorials [5]. Face features can be obtained for makeup image synthesis with different makeup references [8]. The face dataset was constructed to analyze the facial attractiveness for recommend system [12]. A rule-based system was proposed to automatically classify the faces and suggest the makeup approach [1]. iMake proposed a novel eye-makeup design system to extract the color and shape features [15]. A mirror system utilizing augmented reality (AR) was proposed to capture and analyze the facial features in real-time [17]. An interactive mobile application was proposed for makeup tutorials [2]. A makeup instruction with AR was proposed to provide special makeup tools [18]. In this work, we aim to provide the real-time guidance to support users for drawing better eyeliner and improving makeup skills.

3   System Overview

Figure 1 shows the overall workflow of the proposed ILoveEye system which consists of three main parts with deep learning based face detection. First, the proposed system detects eyes using the normal web camera and obtains the geometric information of the eye contours. Second, we design a rule-based model using geometric features to classify eye shapes. Finally, the proposed system generates the eyeliner makeup guidance based on the previously analyzed classification results displayed on the screen in real-time. The proposed ILoveEye system enables users to observe the guidance patterns on the screen while drawing suitable eyeliners.

3.1 Eye Detection

About eye shape detection, we adopted the open-source deep learning based face detector, MediaPipe [13], which contains a pre-trained face recognition model with high efficiency and accuracy. The detection output is a face mesh map as shown in Fig. 2 with 486 landmarks. From this face mesh, 16 feature points are selected to describe the eye contour. Through our practical testing, we found that this detector can effectively identify the shape of the eye contour with high accuracy to meet our research target.

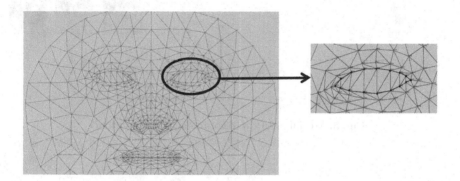

Fig. 2. Eye contour landmarks in human face mesh.

To implement the eye detection module, we first invoke the connected camera (60 fps in our prototype). To get the shape of the eyes accurately, the proposed system asks the user to save the mesh map of the current frame. The proposed system selects the landmarks of the eye parts from the saved mesh map. We found that the camera angle may affect the detection results, so we asked the user to sit at a proper distance and allow the user to adjust the camera angle until the satisfied eye contour is achieved.

3.2 System Workflow

After analyzed the eye features, the proposed system then determines the eyeliner styles in terms of eye aspect ratio, outer corner orientation, and eye distance as shown in Fig. 3. If the eye size is labeled with "small" or "average", the system recommends Style-Basic. If the aspect ratio is labeled with "big", the proposed system recommends Style-Basic with the lower eyeliner thickness. The system can analyze the orientation of the eye outer corner to recommend Style-Winged, Style-Drop and Style-Extend. The system can recommend lower eyeliner styles: Style-Inner and Style-Outer, depending on eye distances. Finally, the proposed system combines all recommended eyeliner styles to generate a polygon as a mask, which is visualized in orange color for makeup guidance (Fig. 3).

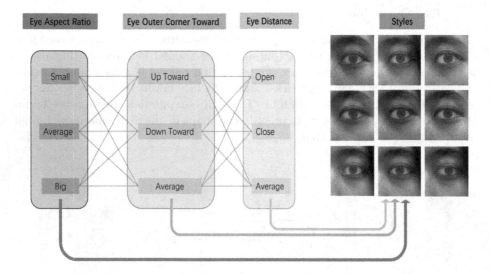

Fig. 3. Workflow of the recommendation function.

3.3   Eye Features Analysis

We reviewed different typical eye types, such as round eyes: larger and more circular; close-set eyes: less space between eyes; down-turned eyes: taper downward at the outer corner [3]. We found that the eye shapes are related to the size of the eye, the angle of the eye outer corner, and the distance between the two eyes. Therefore, we set these three features as the classification conditions for determining eye types.

Aspect Ratio. In the obtained eye contour landmarks (the right eye as an example), we define the leftmost and rightmost points $p0$ and $p8$ as the positions of the head and tail positions. We calculate the lengths of the perpendicular lines from $p1 - p7$, and $p9 - p15$ respectively, then find the maximum values of the upper part and lower part, and sum them as eye height (Fig. 4(a)). The length of $p0 - p8$ line is considered as eye width. The ratio of eye width and height is calculated as aspect ratio a. We obtained a relatively reasonable range for eye

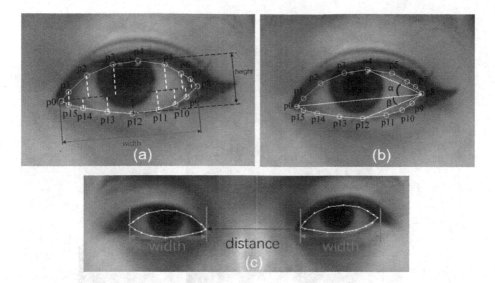

Fig. 4. Eye feature parameters: (a) width and height of eye; (b) outer corner orientation; (c) eye distance.

styles: $a \in [2.75, 3.00]$ for average eye, small long eyes for $a > 3.00$, big round eyes for $a < 2.75$.

Outer Corner Orientation. We found that the angle of the eye's outer corner can be used as a reliable reference, for determining up-turned or down-turned eyes. As shown in Fig. 4(b), $p4$ and $p12$ denote the highest and lowest points. Line $p0 - p8$ divides the outer corner angle into two angles. The upper angle α is defined as included angle of lines $p4 - p8$ and $p0 - p8$, and lower angle β with $p12 - p8$. If $\alpha > \beta$, the outer corner is determined to be down toward and the eye will be labeled as down-turned eye. On the contrary, the eye will be labeled as upturned eye.

Eye Distance. We measure the distance between point most left and right points as eye-distance D_e. The average distance of two eyes D_{avg} (Fig. 4(c)) is calculated. If $D_e > 1.05 \times D_{avg}$, the eyes are considered as open-set type and labeled as "open"; If $D_e < 0.95 \times D_{avg}$, the eyes are considered as close-set type and labeled as "close". Otherwise, the eyes are labeled as "average".

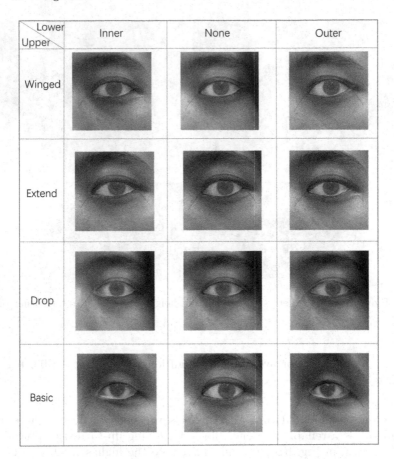

Fig. 5. Eyeliner styles defined in the proposed ILoveEye system.

4   Makeup Guidance

Eye makeup uses eyeliner to make eyes look bigger and modify the eye shape [14]. Generally, there are two typical eyeliner types: outer and inner eyeliners. However, the inner eyeliner is applied in eyelashes roots which is difficult to detect by the proposed detection method. Therefore, we focus on the outer eyeliner in this work. We defined the typical styles of eyeliners related to eye shape features as shown in Fig. 5.

4.1   Eyeliner Styles

Style-Basic. Style-Basic is a thick line along the upper eyelid that starts from the inner corner of the eye to the outer corner (Fig. 6(a)). In this work, a thick line is generated along with the shape of the upper eyelid. We calculate the mid-point for each line segment of the upper eyelid. Then, we set the perpendicular line of each line segments from mid-points to $p0', p1'...p7'$, and the length of each perpendicular line is defined as eyeliner thickness h. By connecting the points of the upper eyelid, the closed polygon is the basic upper eyeliner style as Style-Basic.

Style-Winged. For the upturned-winged eyeliner pattern, we define an outermost point E. As shown in Fig. 6(b), we measure typical samples of winged eyeliner. The angle of the eyeliner tail is about 15° for natural look, so we define the average angle between $E - p8$ and $p0 - p8$ as 15°. The length of the wing is set to be about 12% of the eye-width. The wing length is also suitable for other wing eyeliner styles. The coordinates of E point is calculated by trigonometric functions. By connecting the Style-Basic to the point E to form a closed polygon, we get the upturned wing eyeliner as Style-Winged.

Style-Drop. As shown in Fig. 6(c), we get a droopy eye-line style by defining a point E. $p8 - E$ is the extension of $p4 - p8$, and the length of $p8 - E$ is 12% of eye-width in Style-Winged. The closed polygon formed by connecting Style-Basic to point E is regarded as Style-Drop.

Style-Extend. As shown in Fig. 6(d), Style-Extend is the eyeliner style that the end of an eyeliner extends out horizontally. We first define the outermost point of the eye-liner as point E. Points $p0$, $p8$, and E are on the same horizontal line, and the length of $E - p8$ is 12% of eye-width. Finally, we connect Style-Basic with point E to get Style-Extend.

Basic Lower Eyeliner. The lower eyeliner is always thinner than the upper eyeliner and applied on corner sides. For lower eyeliner in outer corner side (Fig. 6(e)), we connect the wing of upper eyeliner, and select $p8$, $p9$, $p10$, $p11$ to create a lower eyeliner style on the inner corner side. We decrease the value of h from large to small in equal proportion to make the shape appearance from big to small. We connect the upper eyeliner and select $p13$, $p14$, $p15$, $p0$ to create a lower eyeliner style on the inner corner side. We found that the value of h as $h/3$ of the outer side can make this pattern look thinner than the outer side. We defined these two lower eyeliner styles as "Style-Inner" and "Style-Outer".

4.2   Merging Types

For the lower eyeliner styles, the system selects the recommended style and merges it with upper eyeliner styles. In the case of merging Style-Outer with Style-Winged, the system connects point E to lower $p8'$, and gets the closed polygon as the result for guidance as shown in Fig. 7(a).

In the case of merging Style-Inner and Style-Basic (Fig. 7(b)), the system first calculates a fitting curve by $p0 - p8$ and $p0' - p7'$, then finds the cross-point E with line $p0 - p8$. We obtain a closed polygon by connecting the point E, Style-Basic and Style-Inner as shown in Fig. 7(b).

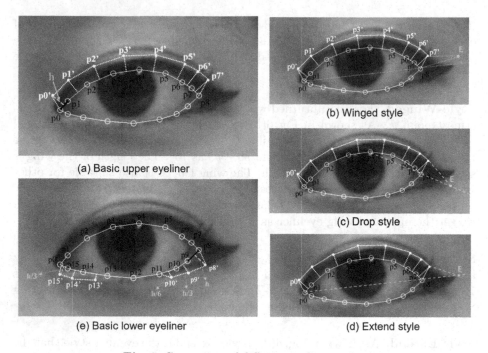

(a) Basic upper eyeliner

(b) Winged style

(c) Drop style

(d) Extend style

(e) Basic lower eyeliner

Fig. 6. Generation of different eyeliner styles.

4.3 User Interface

Figure 8 shows the user interface of the proposed ILoveEye system. The proposed system creates new windows of the left and right eyes separately and enlarges them to make the user observe the guidance and their eyes clearly. For the recognized eye contour, the system can display it that the user can find how well their eyes are recognized.

The user can select one window to save the feature information of the current frame through keyboard input. Because the feature information of both eyes is saved at the same time, the recognized eye shape may be in error with the real shape due to the shooting angle and lens distortion of the camera. Therefore, we asked the user save the eye features at the correct angle while looking straight to the camera.

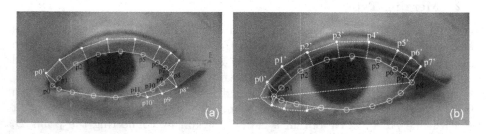

Fig. 7. Merging types of eyeliner styles.

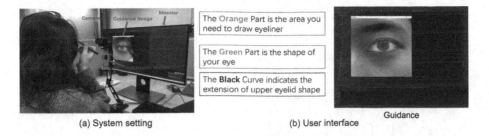

(a) System setting   (b) User interface

Fig. 8. Experiment environment and user interface.

5   System Evaluation

The prototype system was implemented on a desktop computer with Windows 10, Intel i7-9700 CPU, GeForce RTX 2060 SUPER GPU. The proposed system was developed in Python with OpenCV ver4.5.2, MediaPipe ver0.8.6, and we set up Logitech HD Pro camera C920 on the table to capture user's face.

5.1   Experiment Procedure

We conduct a comparison study that participants are asked to draw eye makeup with and without the proposed system. After the eye-makeup is applied, we provided makeup remover for the participants to remove their makeup and start to draw makeup using the proposed system. We first introduced the system workflow and asked users to draw an eyeliner by referring to the makeup guidance in orange color. All participants were asked to look at the camera, the monitor can display the participant's face and independent images of the two eyes. The participants can adjust the angle of the camera until the satisfied eye recognition results were achieved. The analyzed eye features results were prompted to the participant in text. We provided participants with two eyeliners (one gel eyeliner and one liquid eyeliner) to draw eyeliner. When the participant completed the eye makeup, we counted the time cost and archived makeup images.

5.2   Evaluation Method

In the user study, we adopted both subjective evaluation and objective evaluation to verify the proposed ILoveEye system. Regarding to the subjective evaluation, we designed a questionnaire to confirm the user experience and the system usefulness. After collected the basic information of the participants (gender and age), we first investigated the users' knowledge about eye makeup. The question items of the questionnaire are listed as below.

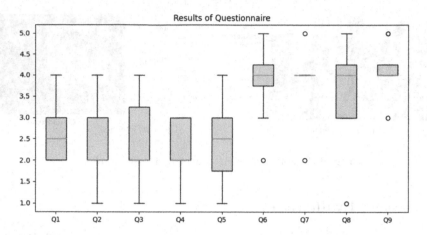

Fig. 9. Results of questionnaire in our user study.

- Q1: Do you wear makeup?
- Q2: Do you watch makeup tutorials (videos, magazines, blogs)?
- Q3: How do you evaluate the level of your makeup skill?
- Q4: How do you evaluate the level of your eye-makeup skill?
- Q5: Do you draw eye-makeup by considering your eye shape?
- Q6: Do you consider this system describe your eye shape correctly?
- Q7: Do you consider this system support you to draw eyeliner?
- Q8: Do you consider this system help you to draw better eyeliner than by yourself
- Q9: Do you think this system improve your drawing eyeliner skill?

All these questions were evaluated with 5-point Likert scale where 1 denotes strongly disagree and 5 for strongly agree. Finally, we collected the free feedback from the participants for any advice or improvement. We invited 8 female graduate students, around 25-year-old. All participants were asked to join without eye makeup, and the few experience of eye makeup was required.

6   Results

6.1   Subjective Evaluation

The results of the questionnaire are shown in Fig. 9. We found that most participants did not wear makeup very often (Q1: average = 2.63) and were not confident in their makeup skills (Q3: average = 2.50; Q4: average = 2.25). The average scores of questions from Q6 to Q9 were Q6: 3.88; Q7: 3.88; Q8: 3.63; Q9: 4.13. We found that the overall score of our system was above 3.0, which means our system can help users to draw better eye makeup and improve their makeup skills. We noticed that one participant rated lower scores on Q6, Q7, and Q8, and confirmed that it was difficult to satisfy experienced users well with

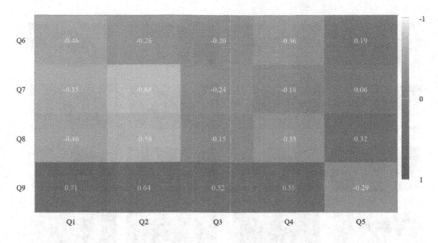

Fig. 10. Pearson coefficients for Q1–Q5 and Q6–Q9.

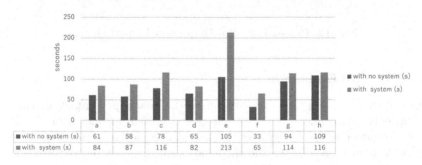

	a	b	c	d	e	f	g	h
■ with no system (s)	61	58	78	65	105	33	94	109
■ with system (s)	84	87	116	82	213	65	114	116

Fig. 11. Comparison of completion time in our user study.

our makeup guidance, who already had a well-tested routine for drawing eye makeup.

We analyzed the correlation between participants' background of makeup knowledge and evaluation of the system by the Pearson correlation coefficients for Q1–Q5 and Q6–Q9 as shown in Fig. 10. We found a significant and positive correlation between Q1 and Q9 (p = 0.71). This means that the participants who wear makeup frequently can feel the skill improvement with the proposed system. We observed that participants who normally wear makeup are acquired with some experience and knowledge of makeup. When they learn the exact contours of their eye features, it can help them to draw eyeliner afterward. In addition, one participant reported that a better experience for improving makeup skills than watching tutorial videos was obtained because the system could describe and show the eye features.

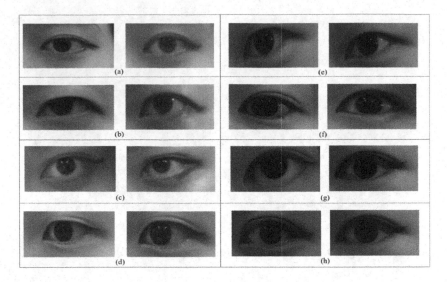

Fig. 12. Comparison of makeup effect by participants.

6.2   Objective Evaluation

In the objective evaluation, we compared the makeup completion time and makeup effect with and without the proposed system as shown in Fig. 11. We found that after using our system, all users' completion time were increased. The users spent more time to find the location of the guidance pattern and consider how to reproduce the guidance pattern while using the system. In addition, we analyzed the correlation between completion time, and the questions in the questionnaire. We found a significant negative correlation between Q2 and completion time without system ($p = -0.723$), which means that the less frequently people watched makeup tutorials, the more time they spent on completing their eyeliner.

From the eye makeup results in both conditions as shown in Fig. 12, we found that the eye makeup which refers to the makeup guidance was generally thicker and more clearly than those without using the system. It is verified that the proposed makeup guidance is easy to understand and could be reproduced by the users. The detailed comparison results are shown in Fig. 13. Observing the comparison results of winged parts, we found that the eyeliners drawn by the participants with the proposed system look thicker and much clearer. After reviewed the evaluation of system interactivity, it is verified that the makeup guidance provided by the proposed system was easy to understand and could be reproduced by users. Basically, these thicker and much clear eyeliners make users' eyes became visually bigger.

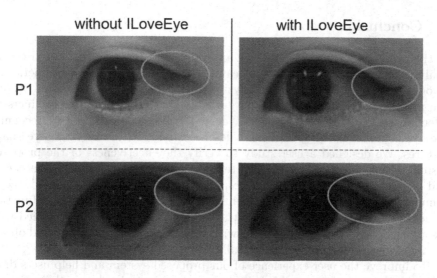

Fig. 13. The difference of eyeliner's wing (Left without-system; Right: with-ILoveEye).

7   Discussion

From the interviews with participants and free comments in questionnaires, we found that the proposed system can effectively help users with less makeup experience to draw a better eyeliner that suits the feature of eye shape. Besides, from the collected participants feedback, we found that the proposed system could help users understand their eye features and improve their eye-makeup skills. However, most recommended eyeliner styles to the participants were Style-Wing, Style-Extend and Style-Outer, which means that they have similar eye shapes. All participants in this study were Chinese and Japanese, it would be better to verify the proposed system for more nationalities. For the eye recognition, it was difficult to achieve accurate recognition results due to camera angle, which was adjust the camera angle manually. Although the proposed system can capture the eye contour accurately, the detection of eyelids was very limited which means that the system may have difficulty for users when eyelids heavily affect the visual effect of the eyes.

In our user study, the users had to keep a distance from the monitor so that the nearsighted users may have difficulty to see the exact location following the guidance. The frame rates were sometimes low for the smoothness of observing the guidance. One participant reported that it would be helpful if the system could provide more recommended solutions for users to choose. For example, a scene choice function can be added for different usages such as the party, business and shopping. It would be promising to provide a face preview with the applied makeup styles.

8   Conclusion

In this work, we proposed ILoveEye, an eyeliner guidance system based on the analysis of eye shape. This system captures the user's eyes by a camera and recognizes the eye contours for analyzing the eye shape. The proposed system can provide a recommended eyeliner guidance that users can draw eyeliners by referring to guidance. We utilized a deep learning based model to recognize the eye contour. We proposed the rule-based model to classify the eye shape features. We designed a user study to verify the effectiveness of the proposed system. From the user study, our system can effectively identify the user's eyes and recognize the contours of the eyes, and the user can follow the guidance to complete the eye makeup. The system provided an effective guidance function to help users draw better eye makeup. We found that users were satisfied with recommended guidance and the user with more makeup experience could obtain the improvement for their makeup skills.

To improve the user experience of our proposed system and help users draw eyeliner, there are some possible future works. In this study, the facial recognition model could not effectively identify the eyelids when the system recognized the user's eye contour, which may decrease the classification accuracy of the user's actual eye shape. To solve this issue, we plan to build a novel dataset containing images of eyes with labeled eyelid shapes and train a supervised learning model to optimize the effect of eye shape recognition. Furthermore, some participants reported that our proposed system provided few eyeliner styles, which may not be suitable for rare eye shapes, it is worth to designing more eyeliner styles as makeup guidance in the future to meet the user diversity. We think that the proposed ILoveEye system can be extended to improve other makeup skills such as eyebrows makeup.

Acknowledgement. The authors thank the anonymous reviewers for their valuable comments. This project has been partially funded by JAIST Research Grant and JSPS KAKENHI grant JP20K19845, Japan.

References

1. Alashkar, T., Jiang, S., Fu, Y.: Rule-based facial makeup recommendation system. In: 2017 12th IEEE International Conference on Automatic Face Gesture Recognition (FG 2017), pp. 325–330 (2017). https://doi.org/10.1109/FG.2017.47
2. Almeida, D.R.O.D., Guedes, P.A., Silva, M.M.O.D., Buarque Vieira e Silva, A.L., Lima, J.P.S.D.M., Teichrieb, V.: Interactive makeup tutorial using face tracking and augmented reality on mobile devices. In: 2015 XVII Symposium on Virtual and Augmented Reality, pp. 220–226 (2015). https://doi.org/10.1109/SVR.2015.39
3. Alzahrani, T., Al-Nuaimy, W., Al-Bander, B.: Integrated multi-model face shape and eye attributes identification for hair style and eyelashes recommendation. Computation **9**(5), 54 (2021). https://doi.org/10.3390/computation9050054, https://www.mdpi.com/2079-3197/9/5/54

4. Bhat, S., Savvides, M.: Evaluating active shape models for eye-shape classification. In: 2008 IEEE International Conference on Acoustics, Speech and Signal Processing, pp. 5228–5231 (2008). https://doi.org/10.1109/ICASSP.2008.4518838

5. Chiocchia, I., Rau, P.-L.P.: Facial feature recognition system development for enhancing customer experience in cosmetics. In: Rau, P.-L.P. (ed.) HCII 2021. LNCS, vol. 12771, pp. 27–40. Springer, Cham (2021). https://doi.org/10.1007/978-3-030-77074-7_3

6. Fang, N., Xie, H., Igarashi, T.: Selfie guidance system in good head postures. In: ACM International Conference on Intelligent User Interfaces Workshops (2018)

7. Fuhl, W., Santini, T., Kasneci, E.: Fast and robust eyelid outline and aperture detection in real-world scenarios. In: 2017 IEEE Winter Conference on Applications of Computer Vision (WACV), pp. 1089–1097 (2017). https://doi.org/10.1109/WACV.2017.126

8. Guo, D., Sim, T.: Digital face makeup by example. In: 2009 IEEE Conference on Computer Vision and Pattern Recognition, pp. 73–79 (2009). https://doi.org/10.1109/CVPR.2009.5206833

9. He, Y., et al.: Interactive dance support system using spatial augmented reality. In: 2021 Nicograph International (NicoInt), pp. 27–33 (2021). https://doi.org/10.1109/NICOINT52941.2021.00013

10. He, Z., Xie, H., Miyata, K.: Interactive projection system for calligraphy practice. In: 2020 Nicograph International (NicoInt), pp. 55–61 (2020). https://doi.org/10.1109/NicoInt50878.2020.00018

11. Ikeda, A., Tanaka, Y., Hwang, D.H., Kon, H., Koike, H.: Golf training system using sonification and virtual shadow. In: ACM SIGGRAPH 2019 Emerging Technologies. SIGGRAPH 2019. Association for Computing Machinery, New York (2019). https://doi.org/10.1145/3305367.3327993

12. Liu, L., Xu, H., Xing, J., Liu, S., Zhou, X., Yan, S.: Wow! you are so beautiful today!. In: Proceedings of the 21st ACM International Conference on Multimedia, MM 2013, pp. 3–12. Association for Computing Machinery, New York (2013). https://doi.org/10.1145/2502081.2502126

13. Lugaresi, C., et al.: MediaPipe: a framework for building perception pipelines. CoRR abs/1906.08172 (2019). http://arxiv.org/abs/1906.08172

14. Matsushita, S., Morikawa, K., Yamanami, H.: Measurement of eye size illusion caused by eyeliner, mascara, and eye shadow. J. Cosmet. Sci. **66**(3), 161–74 (2015)

15. Nishimura, A., Siio, I.: iMake: eye makeup design generator. In: Proceedings of the 11th Conference on Advances in Computer Entertainment Technology, ACE 2014. Association for Computing Machinery, New York (2014). https://doi.org/10.1145/2663806.2663823

16. Peng, Y., et al.: Sketch2Domino: interactive chain reaction design and guidance. In: 2020 Nicograph International (NicoInt), pp. 32–38 (2020). https://doi.org/10.1109/NicoInt50878.2020.00013

17. Rahman, A.S.M.M., Tran, T.T., Hossain, S.A., Saddik, A.E.: Augmented rendering of makeup features in a smart interactive mirror system for decision support in cosmetic products selection. In: 2010 IEEE/ACM 14th International Symposium on Distributed Simulation and Real Time Applications, pp. 203–206 (2010). https://doi.org/10.1109/DS-RT.2010.30

18. Treepong, B., Mitake, H., Hasegawa, S.: Makeup creativity enhancement with an augmented reality face makeup system. Comput. Entertain. **16**(4), 1–17 (2018). https://doi.org/10.1145/3277452

19. Xie, H., Peng, Y., Wang, H., Miyata, K.: SketchMeHow: interactive projection guided task instruction with user sketches. In: Stephanidis, C., et al. (eds.) HCII 2021. LNCS, vol. 13096, pp. 513–527. Springer, Cham (2021). https://doi.org/10.1007/978-3-030-90328-2_35
20. Xie, H., Watatani, A., Miyata, K.: Visual feedback for core training with 3D human shape and pose. In: 2019 Nicograph International (NicoInt), pp. 49–56 (2019). https://doi.org/10.1109/NICOInt.2019.00017

Evaluating the Effectiveness of Adaptive Instructional Systems

Lessons Learned from Creating, Implementing and Evaluating Assisted E-Learning Incorporating Adaptivity, Recommendations and Learning Analytics

Daniela Altun[1](✉), Christopher Krauss[2], Alexander Streicher[3], Christoph Mueller[2], Daniel Atorf[3], Lisa Rerhaye[1], and Dietmar Kunde[4]

[1] Fraunhofer FKIE, Bonn, Germany
Daniela.Altun@fkie.fraunhofer.de
[2] Fraunhofer FOKUS, Berlin, Germany
[3] Fraunhofer IOSB, Karlsruhe, Germany
[4] German Army Headquarters, Strausberg, Germany

Abstract. Applications of adaptive e-learning, recommender systems and learning analytics are typically presented individually, however, their combination poses several challenging requirements ranging from organizational to technical issues. This article presents a technical study from a holistic application of a variety of e-learning assistance technologies, including recommender systems, chatbots, adaptivity, and learning analytics. At its core we operationalize interoperability standards such as the Experience API (xAPI) and Learning Tools Interoperability (LTI), and control the data flow via a standard-encapsulating middleware approach. We report on the challenges regarding organization, methodology, content, didactics, and technology. A systematic evaluation with the target group discusses the users' expectations with the measured interactions.

Keywords: Adaptive instructional systems · Learning management system · Artificial intelligence · Adaptive learning · Adult education · Evaluation

1 Introduction

New learning approaches and the rapid technological development of the last years offer exciting opportunities for education. The digitalization provides access to these educational technologies as Learning Management Systems (LMS) for a growing number of people [1] and the Covid-19 pandemic fuels the need for effective and assisted e-learning applications even further. These factors contribute to the recent increase of LMS usage and suggest a high demand in the future. Conventional LMS are software tools that help manage the entire education process, including preparing, conducting and post-processing classes [2]. As such, they offer an entry point for learners and instructors where learning material can be stored, edited, and processed at any time. Some LMS include additional functionalities, like platforms for communicating between peers and instructors or social media components. While these conventional LMS offer a good basis for assisted e-learning, adding Artificial Intelligence (AI) functionalities such as personalized and adaptive assistance can bring the whole education process to the next level.

The influence of AI on all our lives in varying ways is believed to grow continually [3] and already we observe an increasing empowerment in education through AI [4, 5]. We agree with Chaudhry and Kazim (2021) when we define AI for the context of this paper as a computer system that can achieve a particular task through certain capabilities and intelligent behaviour that was once considered unique to humans [3]. The continuously growing demand for adaptive and personalized education [4] cannot realistically be fulfilled without AI support. Integrating AI supported functionalities into LMS can make their usage even more attractive by offering a broad variety of benefits for all kinds of users, which are involved in the education process, including learners, instructors or training managers.

Prior to implementing assisted learning functions, a couple of requirements on organizational, methodological, didactical, content-related, and technical levels need to be fulfilled. Here, AI supported e-learning functions involve the processing of usage observations to optimize learning and teaching behavior as well as e-learning content. The analysis can be purely machine-driven, for example, the utilization of data for adaptivity (e.g., for educational recommender systems), generation of learning paths or the automatic and dynamic difficulty adjustment in exercises. For this purpose, Learning Analytics (LA) data is commonly used. LA is the collection, aggregation and analysis of data generated by learners, usually generated in specific environment, such as an LMS. The learners and teachers themselves can interpret the data, typically by inspecting aggregated data visualizations in learning analytics dashboards, which, for instance, visualize which tasks have been solved, or for how long the individual tasks took to complete. Processed data can also indicate learning progress, weaknesses or learning needs. Assistive AI functionalities depend on observation data as a necessary basis for their algorithmic decisions. Processed information from LA, e.g., about a learner's progress over multiple courses in an LMS, can be used to further improve AI methods. Additionally, AI approaches from other domains, for instance chatbots as virtual learning assistants may also lead to increased user satisfaction and, thus, to a higher user motivation for e-learning.

The primary research question of this article is (RQ1): which requirements need to be met in order to successfully implement AI supported functions in e-learning environments? When implementing different types of e-learning and assistance systems for different course environments, we often face similar challenges. Our analysis considers the experience from various sources of implementations in professional trainings and formal classes with typically 8 to 15 participants. While the application of such functions in small settings is challenging – e.g., because data is only sparsely available – it represents a large proportion of real-world scenarios for which we deploy assistive functions. Although we assume that our experiences are applicable to several other learning settings, we rely on the experiences of smaller courses.

We present a systematic requirements analysis as a guide for the initial steps when implementing an AI supported LMS for small course environments. Additionally, we report on the specific user requirements from a study, for which we designed a demonstrator software for a small-scale course in higher education. In this study, we accompanied five identical course runs on mathematical topics, each lasting approximately four weeks. We used the first course to gather requirements. During the fourth course, we were able to field test our demonstrator over a period of one week with some of the AI functionalities for the first time and use the feedback to adjust our demonstrator, e.g., control of feedback frequency or data collection granularity. The fifth course used all four assistive functionalities and evaluated the demonstrator in conclusion.

We were interested in what specific AI-enabled features would best support our users and which are generally accepted. This led to the second exploratory research question (RQ2): What AI supported functionalities do our users require? To answer the second research question, we went through a requirements engineering process, as defined by Nuseibeh and Easterbrook (2000) [6]. Based on the user requirements, we developed a software demonstrator, which offers a unique interplay of AI functionalities by using a middleware as a communication mediator and a web-based portal app as entry point for the users. This adds up to an innovative system approach. For the evaluation we mainly focused on the possible benefits of the AI assistance. This led to the third research question (RQ3): Do the implemented AI supported functionalities offer a benefit for the users?

The remainder of this paper is structured as follows. Chapter two discusses the set of core requirements which need to be fulfilled when implementing adaptive functionalities in digital learning environments. In the following, the key AI functions are presented that we have offered to our learners, followed by some general usage statistics and the results of a qualitative evaluation. The paper concludes with our lessons learned and an outlook.

2 Requirements

The first research question of this article concerns how to implement AI supported functionalities for heterogeneous e-learning system landscapes and what the challenges are. In our application context we observed various non-obvious challenges while implementing different types of e-learning and assistance systems. The systems include Learning Management Systems (LMS, e.g., Moodle) and plugins, web portals, adaptive serious games, dashboards, recommender systems and chatbots.

We first report on general requirements that should be met in order to implement AI-functionalities successfully. Then we describe the requirement engineering process tailored to the study that focuses on our users' requirements.

2.1 Organizational Requirements

The introduction of AI supported LMS requires the involvement of different stakeholders. Above all, the responsible organizations must ensure that the application of LMS is in accordance with applicable laws (in the EU, e.g., GDPR, data processing contracts, etc.), all regulatory requirements are met (e.g., naming of responsible persons, such as data owners, anonymization, etc.) and that IT and data security concepts are state-of-the-art. In this context, Flanagan and Ogata (2017) discussed the increasing need for data and privacy protection throughout the entire Learning Analytics (LA) workflow [7]. They find that the privacy of key stakeholders, such as learners, teachers and administrators need to be protected, while still maintaining the usefulness of user interaction data. Renz and Meinel (2018) addressed the requirement to use pseudonymization for the GDPR-compliant collection of xAPI learning records and argue for the use of an appropriate middleware [8]. Moreover, the introduction of LA components needs to be adequately supported. The organization must ensure that the role holders have sufficient resources, even after the initial launch of LA [9]:

- Operators and engineers are responsible for deploying the services, monitoring the components' technical operability, and immediate response if something does not run as expected.
- Data analysts ensure the fulfillment of the objective of LA components through regular evaluations and adjust data and algorithms if necessary.
- Domain experts keep data, content and media updated according to the individual context, evaluate high-level usage as well as the need for optimizations and prepare new content.
- Supporters for learning analytics users introduce the LA functions, motivate its usage, answer questions, help with the use of the system and explain how it works.

Depending on the size of the learning setting and number of people involved, the roles can be taken over by people that are already involved. For instance, instructors in smaller courses communicate with their learners on a regular basis. They are ideal supporters, and, in most cases, can also act as domain experts for specific course topics. Technical staff, such as data analysts as well as operators, can take care of multiple LA instances at the same time. However, it is very important to train the instructors and raise awareness for any LA particularities beforehand.

2.2 Methodological Requirements

The "appropriateness" of LA functionalities is of essential interest for the implementing institutions. However, what needs to be realized and how can it be evaluated? Most LA applications in learning environments aim to optimize learning by making it more efficient and more effective through data analysis. In the context of learning recommender systems, for instance, 'efficiency' describes the way to achieve a personal goal. In a small-scale course setting, a higher efficiency can optimize the process, save efforts and time to reach the course goal. 'Effectiveness', in turn, can directly affect the results achieved, e.g., a better mark in the exam or longer lasting knowledge [10]. Thus, the actual task of an LA function is of essential importance for design choices, development of an appropriate methodology and selecting an optimal evaluation framework [11]. There are numerous approaches and attempts to measure the intelligence of a learning system. Rerhaye et al. (2021) propose to conduct a combination of qualitative and quantitative evaluation methods [12]. This is needed to not only gain deep qualitative insights in the usage of LA functions, but also support the findings with reliable quantifiable values.

The user interface, satisfaction and the user experience are also of enormous importance for an LA supported system to ensure user acceptance of the system.

2.3 Didactical Requirements

In our experience, one of the users' biggest fears about the implementation of an AI supported LMS is that in-person classroom teaching could be replaced by just online learning. An ideal implementation concept, however, should only consider AI supported LMS for recurring and well-specified tasks, which would normally involve several members of the institutional staff. When organizers know how humans accomplish a certain task, such as analyzing learning groups at the beginning of a course or recommending appropriate learning materials, they can then consider having an LA component take on that task. We strongly believe that following a didactical concept instead of blindly replacing all classroom sessions would not only improve the learners' and teachers' overall acceptance of a system, but also result in better learning outcomes. The next step in a didactical concept would be to decide what to do with the information that LA provides. How can we use the results from a learning group analysis in the most beneficial way? How can we improve user motivation, push the self-responsibility in learning, help with useful reflections and keep the learner in an active role? What degree of freedom in an individual learning path can increase efficiency? As many LA applications are aimed at automating didactic activities [3], e.g., selection of learning material, it is necessary to decide on a robust didactic concept as a foundation [3]. As such, the didactic concept should be evaluated as thoroughly as the analytics' functionalities themselves.

2.4 Content Requirements

When well-structured data on content and usage is available, learning analytics can offer added value for various users of learning systems. However, it is not sufficient to only describe the content. Learning Analytics (LA) is based on digital data and hence, content must be available in a digital and compelling format. For example, even if several learning records are collected, the best LA approaches are of little use if the content is not annotated, e.g., PDF documents. To support LA functionalities in a meaningful way, learning content must be digitally edited and enriched with metadata. Digitally edited content means that it is machine-readable and that user interactions can be tracked. Ideally, the learning content should be organized into learning units, which can be linked together, e.g., combination of single multiple-choice tasks into a quiz. The minimum metadata of interest contains the users' activities and their achieved results. For LA, additional metadata should be provided, e.g., targeted learning time, difficulty level or knowledge type. For interoperability purposes, we encourage the use of standards, such as IEEE Learning Object Metadata (LOM) or IMS Common Cartridge. The needed metadata for LA functionalities depends on the respective application's purpose and overall didactic concept. From our experience, the implementation and maintenance of metadata standards for one's own content involves a great deal of effort, which, in the best-case scenario, is automated or already realized during the creation of individual content.

2.5 Technical Requirements

Finally, a successful implementation and integration of learning analytics into corporate learning environments – especially in environments of multi-institutions with distributed services - requires the use of widely accepted interoperability standards [9]. To address typical technical challenges such as IT security and network limitations (e.g., CORS) while still adhering to given data-privacy and data-protection regulations, we identified and implemented multiple core technologies and protocols which adhere to established specifications. Notable standards include the learning record specifications Experience API (xAPI) and CALIPER, which can be persisted in distributed Learning Record Stores [13] or user-controlled Data Wallets [14], LTI (Learning Tools Interoperability) or cmi5 (computer managed instruction, 5th attempt) launch specifications, as well as standards for the exchange of content metadata, such as Common Cartridge or LOM. However, not every service in a complex educational ecosystem follows the same standard, and many direct links between individual adaptive services are difficult to maintain. Therefore, we recommend a middleware architecture for service orchestration [15]. This middleware can either be standards-agnostic and allow communicating services to agree on a particular form of communication, or act as a standards-translator, e.g., between LTI and cmi5 [16].

A specific challenge arises when decentralized storage or replication of learning record data becomes necessary. We observed a typical corporate requirement: the operation of multiple as well as decentralized xAPI Learning Record Stores (LRS) instances. Each subsidiary can have its own data handling constraints, resulting in the need for individual stores. This motivates LRS replication strategies, control of the data flow, and operating dashboards under customer sovereignty.

3 Requirements Engineering Tailored to Our Users Needs

While multiple studies showed that users can benefit from AI supported LMS [17, 18], the research question remains: What specific functionalities in an LMS do our users require? Thus, before we decided on functionalities for a learning management system, we gathered the requirements of different stakeholders. The focus lies on the learners as they are the main users of the learning management system (LMS). Among others, we included teachers, authors that create the content for the LMS, and superiors that deal with education on an organizational level. Due to the pandemic, some requirements methods, like observations had to be excluded. Instead, we raised the requirements in online workshops with one stakeholder group at a time. We prepared discussion rooms where users could work alone or in groups of two people or as a group, depending on the topics and worksheets. The size of 3 to 5 participants for each workshop worked well for us, giving every individual the opportunity to engage and yet leaving enough room for discussion and brainstorming. We derived 89 user stories, which we converted into technical requirements. In consultation with stakeholders and software engineers, we prioritized the requirements and decided which AI functionalities could best support the user's needs. Under the strict observance of the mentioned requirements, such as ethical and regulatory requirements as well as standards and norms, we focused on four AI supported functionalities for the key users learners and teachers.

4 AI Supported Functionalities for Learners and Teachers

During the requirements engineering process users and stakeholders asked for support in varying ways. Some requirements relate to functionalities that are not related to AI and were therefore not within the scope of this study. The remaining requirements were sorted, validated and prioritized with stakeholders and software engineers. To summarize, most of the requirements were related to learner support, e.g., adaptivity, exercises with feedback, learning recommendations, gamification elements, display of own learning deficits and strengths and around the clock support for questions. The second most frequently mentioned requirements concerned the instructors. The system should provide overviews of the students' learning process, e.g., the presentation of the current knowledge and learning status as well as the most frequent errors during the course. Based on these requirements, we decided to implement four core functionalities, that address several of these requirements: a learning recommendation system, a chatbot, adaptive tasks and a learning analytics dashboard for instructors. Figure 1 illustrates the general system design with the various services interconnected by a middleware .

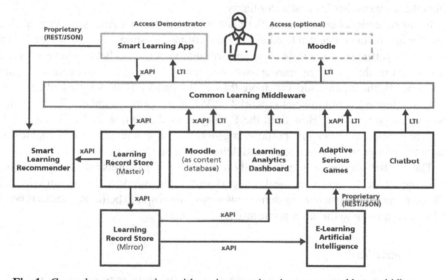

Fig. 1. General system overview with various services interconnected by a middleware.

4.1 Learning Recommender System

The learning recommender system provides learning recommendations, based on observed interactions, self-assessments, and practice successes [10]. Learning recommendations can refer to learning units which have not been worked on, or which indicate an increased learning need. The system can collect and use the following data:

- When and how often a learner opens a specific learning content.
- How long a learner has left a specific learning content page open.

- How learners assessed their knowledge about the content.
- How well the learners perform at learning exercises or assessments.
- Assessing the learner's prior knowledge of a learning content, e.g., by assessing how many underlying learning units have been completed.
- Whether the learning content is relevant for an upcoming face-to-face event.
- How long ago the learners learned the content and account for forgetting over time.
- How well other learners from the same course learned the content.

Learning records are collected during the whole learning process for each student individually and stored in a pseudonymized way. This data is not only the basis for the learning recommender system, but it offers an overview of the student's study progress and learning needs for instructors.

4.2 Chatbot

A chatbot system can help to answer frequently asked questions that have been implemented into the system. The chatbot recognizes the users' question and its intent and offers the answers that fits best accordingly.

In our educational environment, the chatbot lowers system barriers as a central focal point for answering content-related and organizational questions. This offers the benefit for users of getting answers right away and around the clock and helps relieve workload for instructors, that no longer must answer frequently asked questions again and again.

Currently, the chatbot supports more than 250 topics on frequently asked questions such as "How does multiplication work?", "What are prime numbers?", "How do I dissolve parentheses?", "How does the Pythagorean theorem work?", "What are the exam requirements?" For this purpose, a glossary was connected, which enables the chatbot to recognize common terms and define them on request.

The chatbot is optimized by manually reviewing the questions asked, e.g., after the end of the course. On this basis, the developers and editors extended the system to include unknown answers or teach the system to automatically respond better to questions posed and to assign them to the appropriate answer.

4.3 Adaptive Tasks

The adaptive and gamified tasks are primarily for training purposes, e.g., repetition according to the Spaced Repetition Method [19]. In our course setting they are a facultative element because of their adaptive nature. In a mandatory setting, where students would have to complete a given set of tasks in a given sequence, any adaptation would be obsolete. To enable quick, casual and successful learning, suitable task types and motivational incentives are needed. For learning objectives in the domain of natural sciences and general knowledge, classic task types such as multiple choice, hot spot and free text are suitable. Classic gamification methods such as leaderboards, achievements and playing against each other in quiz duels form additional motivational incentives according to the ideas of immersive didactics [20].

For the study, the software has been extended by an adaptivity framework [21] and suitable content has been prepared. The latter contains tasks from mathematics, notably equations. The AI component adapts the exercise tasks, i.e., within a quiz the difficulty of the tasks is individually adjusted to the user. Conceptually, the adaptivity approach follows the 4-phase adaptivity cycle [22]. The adaptivity framework works in the analysis phase and in the phase of generation of user models. The output of the framework transitions to the selection phase and includes a new difficulty level, normalized between 0–100% [23]. The computations are personalized for individual users because the adaptivity framework primarily uses user-specific xAPI tracking data. The game itself produces a performance score by a linear weighted sum of correctness score, completion time and base difficulty category. The adaptivity framework retrieves the most recent xAPI statements (with performance score results) and computes new difficulty levels based on a windowed harmonic sum approach. The effect is a typical dynamic difficulty adjustment or so-called rubber-banding where the difficulty level depends on the users' performance (Fig. 2).

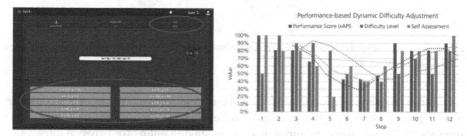

Fig. 2. Dynamic difficulty adjustment for a quiz game; (left) number and type of multiple-choice questions are adjusted as well as the time budget; (right) visualization of the various metrics.

4.4 Learning Analytics Dashboard for Instructors

In learning analytics, the observed interaction data is evaluated with the goal of learning optimization. The analysis can be carried out solely by the computer, for instance by an adaptivity system, or by the users by looking at visualized selected data aggregations in so-called learning analytics dashboards. Based on the requirements analysis with the customer, dashboards were designed primarily for the instructor level. A dashboard was integrated that presents overview statistics about the course, for example on the average performance in exercises or the average usage time. This helps instructors to identify explanation needs and optimize the course or curriculum. Figure 3 depicts how this approach has been realized. Technically the learning analytics dashboard has been implemented using the open-source software Learning Locker which is an xAPI Learning Record Store (LRS) with reporting and dashboard functionalities. The various visualization components (widgets) were developed iteratively, considering the available input data and the customer's requirements. The dashboard was embedded in an LTI wrapper for integration into the superordinate assistance system.

Our technical solution allowed recording implicit and explicit user feedback. Implicit feedback includes the amount of pageviews or the duration of the page visit. Explicit feedback includes given answers to the exercises or self-assessment on how well the

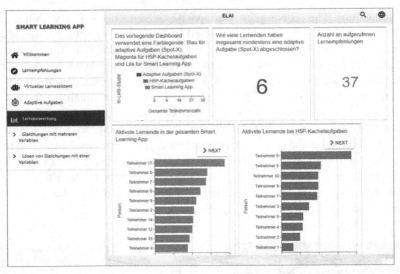

Fig. 3. Example screenshot of the realized software demonstrator, here the learning analytics view for the instructors.

user thinks the learning material was learned on a 1 to 5 scale. This feedback is used to optimize the AI but it also gives insights for the evaluation. During the fifth course the AI supported LMS was used 167 times, with users staying online for between three and a half and eight minutes on average. For a total of 4.814 interactions, 29 on average, 97,6% of the time the LMS was used on the laptop or desktop PC; only 2,4% via smartphone. We also observed a decline in access rates throughout the week. On Monday we observed 51 visits which went down to 16 visits on Friday. The highest access rates were before 12 o'clock AM.

Regarding the AI functionalities, we overserved the following access rates:

- 56 visits for adaptive tasks.
- 50 visits for learning recommendations.
- 35 visits for the chatbot.
- 34 visits of the individual learning indicators.

5 Evaluation

The goal of the evaluation focuses on the third research question (RQ3): Do the AI supported functionalities offer a benefit for the users? As suggested in our previous work [12] we decided on a mixed method approach with quantitative and qualitative methods to evaluate. During a course with mathematical learning content at a higher education institution the participants first used an LMS without AI supported functionalities for a week, and our AI supported LMS the week after. We used an online-survey tool and conducted the surveys on the last day of usage of each LMS. Both LMS were evaluated with the same set of questionnaires, while we added questions specifically for each AI supported functionality, e.g., asking for suggestions of improvement. The set of questionnaires contained the User Experience Questionnaire [24], the Technology Acceptance Model [25], a self-developed questionnaire for learning media, the affinity for technology interaction (ATI) scale [26] as well as tailored questions about the usage of LMS in the course and demographical questions. On a voluntary basis, the participants could create a subject code, allowing for an individual coupling between the survey-answers and the use of the AI supported LMS while guaranteeing anonymity. In the last week of the course, we additionally conducted semi-structured interviews, which took place in person at the institution, with four learners and two instructors.

Due to the low number of participants that filled out the online questionnaires at both times of measurement the data from the quantitative questionnaires were not sufficient for a statistical analysis. Therefore, we mostly rely on the answers from the open questions and the interviews for the evaluation results. For the qualitative analysis we used a structured qualitative approach [27]. We report on the open-answer questions from the online-questionnaire and interpret them by using user's statements from the interviews.

5.1 Evaluation Results

The evaluation finds added value of an AI supported assistive e-learning in a small course setting for learners and instructors. For the evaluation we distinguished between the learning recommender system itself and the integrated learning progress indicator. Six out of Nine learners who answered the open question about the added value of the learning recommender system found that the system had an added value. According to the interviews, this was due to a good clarity in comparison to the alternative system and a high clarity on which exercises had already been solved. Displaying learning deficits helped the learners adjust. Two learners did not find the learning recommender system meaningful enough and one person stated that she/he did not use this functionality. Seven Learners answered the question about the learning recommender system. One person did not find the learning recommender system helpful, as the teachers determine the order of the learning content. Six learners found the learning recommender system helpful. They liked the clarity on which content should be worked on next and reported a high motivation to reach a fully processed learning progress indicator. The functionality was rated as helpful with self-assessment.

Regarding the chatbot opinions differed. Some users found the chatbot very helpful and used it frequently. Others found the chatbot obsolete and preferred to continue using google. Only a few users tried the adaptive tasks. Here, the gamification element along with the alternation and the adaptivity of the difficulty level were praised.

All users – the learners as well as the instructors – emphasized during the interviews that they do not wish for any e-learning system to replace face-to-face teaching.

6 Discussion and Lessons Learned

A field test of the AI supported demonstrator at a higher education institution confirmed a clear benefit of the implemented functionalities, but also showed a necessary need for development and change. One limitation of the study was the small course-sample. Having more participants would enable statistical analysis of the evaluation questionnaires and would highly benefit the AI components – which work better with more input data. Therefore, we believe that having more users would lead to a higher user acceptance in respect to RQ3. While the application of AI supported LMS in small settings is challenging – e.g., because only little data is available – it represents a large proportion of real-world scenarios for which we deploy AI supported functions.

In order to implement an AI supported LMS successfully, many factors must be considered. A didactical concept is the most important prerequisite. Defining the purpose of the AI supported LMS and deciding on how to integrate the system into the course beforehand is indispensable. AI supported learning management systems should not replace face-to-face teaching but support the learning process, including preparing and post processing lessons in an expedient way. According to the authors of this article, communicating the supporting purpose and making clear that the system won't replace face-to-face teaching does help with one of the biggest challenges in implementing any new system: gaining user acceptance. The user's fears and reservations in terms of an AI supported LMS should be addressed, e.g., by explaining how data privacy is handled.

Learning content must be available digitally and should be appealing for students, e.g., by integrating quizzes, animation, interactive graphics, videos or even games. Additionally, the learning content must contain metadata and enable standardized collection and saving of user data, e.g., using xAPI. For an ergonomic usability, the AI supported functionalities should be accessible for users without detours. Hereby, meeting the users' requirements should always prioritize over the technological solution. For some stakeholders, e.g., administers and instructors, training for handling the AI supported LMS can be beneficial.

7 Conclusion and Further Research

Effective support of individual learners seems possible based on this study's experiences. Future studies should widen the scope of this study and raise requirements for team training, mobile learning and learning on the job. Long-term-studies, that accompany learners and their development over several years, could offer insights that help optimize AI supported recommendations and evaluate the learning journey. Implementing an extending range of functions might offer additional benefits for users and should be field tested and evaluated. From a technical perspective, future studies could focus on methods of data maintenance in order to optimize AI supported functionalities. As the usability and user experience (UX) have a tremendous effect on the user acceptance and the intention to use an LMS [28], usability studies should be included in all further research on LMS.

References

1. Farhan, W., Razmak, J.: A comparative study of an assistive e-learning interface among students with and without visual and hearing impairments. Disabil. Rehabil.: Assistive Technol. 1–11 (2020). https://doi.org/10.1080/17483107.2020.1786733

2. Sezer, B., Yilmaz, R.: Learning management system acceptance scale (LMSAS): a validity and reliability study. AJET **35** (2019). https://doi.org/10.14742/ajet.3959
3. Chaudhry, M.A., Kazim, E.: Artificial Intelligence in Education (AIEd): a high-level academic and industry note 2021. AI Ethics **2**, 1–9 (2021). https://doi.org/10.1007/s43681-021-00074-z
4. Ciolacu, M., Tehrani, A.F., Binder, L., Svasta, P.M.: Education 4.0 - artificial intelligence assisted higher education: early recognition system with machine learning to support students' success. In: 2018 IEEE 24th International Symposium for Design and Technology in Electronic Packaging (SIITME). IEEE (2018). https://doi.org/10.1109/siitme.2018.8599203
5. Streicher, A., Smeddinck, J.D.: Personalized and adaptive serious games. In: Dörner, R., Göbel, S., Kickmeier-Rust, M., Masuch, M., Zweig, K. (eds.) Entertainment Computing and Serious Games, vol. 9970, pp. 332–377. Springer, Cham (2015). https://doi.org/10.1007/978-3-319-46152-6_14
6. Nuseibeh, B., Easterbrook, S.: Requirements engineering: a roadmap. In: Proceedings of the Conference on the Future of Software Engineering (2000)
7. Flanagan, B., Ogata, H.: Integration of learning analytics research and production systems while protecting privacy. In: The 25th International Conference on Computers in Education, Christchurch, New Zealand (2017)
8. Renz, J., Meinel, C.: Can pseudonymized xAPI-Tracking solve data privacy issues in german schools? SAILA-ECTEL (2018)
9. Ifenthaler, D., Drachsler, H.: Learning analytics. In: Niegemann, H., Weinberger, A. (eds.) Handbuch Bildungstechnologie, pp. 515–534. Springer, Heidelberg (2020). https://doi.org/10.1007/978-3-662-54368-9_42
10. Krauss, C., Merceron, A., Arbanowski, S.: The timeliness deviation. In: Proceedings of the 9th International Conference on Learning Analytics and Knowledge. ACM (2019). https://doi.org/10.1145/3303772.3303774
11. Wise, A.F., Vytasek, J.: Learning Analytics Implementation Design Handbook of Learning Analytics, pp. 151–160. Society for Learning Analytics Research (SoLAR) (2017). https://doi.org/10.18608/hla17.013
12. Rerhaye, L., Altun, D., Krauss, C., Müller, C.: Evaluation methods for an AI-supported learning management system: quantifying and qualifying added values for teaching and learning. In: Sottilare, R.A., Schwarz, J. (eds.) Adaptive Instructional Systems. Design and Evaluation. LNCS, vol. 12792, pp. 394–411. Springer, Cham (2021). https://doi.org/10.1007/978-3-030-77857-6_28
13. Samuelsen, J., Chen, W., Wasson, B.: Integrating multiple data sources for learning analytics—review of literature. Res. Pract. Technol. Enhanc. Learn. **14**(1), 1–20 (2019). https://doi.org/10.1186/s41039-019-0105-4
14. Kagermann, H., Ulrich, W. (eds.) European Public Sphere: Towards Digital Sovereignty for Europe, acatech IMPULSE (2020)
15. Folsom-Kovarik, J.T., Raybourn, E.M.: Total Learning Architecture (TLA) enables next-generation learning via meta-adaptation. In: Proceedings of the I/ITSEC (2016)
16. Krauss, C., Hauswirth, M.: Interoperable education infrastructures: a middleware that brings together adaptive, social and virtual learning technologies. In: The European Research Consortium for Informatics and Mathematics, ERCIM NEWS. Special Theme: Educational Technology, pp. 9–10 (2020)
17. Ma, W., Adesope, O.O., Nesbit, J.C., Liu, Q.: Intelligent tutoring systems and learning outcomes: a meta-analysis. J. Educ. Psychol. **106**, 901–918 (2014). https://doi.org/10.1037/a0037123
18. Coffin Murray, M., Pérez, J.: Informing and performing: a study comparing adaptive learning to traditional learning. Inform. Sci.: Int. J. Emerg. Transdiscipl. **18**, 111–125 (2015). https://doi.org/10.28945/2165

19. Leitner: So lernt man lernen: Der Weg zum Erfolg. Nikol, Germany (2011)
20. Prensky, M.: Digital game-based learning. Comput. Entertain. (CIE) **1**, 21 (2003). https://doi.org/10.1145/950566.950596
21. Streicher, A., Schönbein, R., Pickl, S.W.: A general framework and control theoretic approach for adaptive interactive learning environments. In: Kotsireas, I.S., Nagurney, A., Pardalos, P.M., Tsokas, A. (eds.) Dynamics of Disasters. SOIA, vol. 169, pp. 243–257. Springer, Cham (2021). https://doi.org/10.1007/978-3-030-64973-9_15
22. Shute, V.J., Zapata-Rivera, D.: Adaptive educational systems. In: Durlach, P.J., Lesgold, A.M. (eds.) Adaptive Technologies for Training and Education. Cambridge University Press (2012)
23. Streicher, A., Roller, W.: Interoperable adaptivity and learning analytics for serious games in image interpretation. In: Lavoué, É., Drachsler, H., Verbert, K., Broisin, J., Pérez-Sanagustín, M. (eds.) Data Driven Approaches in Digital Education. LNCS, vol. 10474, pp. 598–601. Springer, Cham (2017). https://doi.org/10.1007/978-3-319-66610-5_71
24. Schrepp, M., Hinderks, A., Thomaschewski, J.: Applying the user experience questionnaire (UEQ) in different evaluation scenarios. In: Marcus, A. (ed.) Design, User Experience, and Usability. Theories, Methods, and Tools for Designing the User Experience. LNCS, vol. 8517, pp. 383–392. Springer, Cham (2014). https://doi.org/10.1007/978-3-319-07668-3_37
25. Davis, F.D.: A technology acceptance model for empirically testing new end-user information systems: theory and results. Doctorial dissertation, Massachusetts Institute of Technology (1985)
26. Franke, T., Attig, C., Wessel, D.: Affinity for technology interaction (ATI) scale. Int. J. Hum. Comput. Interact **2018** (2018)
27. Mayring, P., Fenzl, T.: Qualitative inhaltsanalyse. In: Baur, N., Blasius, J. (eds.) Handbuch Methoden der empirischen Sozialforschung, pp. 543–556. Springer, Wiesbaden (2014). https://doi.org/10.1007/978-3-531-18939-0_38
28. Eraslan Yalcin, M., Kutlu, B.: Examination of students' acceptance of and intention to use learning management systems using extended TAM. Br. J. Educ. Technol. **50**, 2414–2432 (2019). https://doi.org/10.1111/bjet.12798

Development of the Assessment Toolkit for Leader Adaptability Skills (ATLAS)

Tarah Daly[1]([✉]) [iD], Allison Hancock[1], Jennifer Phillips[1], Marcus Mainz[2], and Breck Perry[2]

[1] Cognitive Performance Group, Independence, OH 44131, USA
tarah@cognitiveperformancegroup.com
[2] Covan Group LLC, Stafford, VA 22556, USA

Abstract. Adaptability, the ability to efficiently adapt to new contexts, is necessary for current and future operating forces to successfully achieve their missions in complex arenas where problems are shifting continuously. The Assessment Toolkit for Leader Adaptability Skills (ATLAS) system was the first training system of its kind to aid operating forces as they build adaptive expertise by providing a validated assessment of adaptability skills targeted at Marine infantry squad leaders. The goals for development of the ATLAS system included: (1) operationalizing adaptability where the definition included the adaptive process within a cognitive framework; (2) assessing adaptability in a simulated tactical context; (3) providing diagnostic assessment outcomes of adaptability proficiency that support training plans and remediation activities where developmental needs are identified; (4) delivering a validated, sustainable measurement tool that does not require human raters; and (5) supporting an adaptive training approach for improving adaptability skills. To date the ATLAS effort has achieved these goals through iterative, resource-conscious methods of design and development; close collaboration with our subject matter expert partners; and field testing with end-user community populations. The validated measures of adaptability provide users with real-time performance feedback and assessment of adaptive proficiency. Future capabilities would extend the benefits of automated adaptability measures to provide comparative analysis, robust formative and summative feedback, and support improved adaptability proficiency through training and remediation recommendations during the Warfighter career continuum.

Keywords: Decision making · Adaptability · Adaptability measurement · Automated assessment · Simulation · Adaptive training

1 Introduction

Adaptability is a critical skill necessary for current and future operating forces to successfully achieve their missions in increasingly complex arenas where problems are shifting continuously. The ability to efficiently adapt to new contexts and use tactical thinking skills increases effectiveness in the execution of any battle plan. Previous work on decision making and adaptability revealed that adaptability is a key performance area

© The Author(s), under exclusive license to Springer Nature Switzerland AG 2022
R. A. Sottilare and J. Schwarz (Eds.): HCII 2022, LNCS 13332, pp. 271–285, 2022.
https://doi.org/10.1007/978-3-031-05887-5_19

and a primary performance goal for Marine infantry squad leaders, thus substantiating the value of objective assessment of adaptability [1]. However, despite decades of research, progress to operationalize the construct of adaptability and assess an individual's adaptability proficiency is limited [2]. The lack of a cohesive research and development program in the Services is also reflected in the range of non-military research approaches and methods undertaken across individual and team settings [3]. Past efforts aimed at measuring adaptability have not produced tools adequate for Marine Corps application. Many of the current assessments are trait-based and do not account for the domain-specific adaptive decisions required of Marine infantry squad leaders [4, 5]. The domain-specific assessments that have been created require human raters to code participant responses and score performance outcomes, rendering them resource-intensive and subjective. Moreover, a challenge in measuring adaptability is that no agreed upon operational definition exists, and none of the definitions break down the adaptive process in a manner that allows users of assessment outcomes to identify training activities to remediate inadequate skills or knowledge. To address these needs and gaps posed in previous research, we created the Assessment Toolkit for Leader Adaptability Skills (ATLAS) system. ATLAS is the first of its kind training system to support operating forces as they build adaptive expertise by providing a validated, automated assessment of adaptability skills targeted toward developing Marine infantry squad leaders [6].[1]

In this paper, we describe how we automated adaptability measurement and assessment to extend the research area with validated results and to provide a future landscape for the use of automated adaptability measures. The measurement of learner attributes, specifically adaptability, continues to remain an area of ongoing research. To that end, we detail our design, development, and research efforts on the ATLAS system. We provide our research process, from the conception of the hypothesized cognitive framework of adaptability, to the identification of adaptability measures within that framework, and how those measures map onto military-relevant contexts within ATLAS. To address how we were able measure adaptive performance, we provide a detailed description of the ATLAS simulation environment, how the system was designed to capture user responses for adaptability measurement, and how we used automated scoring algorithms to provide adaptability metrics. Further, we address verification and validation concerns of the adaptability measures. To date, the adaptability measures have been validated through a series of field testing with end-user community groups. A final field test will provide us with verification of our processes and the results of our adaptability measure evaluation. Finally, we outline how ATLAS meets the training needs of user communities through our current efforts with ONR—providing automated validated measures of adaptability within a simulated, tactically-relevant environment. We also detail where we see great benefit in expanding ATLAS capabilities in future efforts, specifically the automated measurement and assessment of adaptive expertise within training and education contexts, and how this could be applied across varying stakeholder groups.

---

[1] The authors have published an earlier version of this work under technical reports delivered to the Office of Naval Research.

1.1 Cognitive Framework of Adaptability

We developed an operational definition of squad leader adaptability based on a cognitive framework of change and anomaly detection, cognitive flexibility, and decision making [7]. The cognitive framework begins with a Warfighter in a mission context with certain assumptions and expectations. First, a change in the situation or an anomaly must be perceived. Either a change or an anomaly can trigger the adaptation. Second, the implication(s) for the mission must be recognized. The Warfighter then compares their existing knowledge and schema, or mental models (schema comparison), to arrive at a new interpretation of what is happening in the current situation (new schema assembly) and to generate a basis for deciding how to adapt [8]. Finally, a decision is made as a result of generating the new schema. That decision is demonstrated by actions taken to achieve the mission.

The ATLAS system was designed to measure adaptive performance along each of the discrete cognitive processes (perception, recognition, schema comparison and new assembly, and adaptive decision making [8]; to provide diagnostic feedback on strengths or developmental needs that is instructive for training. We identified six measures of adaptability that align with the discrete steps of the cognitive framework:

1. *Speed of reaction to the event.* A time-based measure indicating how quickly the user perceived a change and recognized that the change has an impact on mission accomplishment.
2. *Duration of situation assessment.* A time-based measure of how quickly a user recognized the importance of a new situation as well as how long it took them to make that assessment.
3. *Quality of situation assessment.* A comparative measure of one's ability to produce an accurate new situation assessment given the perception of situational cues and detection of a change or anomaly based on one's knowledge and experience.
4. *Duration of adaptive activity.* A time-based measure of how quickly a user can compare schemas and assemble new schemas, or in other words, mentally retrieve relevant experience to assess the new situation and to re-plan, which is an element of cognitive flexibility.
5. *Quality of adaptive action.* A comparative measure of one's ability to accurately adapt schema and re-plan based on one's knowledge and experience, which is another element of cognitive flexibility.
6. *Efficiency of adaptive process.* A measure of the ratio between duration and quality of actions as a means of discriminating levels of proficiency.

Figure 1 illustrates the link between the measures of adaptive performance, the discrete cognitive elements, and the performance within the ATLAS system that produces the data to score the measures [6].[2]

---

[2] The authors have published an earlier version of this work under technical reports delivered to the Office of Naval Research.

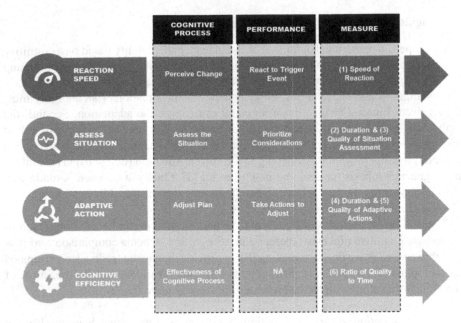

Fig. 1. Measures of adaptability embedded in the ATLAS system.

1.2 ATLAS Simulation Environment

ATLAS leverages a Unity-based simulation environment where users are placed into squad-level tactical missions and must adapt or change their mission plan based on events that occur throughout the scenario. Scenarios incorporate a variety of mission types, environment types, and adaptation types to test Marines in different contexts where adaptations are necessary. The ATLAS system currently includes two offensive missions set in wooded terrain and two defensive missions set in dessert terrain. Each scenario is comprised of phases, which vary in number and represent the progress of the mission. Each phase contains distinct "item sequences" which are the test items, including trigger events, that present a challenge to the user eliciting response data to calculate each of the six adaptability measures. Trigger events include communications, events, or cues that indicate a change in the situation and are presented through a pop-up video during scenario play (Fig. 2). Trigger events vary in their levels of subtlety, with a series of additional cues following the first, most subtle, trigger event.

Fig. 2. Event pop-up vignette.

The user decides when to adapt in response to the trigger events presented in a phase, what adaptations or changes to their mission plan to make in response to the new situation created by the trigger events, and what their priorities are based on the new situation. Adaptive action selections are grouped into three categories: *Exchange Information, Provide Leadership,* and *Request Assets* (Fig. 3). The *Change Priorities* screen captures users' top four priorities used for scoring the situation assessment measures (Fig. 4).

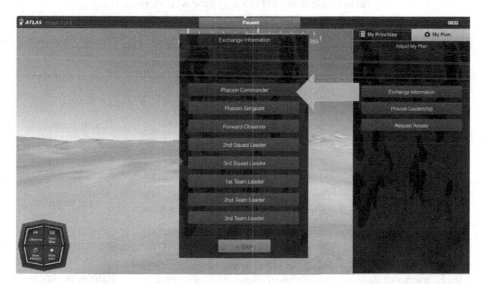

Fig. 3. Adaptive action menu and sample list of options.

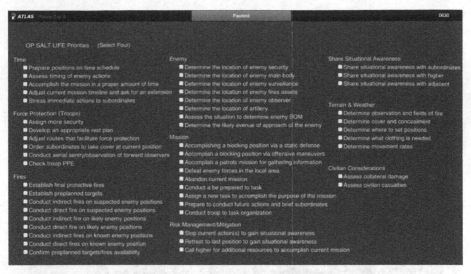

Fig. 4. Change priorities ATLAS user interface.

1.3   ATLAS Measure Scoring

To score the six adaptability measures, we developed automated algorithms, which use a combination of time-based calculations and comparative assessment (accuracy) of user responses to expert referent profiles for the quality calculations (Table 1).

Table 1.   Measures and scoring in the ATLAS system.

Measure	User Performance Captured by System	Score Calculation
(1) Speed of Reaction	Time from trigger presentation to time user pauses scenario to make an adaptation; weighted by trigger.	Time in seconds
(2) Duration of Situation Assessment	Time it takes user to select or update priorities in each phase; weighted by trigger.	Time in seconds
(3) Quality of Situation Assessment	Selection of priorities as compared to expert referent priorities.	Number of quality points earned
(4) Duration of Adaptive Action	Time it takes user to make adaptations in each phase divided by number of actions.	Time in seconds
(5) Quality of Adaptive Action	Selection of adaptive actions compared to expert referent adaptive action profiles.	Number of quality points earned
(6) Efficiency of Adaptive Process	Quality of situation assessment (3) divided by duration of situation assessment (2); quality of adaptive activity (5) divided by duration of adaptive activity (4)	Ratio of quality to duration

Expert referent profiles detail adaptive actions across three levels of squad leader experience (adequate, skilled, and expert), assigning point values to adaptive actions based on which experience level the subject matter experts (SMEs) determined would take that action at that point in the scenario. For example, in first phase after the first trigger, the SME panel may have determined an expert squad leader would have sent up a small Unmanned Aircraft System so if a user made that selection at that point in the scenario, they would receive the highest point value. Not all adaptive action selections were included in every profile for a scenario, only those identified by the SME panel for that scenario. Adaptive action quality scores are a total of points earned based on the value associated with the selections within the expert profile. Expert referent profiles for priority selections are a total of priority matches to the list of acceptable priorities identified by the SME panel [9].

Scores are provided for every measure for each individual phase and summarized across the entire scenario (Fig. 5). This allows users to see their performance summary for the entire scenario and for each phase within a scenario. The scores are presented alongside a legend with detailed information on the definition of the adaptability measure and how the measure is scored. Additionally, we include an interpretation of what that score means for the user. For example, lower adaptive action duration scores (i.e., taking less time to make adaptive action selections) indicate more decisiveness or higher adaptive action count percentages (i.e., users selecting a high number of adaptive actions from the expert referent profile) within a certain experience category (i.e., adequate, skilled, or expert) indicates which experience profile the user is most aligned.

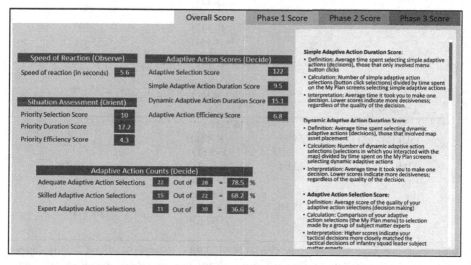

Fig. 5. Example screen shot of ATLAS summary score screen.

2 Field Test Validation

To date, three rounds of field testing have been conducted to evaluate the adaptability measures with participants from The Basic School (TBS) in Quantico, VA and the School of Infantry-East (SOI-E) in Jacksonville, NC suggesting ATLAS is a valid measure of adaptability. Participants ranged in their military occupational specialty (MOS) experience level, time in service (TiS), time spent as a team leader, and time spent as a squad leader (see Table 2 for a summary of participants included in field testing by schoolhouse and course).

Table 2. Participants included in field testing.

Group	Age (*M*)	Grade (n)	Time in Service (*M*)
		Field Test 1[a]	
FISULC (*n*=3)[*]	26.70	E-5 (3)	6.50 yrs
ISULC (*n*=39)	24.40	E-4 (8)	5.08 yrs
		E-5 (31)	
AIMC (*n*=63)	22.70	E-3 (9)	
		E-4 (41)	3.75 yrs
		E-5 (13)	
All (*n*=105)			
		Field Test 2[b]	
CIB (*n*=16)	28.90	O-2 (8)	
		O-3 (6)	6.58 yrs
		O-4 (1)	
WIC (*n*=9)[*]	23.60	E-2 (1)	
		E-3 (3)	
		E-4 (1)	4.33 yrs
		E-5 (3)	
		E-6 (2)	
M Co (*n*=31)	26.10	O-1 (30)	
		UKN (1)	2. 75 yrs
All (*n*=56)			
		Field Test 3	
ISULC (*n*=21)	24.95	E-4 (8)	6.15 yrs
		E-5 (13)	
AIMC (*n*=27)	22.04	E-3 (11)	
		E-4 (12)	3.10 yrs
		E-5 (4)	
M Co (*n*=26)	25.19	O-1 (24)	1.62 yrs
		O-2 (2)	
All (*n*=74)			

Note. [a]AIMC (Advanced Infantry Marine Course) SOI-E, ISULC (Infantry Small Unit Leaders Course) SOI-E, FISULC (Former ISULC students) SOI-E. [b]CIB (Combat Infantry Battalion) TBS, WIC (Warfighting Instructor Company) TBS, M Co (Mike Company – Marines awaiting entry to Basic Officer Course) TBS.

[*]Not included in analyses.

For the analyses, we correlated the adaptability measures with experience criterion variables appropriate for the various participant populations, including single demographic variables (e.g., TiS, time spent as either a squad or team leader) and a combination of demographic variables that resulted in a weighted TiS (wTiS) variable. We also compared mean scores on the adaptability measures across experience levels. In Field Test 1 and Field Test 2, cases in which only two groups were used in the analyses, t-tests were conducted. In Field Test 3, all three experience levels were included in a regression analysis.

For validation of the adaptability measures, we expected that those with high experience would perform better than those with low experience across each measure. For quality-based measures (quality of situation assessment, quality of adaptive action), we expected that those with high experience would obtain higher situation assessment scores and higher adaptive action scores compared to those with less experience. For reaction time and duration of situation assessment, we expected participants with higher experience to have faster reaction times and quicker duration scores compared to less experienced participants. For duration of adaptive action, we expected to see evidence of a curvilinear relationship among the three experience groups. Those with low and high experience would make selections quicker than those in the moderate experience group.

We hypothesized that less experienced participants would make selections more quickly because they do not have the extent of knowledge or experience to spend time considering when making decisions or they may be unable to recognize when an event requires plan modifications thus only selecting familiar options in a reactive manner [10, 11]. Those with high levels of experience would also make faster selections because their decision-making process is more intuitive and requires less deliberate cognitive attention. Additionally, those with higher levels of experience may be thinking in an anticipatory manner, planning ahead, and responding to changes quickly. Those with moderate experience would take the longest to make selections because they are more deliberate about thinking through their decisions, taking the time to consider options based on their experience and knowledge. They recognize the need for change but must think through their execution of the adaptation to it.

Results across the three field tests provided evidence for criterion validity of all six measures of infantry squad leader adaptability, that adaptive performance was able to be individually assessed at each of these discrete measures, and that ATLAS was able to effectively discriminate squad leader experience levels [6].

As hypothesized, participants with higher levels of experience were associated with significantly better assessments of the new situation, which was reflected in higher quality of situation assessment scores compared to participants with less experience, $r(75) = .30$, $p < .01$. Similarly, those with higher levels of experience were associated with significantly better decisions about which actions to take, resulting in higher quality of adaptive action scores compared to participants with less experience, $r(72) = .25$, $p < .05$. These findings suggest that those with more experience were able to better identify important aspects of the situation that required change and made better decisions regarding how to react to the new situation.

Time-based measures also showed evidence in support of our hypotheses. For speed of reaction and duration of situation assessment, more experienced participants were associated with quicker reactions to the new situation, $r(68) = -.25, p < .05$, and quicker assessments of the new situation compared to those with less experience, $r(69) = -.25$, $p < .05$. In previous field tests comparing only our high and low experience groups, participants with more experience showed evidence of quicker duration of adaptive action scores, which points to their ability to draw from relevant experience and re-plan efficiently compared to those with less experience, $t(59, 38) = 2.10, p < .05$. We were unable to fully validate our hypothesis of a curvilinear relationship for duration of adaptive action among our three experience groups, however. Users who make rapid decisions because they have a small repertoire of scripts and schema from which to choose should be differentiated from those who make rapid decisions because they intuitively recognize an effective course of action.

In summary, the field tests to validate the adaptability measures showed that com-pared to participants with less experience, participants with more experience were quicker to identify a change in the situation that would require an adaptation (*speed of* reaction), quicker to make sense of the new situation created by the change (*dura-tion of situation assessment*), better (*quality of situation assessment*) and more efficient (*efficiency of situation assessment*) at making their assessment of the new situation, and better at making decisions about how to adapt to the new situation (*quality of adaptive action*). Further, we attribute low sample sizes to our results that showed an expected but non-significant curvilinear relationship between experience and time spent making decisions (*duration of adaptive action*) such that those with lower and higher experience were quicker to make decisions than those with moderate experience. In our fourth field test, we will verify the validation of the measures with all three experience groups (i.e., high, moderate, and low experience) testing the hypothesized curvilinear relationship for duration of adaptive action with larger sample sizes in the experience groups.

3   Current and Future Implications with Automated Adaptability Measures

With validated measures of adaptive performance, we are able to provide solutions for current user community needs, such as providing students exposure to squad leader missions and decisions without added resource requirements from instructors; pre-post testing in low to moderate experience student populations to evaluate student perfor-mance; or instructor access to raw user data (e.g., actions taken, order of actions) to support detailed, student-centered feedback with targeted remediation. However, we must also plan for and expand on future capabilities that can benefit from automated adaptability measures. The following sections detail how current ATLAS capabilities could be expanded to support more robust training and education needs.

3.1   Adaptability Performance Feedback

Currently, feedback provided by ATLAS is diagnostic across the discrete cognitive ele-ments within the adaptability framework. The information presented to students high-lights areas of adaptive performance strength, areas for improvement, and adaptability insights that can be leveraged in training activities or after-action reviews during reme-diation. The scoring feedback details time- and quality-based scores, which can be interpreted in order to make judgements about how each student perceived change in the scenario, assessed the situation, and adjusted their plan. Within each of these elements, ATLAS provides granular assessment of the duration and quality of adaptation, a novel approach supporting the improvement of adaptive expertise. The scoring feedback for adaptive actions currently only leverages the integrated expert referent profiles for qual-ity assessment within the adequate, skilled, or expert experience groups. However, the ATLAS system can parse adaptive actions by category (i.e., exchange information, pro-vide leadership, and request assets) and compare user selections to the specified adaptive actions in each category within the scenario's scoring profile.

Future score screens could include feedback about adaptive action categories that were used the most (or not) in comparison with expert referent profiles. For example, users could see diagnostic information about the number of opportunities to exchange information and that they chose that adaptive action category 20% of the time. Similarly, the same approach could be taken with priority selections, providing insight into areas commonly overlooked. This information could then be paired with an analysis or expert interpretation for increased awareness of holistic adaptive proficiency.

Formative and summative feedback on the measures of adaptability could be used to support an adaptive training approach, providing students with targeted feedback to support a self-guided, student-centered learning experience [12, 13]. Formative feedback would provide the student with contextual information about their performance [14]. Specifically, detailed comparisons between student decisions alongside expert decisions would be presented at the conclusion of each phase. Presenting feedback in this manner would produce prescriptive knowledge of performance or knowledge of results of expert referent profiles such that students would be able to self-reflect, learn from, and use the feedback to support their decision making in subsequent phases of the scenario. Figure 6 provides an example of student performance feedback compared with an expert referent on an *Adjust Route* adaptive action. If formative feedback were included on the map interface, an explanation on expert route choice could be presented to help the student better understand terrain reasoning.

Providing characteristics of successful performance allows students to make com-parative judgements between their own performance and actions that are preferable and will achieve desired outcomes (e.g., mission success). This is commonly referred to as knowledge of results (KR). During training events, KR has been shown to be effective in improving performance, specifically within cognitive domains. Studies have found that those who were provided KR during training or a pre-task event improved their performance as time on task progressed or during a transfer task compared to those who did not receive KR [15–18]. The ATLAS system already captures adaptive performance data in real-time; however, future capabilities could leverage the automated adaptability measures to produce formative feedback for users in-situ through KR of expert referent profiles to create a more guided, step-by-step walkthrough of scenario play. This scaf-folded approach would provide entry-level and low experience student populations with an interactive introduction to adaptive decision making in squad level missions.

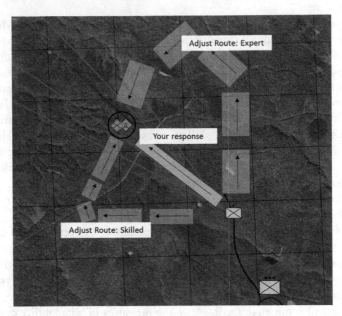

Fig. 6. Formative feedback example.

Summative feedback could provide diagnostic information on each of the discrete cognitive elements of adaptive performance (e.g., situation assessment) that could inform recommendations for remediation and support adaptive training of those cognitive elements. This information, when compiled for instructors, could enable them to facilitate decision games or other experiential learning exercises tailored to strengthen components of the cognitive process of adaptability identified as student developmental needs. For example, based on results that indicate a deficit in reaction speed, the instructor could facilitate a decision game that provides less time for students to engage in scenario decision points with peers or individually, thus providing an opportunity to practice expedited decision making. For results that show situational assessment could be improved, multiple competing factors (e.g., ethical values, cultural norms, resource limitations) could be introduced in the scenario to increase complexity for decision making. The significance of ATLAS supporting an adaptive training capability is that it would allow students and instructors alike to leverage assessment outcomes of these hard-to-observe learner characteristics in contextually relevant applications, to improve individual components of complex decision making to strengthen overall adaptive performance.

Recommendations for future efforts include strengthening the scoring feedback to allow for an array of robust assessment and analysis capabilities, such as establishing norms of time-based measures, creating a repository for stored performance metrics or student profiles, and linkages to doctrinally-sourced remediation materials for targeted support at specific steps in the adaptive process. The establishment of peer group data, in concert with a criterion variable(s), would provide a basis for comparative assessment of student performance. For example, comparisons could be made to peers from previous courses or across students within the same course, performance trends over time, or appropriate performance parameters associated with different skill levels in expert referent profiles. With enough data points, peer group profiles that self-form and provide adaptive performance characteristics of a course, an experience group, or a geographic location that can be leveraged by training communities on a larger scale could emerge.

3.2 Adaptability Throughout Training and Education

Automated adaptability measures from ATLAS can apply throughout the entire training and education continuum for any individual Warfighter and military unit in garrison and while forward-deployed. For example, a Marine recruit or officer candidate can be given a diagnostic adaptability test prior to recruit training or Officer Candidate School, conduct a pre and post-test at the School of Infantry and TBS, improve upon adaptability deficiencies during pre-deployment training, and apply instilled adaptability decision-making skills while being deployed. Comparably, at the unit level, automated adaptability assessment and development can enable military units to create tempo, fluidity, and faster decision-cycles in planning and execution. As units reconstitute in the beginning of the deployment lifecycle, diagnostic adaptability tests can be used to assess the current state of a unit's combat effectiveness and provide feedback on where adaptability improvements are needed. As the unit progresses through pre-deployment training, evaluation teams can incorporate automated adaptability measures into after-action reviews, debriefs, and readiness reports, providing essential feedback to both unit leaders and higher-echelon commanders that certify a unit's combat readiness prior to deployment. The overarching goal would be to create redundancies, efficiencies, and overlap for enhancing individual and unit adaptability capability, fostering habit of thought and habit of action during combat operations.

In addition to comprehensively applying automated adaptability measures for individual Warfighters and military units, career counselors, senior enlisted advisors, and officer monitors can incorporate adaptability metrics into MOS selection and individual assignment suitability criteria. Servicemembers can be screened and surveyed for high stress, high demand occupations to ensure the individual is deemed adaptable enough to be a good fit in a particular job or organization. Moreover, these counselors and mentors can more effectively provide feedback to professionally develop individual servicemembers to improve upon deficiencies discovered through adaptability training.

Due to the expeditionary nature of automated adaptability capabilities, training and education events can be conducted anywhere at any time. Automated adaptability training is not resource-intensive and can be incorporated into individual professional development plans. As individual servicemembers receive feedback on their deficiencies, ATLAS will include recommended references, activities, and techniques that can be applied to enhance overall adaptability, ultimately improving the lethality and survivability of the servicemember and their unit.

4 Conclusion

The ATLAS system provides a novel approach to automated measurement and assessment of adaptive expertise among Marine infantry squad leaders. With validated measures of adaptability, ATLAS can give users a detailed window into their adaptive performance with diagnostic feedback, assessment, and comparative analysis of quality scores with expert referent profiles. Through the field testing that has been conducted, we have validated the adaptability measures. However, we seek to stipulate extended capabilities with automated adaptability measures. We see added benefits to students and instructors with the development of robust formative and summative feedback on adaptive performance. Extending beyond educational activities, a holistic view of an individual's adaptability proficiency, whether that be over time or compared against benchmark peer group data, can enable other stakeholders to apply adaptability measures throughout the military lifecycle. Future efforts of this kind directly support improvements to and the building of adaptive expertise among our operating forces.

Acknowledgements. This work was supported by the Office of Naval Research (Contract # N00014-21-C-2031). The authors wish to thank the entire Assessment Toolkit for Leader Adaptability Skills project team, Cheryl Johnson, Natalie Steinhauser, Peter Squire, and the instructors, staff, and students from SOI-E and TBS for their support of this research.

The views of the authors expressed herein do not necessarily represent those of the U.S. Marine Corps, U.S. Navy, or Department of Defense (DoD). Presentation of this material does not constitute or imply its endorsement, recommendation, or favoring by the DoD or its components.

References

1. Hancock, A.K., Phillips, J.K., Steinhauser, N.B., Niehaus, J.: Cognitive skill assessment in a virtual environment. In: Proceedings of the Interservice/Industry Training, Simulation and Education Conference. National Training Systems Association, Orlando (2019)
2. Sangwan, D., Raj, P.: The philosophy of be, know, and do in forming the 21st-century military war-front competencies: a systematic review. Def. Stud. **21**, 375–424 (2021)
3. Baard, S.K., Rench, T.A., Kozlowski, S.W.J.: Performance adaptation: a theoretical integration and review. J. Manag. **40**, 48–99 (2014)
4. Kozlowski, S.W.J., Rench, T.: Individual differences, adaptability, and adaptive performance: a conceptual analysis and research summary (Report No. 08146). Battelle Scientific Services (2009)
5. Ployhart, R.E., Bliese, P.D.: Individual adaptability (I-ADAPT) theory: conceptualizing the antecedents, consequences, and measurement of individual differences in adaptability. In: Burke, S., Pierce, L., Salas, E. (eds.) Understanding Adaptability: A Prerequisite for Effective Performance Within Complex Environments, pp. 3–39. Elsevier, San Diego (2006)
6. Hancock, A.K., Daly, T.N.: Assessment toolkit for leader adaptability skills (ATLAS) phase II option period II final technical report (ATLAS Report No. A006 – Not approved for public release). Office of Naval Research, June 2021
7. Ross, K.G., Phillips, J.K.: A cognitive framework and approach for adaptability assessment. In: Proceedings of the 2017 Interservice/Industry Training, Simulation, and Education Conference. National Training Systems Association, Orlando (2017)

8. Morison, J.E., Fletcher, J.D.: Cognitive readiness (IDA Paper P-3735). Institute for Defense Analyses (2002)

9. Hancock, A.K., Daly, T.N., Borders, M.R.: Assessment toolkit for leader adaptability skills (ATLAS) phase II option period II interim technical report (ATLAS Report No. A005 – Not approved for public release). Office of Naval Research, November 2020

10. Phillips, J.K., Ross, K.G., Rivera, I.D., Knarr, K.A.: Squad leader mastery: a model underlying cognitive readiness interventions. In: Proceedings of the Interservice/Industry Training, Simulation and Education Conference. National Training Systems Association, Orlando (2013)

11. Ross, K., Phillips, J., Cohn, J.: Creating expertise with technology based training. In: Schmorrow, D., Cohn, J., Nicholson, D. (eds.) The PSI Handbook of Virtual Environments for Training and Education, vol. 1, pp. 66–79. Praeger Security International, Westport (2009)

12. Chen, K., et al.: Academic outcomes of flipped classroom learning: a meta-analysis. Med. Educ. **52**, 910–924 (2018)

13. Tannenbaum, S.I., Cerasoli, C.P.: Do team and individual debriefs enhance performance? A meta-analysis. Hum. Factors **55**(1), 231–245 (2013)

14. Shute, V.J.: Focus on formative feedback. Rev. Educ. Res. **78**(1), 152–189 (2008)

15. Butler, A.C., Godbole, N., Marsh, E.J.: Explanation feedback is better than correct answer feedback for promoting transfer of learning. J. Educ. Psychol. **105**(2), 290–298 (2013)

16. Kluger, A.N., DeNisi, A.: The effects of feedback interventions on performance: a historical review, a meta-analysis, and a preliminary feedback intervention theory. Psychol. Bull. **119**(2), 254–284 (1996)

17. Teo, G.W.L., Schmidt, T.N., Szalma, J.L., Hancock, G.M., Hancock, P.A.: The effect of knowledge of results for training vigilance in a video game-based environment. In: 56th Annual Meeting of the Human Factors and Ergonomics Society, Boston, MA, October 2012

18. Teo, G.W.L., Schmidt, T.N., Szalma, J.L., Hancock, G.M., Hancock, P.A.: The effects of feedback in vigilance training on performance, workload, stress and coping. In: Proceedings of the Human Factors and Ergonomics Society, vol. 57(1), pp. 1119–1123. Sage Publications, Thousand Oaks (2013)

Measurements and Interventions to Improve Student Engagement Through Facial Expression Recognition

Will Lee[1]([✉]), Danielle Allessio[1], William Rebelsky[1], Sai Satish Gattupalli[1], Hao Yu[2], Ivon Arroyo[1], Margrit Betke[2], Sarah Bargal[2], Tom Murray[1], Frank Sylvia[1], and Beverly P. Woolf[1]

[1] University of Massachusetts-Amherst, Amherst, MA 01003, USA
{williamlee,wrebelsky,ivon,tmurray}@cs.umass.edu, {allessio, sgattupalli,bev}@umass.edu
[2] Boston University, Boston, MA 02215, USA
{haoyu,betke,sbargal}@bu.edu

Abstract. A major challenge for online learning systems is supporting students' engagement. Online systems are sometimes boring, repetitive, and unappealing; external distractions often lead to off-task behavior, and a decline in learning. Student engagement and emotion are also tightly correlated with learning gains because emotion drives attention and attention drives learning. In response to this challenge, we present an exploratory study using computer vision techniques and several engagement strategies to help students re-engage once their attention wanders and their head turns away, within the context of an intelligent tutoring system for mathematics. Initial results indicate that students exposed to our re-engagement strategies were more confident and more persistent. They also responded positively to our avatar or learning companion. We discuss design implications and future work.

Keywords: Facial expression recognition · Engagement · Emotions · Gaze · Intelligent tutoring systems

1 Introduction

Student engagement and motivation is important in learning. Given today's variety of distractions – ranging from cell phones, social media, and online content – students are often distracted, their attention wanders, and engage in off-task behavior. We informally define on-task behavior as sustained attentional involvement throughout an academic task. Monkaresi et al. [7] discussed that attention, and thus on-task academic behavior, cannot always be sustained over a long period of time. In fact, attention fluctuates and goes through stages in which an individual might disengage and re-engage with a task. Additionally, engagement doesn't just focus on visual attention. It could be influenced by emotions, e.g., an individual shows a degree of interest or boredom over a task, that will make the student be on/off-task more often [12]. In academic settings, studies found

R. A. Sottilare and J. Schwarz (Eds.): HCII 2022, LNCS 13332, pp. 286–301, 2022.
https://doi.org/10.1007/978-3-031-05887-5_20

correlations between engagement and test scores [4, 13]. This suggests that detection of students' level of engagement can inform educators about how to adjust teaching plans and contents accordingly. Similarly, students show high on-task behavior, less boredom, and almost non-existent gaming the system with tutoring systems that are capable of intelligently responding to students' engagement [8].

We describe an exploratory study using the MathSpring.org system integrated with a machine learning-based off-task detection system using live videos of students, which assesses whether a student is looking away (head wandering). MathSpring.org is a web-based tutoring system, designed to aid middle school students' practice of mathematics problem-solving [3]. We developed computer vision software to detect students' off- task behavior (head wandering) as they use MathSpring.org, with two main goals: to respond "just-in-time" to students' states as they look away; and to provide teachers with real-time information to gauge the "pulse" of their students, so as to better assist them. Ultimately, teachers would gain insights into their students' motivation, interests, attention, and effort. By monitoring and capturing students' engagement with math problems, these constructs can help teachers realize whether their lesson plans are helping students to get involved with the material and learn.

We piloted a field formative evaluation of the tutoring system with underrepresented K-12 students and introduced a set of dynamic real-time interventions aimed at re-engaging students when the tutor detects distraction, i.e. head wandering. One goal was to better understand how students receive the engagement detection system. The contributions of this research include:

- Created a set of re-engagement strategies for informing students when they wander away while solving mathematical problems;
- Demonstrated the feasibility and scalability of real-time detection of head wandering using a deep learning based approach and off-the-shelf hardware;
- Performed a field formative evaluation with 5th-8th grade students to gather feedback about the system and strategies.

2 Related Work

Educators who utilize dashboards defining student performance gain more insight about their class [14]. This also helps teachers to learn about their students' level of engagement and attention. Gupta et al. [5] created a dashboard for K-12 educators to monitor and assess students' mastery level, emotion, and progression in a mathematics tutoring system. The dashboard was evaluated by educators in a live classroom and human-computer interaction subject matter experts. Arroyo et al. [1] discussed a study to understand and evaluate interventions of a meta-cognitive based student disengagement in which students lose interest due to possible frustration and "speed run" through exercises in order to "game" the system. By showing students their current performance and tips for improvement, the authors demonstrated that students could become re-engaged. Yu et al. [15] trained a convolutional neural network to detect student engagement based on computer vision of their head orientations. The model was integrated into the MathSpring tutoring system [2] with decision support to alert students if they were "looking away" too long. MathSpring is a web-based online tutoring system designed to aid students' learning of mathematical problem-solving strategies, aligned to the Common Core (CC) Standards, which constitute the national standardized mathematics curricula in the USA. The present work is an extension of Yu's work with interventions for re-engaging the students and evaluation in a live classroom setting with underrepresented students.

3   Research Questions

In this paper, we describe efforts to use computer vision interventions aimed at re-engaging students found to be distracted when using an online math intelligent tutoring system. We set out to address three specific research questions (RQs). For this exploratory study, we focus only on two RQs:

- **RQ1:** What is the impact of head wandering-aware interventional strategies on student's attention and engagement with math problems?
- **RQ2:** How are these head wandering-aware interventional strategies perceived by the students?

In order to address these RQs, we rely on individual student engagement with math problems, categorized into Engaged/Disengaged Interactions with Problems, which are specific indications of how much effort and what kind of way the student has been excerpted by students, see Table 1. These indicators of how students engage with math problems are recorded by the intelligent tutor. Positive Efforts are indicative of positive learning activities, with students paying attention. They include behaviors such as SOF (solved on first attempt), ATT (attempted a problem incorrectly and then self-corrected their mistake), SHINT (eventually got the correct answer after seeing one or more hints) and SHELP (solved after reviewing the example problem). Negative Efforts are indicative of negative learning activities, with inattentive states and not paying attention. These negative efforts include behaviors such as GIVEUP (learner performed some action but did not solve the problem at all), GUESS (learner did not see hints, but solved the question after more than 1 incorrect attempts), SKIP (learner skipped the problem with no action) and NOTR (learner performed some action, but the first action was too rapid for them to have read the problem).

While we consider this is a very proper way to comprehend students' actual engagement or disengagement with a math problem solving activity, the main difficulty is that this classification label of the student-problem interaction is unknown until the math problem has finished and the student clicks on "next problem" to request a new math problem.

We believe it would be much more beneficial for the students' learning experience if we were able to detect a students' disengagement (such as being able to understand that the student is looking away, and their attention has been lost) before the math problem solving activity has finished, so that MathSpring could intervene, and potentially even change the effort excerpted on the problem, and thus the kind of interaction with the math problem. This is the ultimate motivation for this paper and the inspiration for the research questions presented.

Table 1. Categorization of student activity/math performance metrics in terms of student *effort* and *activity types*.

Effort type	Learning activity	Definition
Student Behaviors with a Math Problem that indicate **Engagement**	SHINT	The student SOLVED the problem correctly after seeing HINTS
	ATT	The student ATTEMPTED once incorrectly, but self-corrected (answered correctly) in the second attempt
	SOF	The student SOLVED the problem correctly on the FIRST attempt without any help
	SHELP	The student got the problem correctly but saw at least one video
Student Behaviors with a Math Problem that indicate **Disengagement**	SKIP	The student SKIPPED the problem and did not do anything on the problem
	NOTR	The student did NOT even READ the problem and answered too fast in less than 4 s
	GIVEUP	The student started working on the problem but then GAVE UP and moved on without solving it correctly
	GUESS	The student apparently GUESSED by clicking through 3 to 5 answers until getting the correct one

4 Methods

4.1 Intelligent Tutors

Intelligent tutoring systems now perform nearly at the level of human tutors with a median effect of raising test scores over conventional levels from the 50th to the 75th percentile [6]. We collected videos of students solving math problems on MathSpring.com, a game-like intelligent math tutor that offers a personalized approach to online learning and tracks and responds to student performance and engagement, see Fig. 1. The platform adapts problems for students and shifts instruction from one-size-fits-none to completely individualized learning. It dynamically pinpoints gaps in student knowledge and seamlessly moves among topics to address students' knowledge. It builds an internal model of student skills and provides personalized content and remedial tutoring when needed. MathSpring is a well vetted, federally funded, personalized intelligent tutor for grades 5–8, developed at UMass and Worcester Polytechnic Institute.

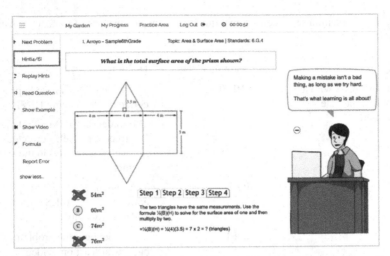

Fig. 1. MathSpring interface. Students solve online problems with support from tools (left) and a companion (right). For example, "Hints" speak and type problem-solving steps; "Show Example" provides worked-out examples, and the character (Jake) talks to students in either English or Spanish and emphasizes the importance of effort and perseverance. Answers are either open ended, short answer, multiple choice, or check-all-that-apply.

4.2 Facial Expression Recognition (FER)

The platform detects students' "head wandering", which is an indicator of possible off-task behavior and disengagement with the tutoring system in general. We know that the head follows the eye gaze [10], so moving the head away from a font-screen focus also very likely means that their gaze is looking away, and likely that their attention is fading. We use deep learning models trained to recognize head orientation and direction. Using muscle points on students' faces, log data of students' mathematics performance, and answers to emotion questionnaires, FER software measures students' emotions. The platform alerts students to bring them back if their attention seems off track. If a student's attention is chronically reduced and they're repeatedly distracted, fatigued, or struggling, they might be provided strong reminders.

We utilized Amazon Mechanical Turk (MTurk), a crowdsourcing marketplace that outsources tasks to a distributed workforce. Each of the 18721 frames of a previous group of students working on online math problems was assigned to one of three labels: 'looking at their paper', 'looking at their screen', or 'wandering' by three different workers, Fig. 2 shows the layout of the students working on MathSpring. These human labeling annotations were used to train computer vision deep learning models that automatically detected wandering and student engagement using state-of-the-art deep learning architecture that classifies a student's gesture into those labels. We then compared the deep learning models to baselines that rely on head pose estimation.

Fig. 2. Data capture setup for the experiments. Students worked on problems on a laptop with a webcam. They had a notebook and pencil available. The webcam captured the student's upper body and face as shown in Fig. 3 below.

Fig. 3. Student engagement dataset. The videos captured student faces and gestures as students solve math problems. Video frames are then annotated to indicate whether a student is on-task ('looking at the screen' or 'looking at their paper') or their head is wandering, thus the student is looking away. Annotations are then used to train computer vision models that automatically detect head wandering. A computer vision model is then integrated into MathSpring, an online tutor system, in order to trigger interventions whenever students are predicted to be head wandering.

4.3 Design

During the Summer of 2021, N = 30 students from a summer camp program in Massachusetts participated in this study by engaging with MathSpring for 5 Monday sessions of at least 30 min each session. The students ranged from 5th (n = 13), 6th (n = 11), 7th (n = 4), and 8th (n = 2) grade and were from underrepresented groups (including bi-lingual Hispanic students). We decided to discard data from the 7th and 8th grades because those students did not have any wandering recorded during the study, and the study focuses on analyzing strategies about what to do when students are distracted (wandering). All other students had at least one instance recorded of head wandering.

Students accessed the web-based MathSpring.org software through a laptop and solved mathematical problems according to their grade level. Before students began using MathSpring, one of the experimenters provided a tutorial of the software, see Fig. 4. Figure 5 illustrates the resulting video capture of students working with the system. Students completed a mixed-methods post-survey at the end of day five.

Fig. 4. Experimenter is shown here demonstrating the MathSpring software in a tutorial.

Fig. 5. Students working online in their camp classroom. The layout of the students working on MathSpring during a summer urban camp.

Each student was randomly assigned to either an experimental or control condition. The control condition did not react when head wandering was detected. The experimental condition reacted to each student's head wondering, in one of four (4) possible ways, described in Table 2. This enabled us to compare the possible impact of responding in "some" way to looking away compared to not responding at all, a between-subject design. Furthermore, we defined a variety of measures to be analyzed that captured how students engaged with math problems. As mentioned before, MathSpring automatically codes student overall interaction with a math problem as engaged or disengaged (see Table 1). Specifically, when the online tutor classified student behavior in a math problem as SHINT, ATT, SOF, and SHELP, the student was coded as "engaged" while student-problem interactions classified as SKIP, NOTR, GIVEUP, and GUESS were coded as a "disengaged" student-problem interaction.

Table 2. Intervention strategies for evaluation conditions. The interventional strategies for re-engaging students once the system detected that the student seemed to be looking away. The experimental condition reacted to students' head wandering in one of four possible ways: flashing screen, text, alert, and companion message. This enabled a comparison of the possible impact of responding and not responding at all.

Condition	Sub-condition	1 Flashing screen	2 Text message	3 Alert sound	4 Companion message
Control	Control				
Experimental (treatment)	Flash	X		X	
	Text		X	X	
	Learning companion				X

Figure 6, Fig. 7, and Fig. 8 show re-engagement interventions used by the online tutor as a part of the treatment condition. Figure 6 shows a screenshot of a *Flashing Screen*, in which the screen flashed and then produced a tone to re-engage the learner. Similarly, Fig. 7 shows a Text Message alerting the learner followed by a tone. Figure 8 shows a Companion Message, in which the MathSpring learning companion was used as a medium to re-engage a distracted learner. The Alert Sound (not shown) displays a text message and plays a sound.

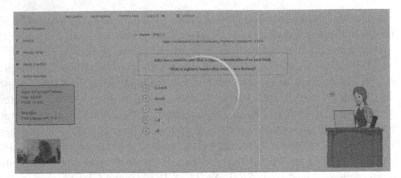

Fig. 6. Screen flash. Visual showing a distracted learner (bottom left) being alerted through screen flashing followed by a tone (Condition "Flash").

Fig. 7. Text prompt. Visual showing Condition "Text," where a text prompt is displayed on the screen followed by an alert sound.

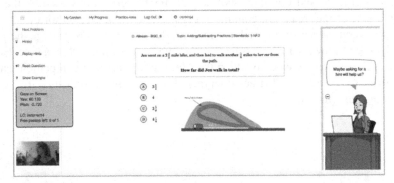

Fig. 8. Avatar learning companion. Visual showing Condition "Learning Companion", in which the learning companion gently reminds a learner who is looking away in an attempt to re-engage with the task.

4.4 Data

All data collected during this study were stored within a MySQL database and categorized into two sets, consisting of the students' level of engagement (expressed as head down, up, right, left, and tilt in degrees), timestamps, interventions, and identifiers for correlating with other data. The second set of data includes the students' performance (e.g., number of mistakes, number of hints requested, mastery level, problem difficulty, number of attempts). Post-test survey included the following questions:

- Did you notice the system was trying to get your attention when you were looking away?
- Do you think the MathSpring software made any mistakes and thought you were looking away when you were actually focused on your work? If so, why?
- Did you find the character, sound, text, and whirling gray blob reminder actions helpful?
- Did you find the reminder actions distracting or annoying?
- What do you think would make the reminder actions better?
- Do you think that giving some type of reminder like this when students look away from the screen for too long is helpful for learning? If so, why?

5 Results

5.1 Exploratory Analysis

We first performed exploratory frequency analyses to identify potential patterns. Specifically, we used the above grouping of student-problem engagement interaction types in Table 1 (engaged and disengaged with problem) and identified their frequency distributions. We identified which student activities (e.g., SOF, ATT, SHINT, SHELP) students excerpted in overall, regardless of condition, see Fig. 9. Solving the problem correctly on the FIRST attempt without any help was the most frequent student activity amongst students. It is also interesting to observe that SHELP – the student got the problem correct and saw at least one video – occurred the least, overall.

Similarly, the student behavior of SKIP was the most frequent kind of "disengaged" student interaction pattern with a problem. Thus skipping the problem and not doing anything was the most frequent student activity when disengaged. NOTR - the student did NOT even READ the problem and answered too fast in less than 4 s - was the least frequently occurring.

The breakdown of the number of students between the treatment and control groups is shown in Table 3 while Table 4 represents the number of problems solved by students according to the experimental and engagement conditions.

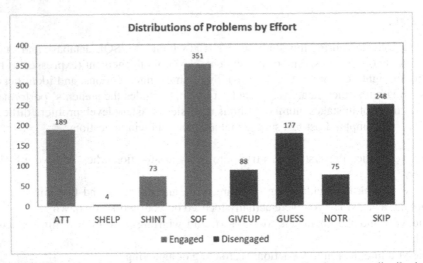

Fig. 9. Distribution of problems by effort. This Chart shows the frequency distribution of student-problem engagement behaviors overall, regardless of condition.

Table 3. Count of students by condition. Number of students who were detected head wandering at least once, during the study.

	Treatment	Control
Number of students	14/15	5/7

Table 4. Student-problem interaction behaviors per condition. The total number of cases where students in each condition were classified as engaged (SOF, ATT, SHINT, SHELP) or disengaged (GUESS, SKIP, NOTR, GIVEUP).

	Engaged	Disengaged
Treatment	448	389
Control	169	199

Note: Only students who experienced wandering during the study are included in this table

5.2 Re-engagement Interventions Analysis

We measured the levels of engagement (disengaged vs. engaged) associated with levels of experimental conditions (control vs treatment): $\chi^2 = 5.91$; $p < 0.02$ and non-directional Fisher's Exact Test ($p < 0.02$). We detected more engaged students in the treatment group, and more disengaged students in the control group, see Fig. 10. Non-directional Fisher's Exact Test also showed that the treatment group used more positive learning strategies than did the control group.

Given emotions could be induced by external factors and emotions could in turn impact learning behaviors, we studied whether levels of engagement could be associated with students' reported emotions. $\chi^2 = 84.08$; $p < 0.0001$ and Fisher's Exact Test ($p < 0.0001$) show a significant association, see Fig. 11. The treatment group also indicates a higher level of confidence compared to the control group. It is also interesting to note that the students in the treatment group also experienced a higher level of frustration.

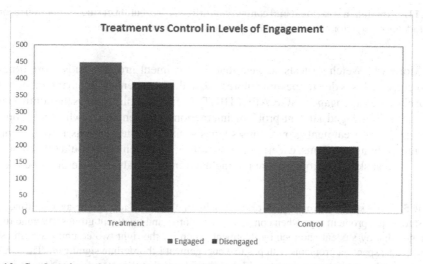

Fig. 10. Students' response to treatment and control conditions. Distribution of instances in which student-problem interactions were classified as engaged (SOF, ATT, SHINT, SHELP) or disengaged (GUESS, SKIP, NOTR, GIVEUP), in each condition.

We conducted a One-Way Analysis of Variance (ANOVA) to identify potential effects by re-engagement interventions. We identified a significant difference between the levels of interventions (control vs treatment) and the level of mastery (F = 19.24; $p < 0.0001$). We studied whether re-engagement interventions would have an effect on the number of problems completed by the students and found no significant difference.

When it comes to requesting hints when solving MathSpring problems, we observed no significant difference with the levels of interventions. Similarly, we also found no significant differences in the number of mistakes made, number of hints requested before solving problems, or the number of attempts tried before solving problems with levels of interventions.

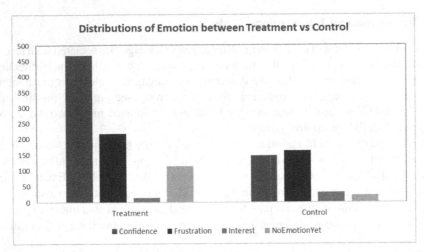

Fig. 11. Emotion in treatment and control conditions. Distribution of student emotion recorded by students in the tutor and classified by treatment and control condition.

Moreover, Welch's t-tests suggest that the treatment group spends more time per problem, and this is due to spending more time in those student-problem interactions that were classified as engaged (SOF, ATT, SHINT, SHELP). Table 5 shows the control group has more disengaged student-problem interactions than engaged, while the opposite happens in the treatment group. This suggests that the interventions lead students to continue solving problems, excerpting more effort, on which they would otherwise give up. Note that skipped problems are not included in this analysis since the time spent is

Table 5. Time spent - two sided T-tests. Results of Welch's T-test on the average time spent per student per problem type when comparing the control and treatment groups. Average time is averaged first by student and then by treatment group for the right two columns. Results show that the time spent on problems with positive engagement is borderline significant. These results and those in Fig. 10 (engagement) provide some evidence that the interventions may be helping students finish problems on which they would otherwise give up.

Average time spent on problems	p value	Control group average time (s)	Treatment group average time (s)
All problems	0.57	56.80	62.53
Problems resulting in engaged student-problem interactions	0.063	46.60	65.00
Problems resulting in disengaged student-problem interactions	0.31	72.74	56.48

recorded as -1, and math problems where the students give up are included with the time of their last guess before giving up.

6 Discussions

Students were attending an urban summer camp and most were excited to be there and to have the opportunity to socialize with their peers after being isolated due to the COVID-19. We observed that some students were chatting, playing, and occasionally arguing with their peers. This tells us that unlike in a regular and structured classroom environment, these summer campers were more "hyperaware." Because of such an informal summer camp setting, the stationary built-in camera on the laptops made it difficult for tracking students' head directions and orientations accurately and continuously.

Yet, in spite of the difficulties of tracking students, we are encouraged by the initial findings since this was an exploratory study of student engagement in MathSpring. As seen from the Chi-Square analysis, students exposed to the treatment group were more engaged overall, as indicated by them engaging in behaviors such as SHINT, ATT, SOF, and SHELP (see Table 1 above) while solving problems in MathSpring. We speculate that the audio component and visual animations used in the treatment group were able to retrieve students' attention and essentially bring them back within the context of MathSpring. One positive side effect is that once students are back on course, their accomplishments as indicated by high frequency of ATT and SOF (see Fig. 9 above) also led them potentially to a higher level of reported confidence. We also saw similar results when looking at the amount of time students spent on solving problems in the treatment group – hence, students exposed to re-engagement interventions tended to spend more time compared to those in the control group. We believe that the re-engagement strategies helped students to not only re-focus but also motivated them to persist through tough problems. This might also explain our significant finding of increased level of mastery of learning in the treatment group. In fact, Immordino-Yang [9] suggested that emotion is one factor that could relate to how students learn, self-regulate, and perform.

Another interesting observation is that both the treatment and control groups show a similar level of frustration experienced by the students. As Pekrun et al. [11] discussed in the Control Value Theory framework, students' ability to control the current situations could determine how much frustration they experience. Factors such as being monitored, using a new software, being in a new environment, or interacting with new peers might all lead to a high level of uncontrollability.

The qualitative results also provided insights about students' motivation. Particularly, we found comments such as "I love jake from state farm" and "I like the [learning] companion but wish you could make you[r] own companion" when we asked the students their perception and preference over MathSpring's learning companion. Students also told us that the engagement detection was useful "[b]ecause it tells people that they [aren't] paying attention" and allowed them to "...do [their] work". We further learned that more subtle interventional strategies might be needed since some of the students told us they were startled or were made nervous. Likewise, some students also commented that they might not like being monitored through a camera. An implication from a human-computer interaction and education perspective is that the design of re-engagement strategies should incorporate multiple modalities and should involve the students and educators themselves.

Hence, for RQ1, the impact of gaze interventions on learning, results from the exploratory analyses illustrate that students were more engaged, confident, persistent, and masterful when exposed to re-engagement interventional strategies once head wandering was detected. For RQ2, how were interventions perceived by students, we found that students enjoyed interacting with the avatar learning companion – one of the re-engagement interventional strategies. Students related the learning companion to real life characters and provided feedback about the strategies that were encouraging; students wanted to see improved versions and to personalize the strategies.

7 Conclusions

We described an exploratory study in which a computer vision detection system using deep learning techniques was used to identify students' mind wandering and a set of re-engagement interventional strategies were applied to bring students "back on course". This exploratory analysis showed that students were more engaged, confident, persistent, and masterful when exposed to re-engagement interventional strategies once head wandering was detected and that students enjoyed interacting with the avatar learning companion (a re-engagement interventional strategy) and related the learning companion to real life characters. Results showed that time spent on problems with positive engagement when compared between control and treatment groups is statistically significant and some evidence exists that the interventions may be helping students finish problems on which they would otherwise give up.

We noticed that 7th and 8th grade students were engaged and did not demonstrate head wandering. We would like to investigate whether this pattern holds more generally (that older students don't wander as much/at all) or if this was just an anomaly due to the structure of the summer camp. We plan to conduct a future experiment in a typical classroom to determine this.

In future research, we plan to understand how several student characteristics might account for head wandering and the impact of the intervention strategies. Thus we will explore the impact of age, gender, time of the day (before/after lunch), fatigue (being tired), and boredom (beginning vs, end of session) on student learning after interventions are applied. Also, in future work, we plan to further understand and identify the most effective engagement interventional strategies.

Acknowledgments. We thank the participants of our experimental study and acknowledge partial funding for this work by the National Science Foundation, grants 1551572, 1551589, 1551590, and 1551594.

References

1. Arroyo, I., et al.: Repairing disengagement with non-invasive interventions. In: Proceedings of the 2007 Conference on Artificial Intelligence in Education: Building Technology Rich Learning Contexts That Work (2007)

2. Arroyo, I., Wixon, N., Allessio, D., Woolf, B., Muldner, K., Burleson, W.: Collaboration improves student interest in online tutoring. In: André, E., Baker, R., Hu, X., Rodrigo, M.M.T., du Boulay, B. (eds.) AIED 2017. LNCS (LNAI), vol. 10331, pp. 28–39. Springer, Cham (2017). https://doi.org/10.1007/978-3-319-61425-0_3

3. Arroyo, I., Woolf, B.P., Burelson, W., Muldner, K., Rai, D., Tai, M.: A multimedia adaptive tutoring system for mathematics that addresses cognition, metacognition and affect. Int. J. Artif. Intell. Educ. **24**(4), 387–426 (2014)

4. Carini, R.M., Kuh, G.D., Klein, S.P.: Student engagement and student learning: testing the linkages. Res. High. Educ. **47**(1), 1–32 (2006)

5. Gupta, A., et al.: Affective teacher tools: affective class report card and dashboard. In: Roll, I., McNamara, D., Sosnovsky, S., Luckin, R., Dimitrova, V. (eds.) AIED 2021. LNCS (LNAI), vol. 12748, pp. 178–189. Springer, Cham (2021). https://doi.org/10.1007/978-3-030-78292-4_15

6. Kulik, J.A., Fletcher, J.D.: Effectiveness of intelligent tutoring systems: a meta-analytic review. Rev. Educ. Res. **86**(1), 42–78 (2016)

7. Monkaresi, H., Bosch, N., Calvo, R.A., D'Mello, S.K.: Automated detection of engagement using video-based estimation of facial expressions and heart rate. IEEE Trans. Affect. Comput. **8**(1), 15–28 (2017)

8. Mulqueeny, K., Kostyuk, V., Baker, R.S., Ocumpaugh, J.: Incorporating effective e-learning principles to improve student engagement in middle-school mathematics. Int. J. STEM Educ. **2**(1), 1–14 (2015)

9. Immordino-Yang, M.H.: Emotions, Learning and the Brain: Exploring the Educational Implications of Affective Neuroscience. W.W. Norton & Company (2016)

10. Khan, A.Z., Blohm, G., McPeek, R.M., Lefevre, P.: Differential influence of attention on gaze and head movements. J. Neurophysiol. **101**(1), 198–206 (2009)

11. Pekrun, R., Frenzel, A.C., Goetz, T., Perry, R.P.: The control-value theory of achievement emotions: an integrative approach to emotions in education. In: Schutz, P.A., Pekrun, R. (eds.) Chapter 2. Emotion in Education, pp. 13–36. Academic Press (2007)

12. Sümer, Ö., Goldberg, P., D'Mello, S., Gerjets P., Trautwein, U., Kasneci, E.: Multimodal engagement analysis from facial videos in the classroom. IEEE Trans. Affect. Comput. (2021)

13. Taylor, L., Parsons, J.: Improving student engagement. Curr. Issues Educ. **14**(1) (2011)

14. Xhakaj, F., Aleven, V., McLaren, B.M.: Effects of a dashboard for an intelligent tutoring system on teacher knowledge, lesson plans and class sessions. In: André, E., Baker, R., Hu, X., Rodrigo, M., du Boulay, B. (eds.) AIED 2017. LNCS, vol. 10331, pp. 582–585. Springer, Cham (2017). https://doi.org/10.1007/978-3-319-61425-0_69

15. Yu, H., et al.: Measuring and integrating facial expressions and head pose as indicators of engagement and affect in tutoring systems. In: Sottilare, R.A., Schwarz, J. (eds.) HCII 2021. LNCS, vol. 12793, pp. 219–233. Springer, Cham (2021). https://doi.org/10.1007/978-3-030-77873-6_16

Assessing the Social Agency of Pedagogical Agents in Adaptive Training Systems

Bradford L. Schroeder[1]([✉]), Nicholas W. Fraulini[2], Wendi L. Van Buskirk[1], and Reganne M. Miller[3]

[1] Naval Air Warfare Center Training Systems Division, Orlando, FL 32826, USA
{bradford.l.schroeder.civ,wendi.l.vanbuskirk.civ}@us.navy.mil
[2] StraCon Services Group, Fort Worth, TX 76109, USA
nicholas.w.fraulini.ctr@us.navy.mil
[3] Morehead State University, Morehead, KY 40351, USA
rmiller4@moreheadstate.edu

Abstract. Pedagogical agents (PAs) could be used to mimic a one-on-one human tutoring experience in adaptive training systems. Social agency theory (SAT) is one perspective that describes how humans learn from PAs. However, there are no measures to test the components of SAT and their effects on learning. Therefore, we discuss the development of a subjective measure, the Social Agency Theory Questionnaire (SATQ), to assess components of the social agency theory framework: social cues, the cooperation principle, and deep cognitive processing. Next, we present a study that investigates the effectiveness of a PA instructor who provides error-sensitive human-voiced feedback to learners in an adaptive training system (PA-Present group), compared against an instructionally equivalent text-based instructor system (PA-Absent group). We hypothesize that the PA-Present group will exhibit higher learning outcomes than the PA-Absent group. Additionally, we hypothesize that participants in the PA-Present group will rate the "instructor" higher on all three subscales of the SATQ than participants in the PA-Absent group. Data collection for this project is currently underway. Future analyses will examine between-group differences on performance and learning outcomes, and their association with subjective ratings on SATQ subscales. In addition, we will present psychometric analyses of the SATQ and proposed revisions to the scale.

Keywords: Social agency theory · Pedagogical agents · Adaptive training · Feedback

1 Introduction

1.1 Adaptive Training – Feedback Strategies

Adaptive training (AT) systems are computer-based instructional systems that provide tailored training to a learner based on variables such as in-situ performance or learner attributes [1]. AT systems are supported by algorithms that adjust some aspect of the training (e.g., difficulty, feedback, content) based on the aforementioned factors. Previous

© The Author(s), under exclusive license to Springer Nature Switzerland AG 2022
R. A. Sottilare and J. Schwarz (Eds.): HCII 2022, LNCS 13332, pp. 302–313, 2022.
https://doi.org/10.1007/978-3-031-05887-5_21

research suggests that such adaptations are beneficial for learners when compared to informationally equivalent non-adaptive systems [2–4]. A goal of AT is to achieve the tutoring effect where students learn as much from an AT system as they would one-on-one from a human tutor [5, 6].

The goal of our research is to investigate characteristics of human tutors that make them effective and incorporate those strategies into AT systems. Providing feedback to students is one of the main strategies human tutors use. Indeed, AT system designers have used a variety of presentation methods for delivering feedback to trainees, such as feedback that is provided in text form [7], with text and images [2], or contextual animations [8]. However, we wanted to investigate the use of pedagogical agents (PAs) as a mechanism to deliver feedback much like a human tutor would. The following sections provide the rationale for the use of PAs for delivering feedback in adaptive training systems and a questionnaire for assessing their effectiveness.

1.2 Learning with Pedagogical Agents

In recent decades, the roles that PAs have played in learning environments have expanded greatly [9]. PAs can be defined as agents in the form of a virtual character equipped with artificial intelligence that can support the students' learning process and use various instructional strategies in an interactive learning environment [10]. A related term is a Non-Player Character (NPC), which often takes the form of an instructor [11–13] that guides learners toward learning objectives by providing them with narrative instruction or feedback. Much like PAs, NPCs are considered effective when they are believable to the user and should be designed to maximize believability for optimal effectiveness [14–16].

When designing instructional systems, PAs are often chosen because they are believed to confer benefits to the learner and the learning experience. Previous research suggests that they can foster motivation [17], enjoyment [18], engagement [9], flow (a state of deep involvement) [19], and immersion [20]. In their meta-analysis examining the effectiveness of PAs during learning, Schroeder, Adesope, and Gilbert found benefits when PAs were utilized rather than on-screen text [21]. Researchers have sought to uncover properties of PAs that facilitate learning, studying their voice [22], likeability [23], and politeness [24]. Critical to the success of these PAs, though, appears to be the role the agent plays during instruction [25]. Baylor and Kim discussed how research since their previous work [26] has affirmed and expanded their claim that personalized roles help PAs to facilitate learning. Similarly, Veletsianos advises that PAs be contextually relevant during the learning process [27]. In the following section, we will describe theoretical perspectives explaining how pedagogical agents influence the learning process.

1.3 Theoretical Background: How Do Pedagogical Agents Influence Learning?

Social agency theory (SAT) is a framework used to explain how a human learns from a PA [28] (see Fig. 1). This theory proposes that a PA presents social cues within its interactions with a human, which lead to improved learning via the cooperation principle and deep cognitive processing. For example, the PA's voice [28], gestures, facial expression, and eye gaze may serve as social cues [29]. These cues prime the learner's social conversation schema, causing the learner to behave as if they are conversing with another human and thereby prompting certain social rules found in human-to-human communication. Mayer and colleagues emphasized Grice's cooperation principle [30] as a framework for the social rules that may come into play in human-PA communication (see Table 1 for descriptions of the four maxims of the cooperation principle). Due to the elicitation of these human-to-human social rules, the learner attempts to make deeper sense of what the PA is communicating to them, thus resulting in deep cognitive processing and more meaningful learning [28].

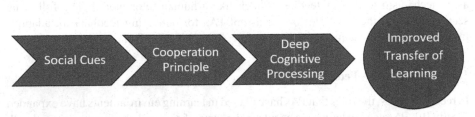

Fig. 1. The process of social agency theory according to Mayer and colleagues [28]

Lin and colleagues suggested that, when a conversation primes the cooperation principle, the learner may be more determined to understand and process what the PA is communicating, leading to deep cognitive processing [31]. Mayer and colleagues defined deep cognitive processing as meaningful learning in which the learner identifies important information and organizes this information into logical verbal and visual cognitive models [28]. The learner then integrates these models with each other and with pre-existing knowledge. This process allows the learner to apply the knowledge they have gained to other situations, which is reflected in transfer tests [28]. Alternatively, Lin and colleagues hypothesized if the PA does not act in accordance with the learner's pre-established social rules, that the cooperation principle may not be provoked. This may cause the learner to withdraw from the conversation, limiting deep cognitive processing, immersion, and meaningful learning outcomes [31].

Another framework that researchers have used to study how PAs influence the learning process is the Persona Effect [32]. This effect states that when a personified educational agent is presented in the learning environment, the learner may perceive conversations with the PA as human-to-human interactions, leading to a positive view of the learning environment and enhanced motivation [32, 33]. In other words, this effect proposes that the mere presence of a lifelike PA may improve the learning experience. Like SAT, enhancing the social aspect of the human-computer interaction improves learning. However, whereas SAT seems to be more focused on priming the social conversation schema, the Persona Effect appears to be centered around altering the learner's perceptions of the learning environment using a personified agent [32]. Many studies only support the subjective benefits of the persona effect, such as supporting heightened entertainment levels, engagement, trustworthiness, believability, and perception [32, 34–36]. Despite the lack of data on objective benefits, much of the literature assumes that the

Table 1. Four maxims of the cooperation principle

Maxim	Rule
Quantity	Speaker is expected to make their involvement in the conversation exactly as informative as required. They are neither under-informative nor over-informative
Quality	Speaker is expected to provide accurate information. They only say what they believe is true and that for which they have sufficient evidence
Relation	Speaker is expected to provide information relevant to the conversation. Relevancy may shift throughout the conversation
Manner	Speaker should be concise. They should maintain organization and avoid ambiguity, obscurity of expression, and unnecessary wordiness

persona effect improves learning. However, there are studies that suggest that actual learning gains from the persona effect alone are small to nonexistent [17, 35, 37].

1.4 Experimental Extensions of Social Agency Theory

Utilizing SAT, previous research tested a "social-cue hypothesis" arguing that PAs that are designed with more social cues (i.e., human-like face, natural voice) would be more likely to improve learning as predicted by SAT [38]. Louwerse and colleagues found mixed interacting social cue effects with two experiments, ultimately concluding that the quantity of social cues may not matter. The clearest example of this is exhibited in their second experiment, which did not identify learning differences between an animated and narrated PA and a PA that was narrated but absent from view. In two experiments examining learning gains using worked-out examples, Atkinson, Mayer, and Merrill showed that high school and college students displayed better performance in both near and far transfer when learning from a PA with a human voice compared to a PA with a machine-like voice [39]. However, Moon and Ryu presented evidence that social cues in the form of conversational gestures (e.g., hand-waving and pointing) can be detrimental to learning comprehension [40]. Due to the variety of possible social cues one could manipulate, Louwerse and colleagues acknowledged that comparing social cues across experiments is challenging in this body of research [40].

If the human conversational schema is the primary mechanism by which PAs improve learning, how do designers know their PAs have elicited it? Previous research has made this inference when observing improved learning outcomes, but improved learning is only indirect evidence. Considering the findings for learning benefits in this literature are mixed, it could be that some PAs do not elicit the human social conversational schema. Or, they may elicit this schema without fostering the cooperation principle or deeper cognitive processing of the instructional material. In the proceeding sections, we describe the theoretical development of a proposed Social Agency Theory Questionnaire (the SATQ) to address this gap in the literature.

2 Scale Development

In previous research, social agency effects are assumed to have occurred when a learning benefit is observed. In order to more fully assess the mechanisms of SAT and their roles in learning, we set out to repurpose items from existing scales or research when possible to assess attributes of social cues, the cooperation principle, and deep cognitive processing. When that was not possible (due to lack of existing social agency measures), we created our own items based on the component processes of SAT. For the SATQ, we treated these categories as subscales, as they are the primary theoretical components of SAT (see Fig. 1).

A relevant scale exists called the Agent Persona Instrument [41, 42] (API; or API-R after revisions suggested by Schroeder, Romine, & Craig), which addresses the Persona Effect [32]. In developing the SATQ, we adapted several items from the API-R that we considered relevant to the processes of SAT. At the time of writing, the most recent published psychometric analysis [42] indicated good reliability and fit indices for the API-R, but these researchers noted that there were only small correlations between subscales of the API-R and learning outcomes. When examining this measure in the context of SAT, the API-R's "Facilitating Learning" subscale contains items that mostly relate to deep cognitive processing, but few items relate to social cues or the cooperation principle. As mentioned previously, the Persona Effect is assumed to enhance the learning experience but does not specifically predict improved learning outcomes. SAT predicts improved learning outcomes, so a measure that addresses the mechanisms of SAT should be more directly related to objective measures of learning.

For the social cues subscale, we adapted questionnaire items that Atkinson and colleagues suggested [39]. In their paper, they argued that learners would rate PAs with more social cues higher on these items as compared to PAs with fewer or no social cues. We also included one item from the API-R's "Engaging" subscale ("The instructor was easy to connect with"). For the items we generated, our goal was to have learners focus on their perceptions of their interaction with the PA during learning.

Next, we considered Lin and colleagues' argument that the cooperation principle is a prerequisite for deep cognitive processing [31], and generated items from Grice's four maxims of the cooperation principle [30] (see Table 1). We generated at least two items for each of the four maxims of Quantity, Quality, Relation, and Manner. We also adapted two items from the Facilitating Learning subscale of the API-R: "The instructor focused me on the relevant information" and "The instructor communicated the main ideas clearly." These items are associated with the relation and manner maxims, respectively. The remaining items were generated for this scale with the intent to gauge subjective perceptions of the PA's cooperation in communicating information to the learner.

Lastly, to generate items for the SATQ's deep cognitive processing subscale, we utilized some of the "Facilitating Learning" items from the API-R, and we also generated items based on Mayer's definition for deep cognitive processing (i.e., assimilating multimedia models into prior knowledge) [28]. The items in this subscale are intended to assess learners' subjective perceptions of their own deep cognitive processing of the material they received. The proposed measure can be seen in Table 2.

The primary use of the proposed scale is to assess learners' subjective perceptions of their interactions with a PA. The goal is to ascertain whether the PA elicited key aspects of

SAT that are integral for learning. We suggest that subjective responses from this measure should correlate with objective learning outcomes, in line with SAT's predictions that PA social cues enhance learning. In addition, another goal is to offer a standardized instrument for examining differences in subjective perceptions of social agency among different PAs or experimental paradigms, addressing a concern of previous research [38]. In light of disparate results in the research literature in terms of learning gains, this scale could distinguish among PAs that foster deeper cognitive processing and those that only trigger the social conversation schema or cooperation principle. SAT predicts that an agent that does not foster deep cognitive processing may not be beneficial for learning. Therefore, a PA that only elicits the human conversational schema may be a PA that enhances the subjective learning experience without yielding learning gains. In the following section, we describe current research with the goal to use and validate the SATQ in an adaptive training system.

3   The Present Study

The present study investigates the effectiveness of a PA instructor who provides error-sensitive feedback to participants in real-time on a radar detection task during simulated electronic warfare training missions. We have approached this differently than other PA or virtual instructor studies, which typically involve the delivery of a lesson and subsequent question-based testing [43]. Our PA delivers feedback in situ during scenario-based training, and learners will be tested within a similar task-based context. The PA instructor will be compared to an informationally equivalent text-based feedback system, which will be driven by the same error-sensitive feedback algorithm. The present study will use a 2-group between-subjects design to compare learning performance and subjective ratings on the SATQ. Our hypotheses for the study are:

H1: Participants who learn with the PA instructor will exhibit better learning outcomes compared to those learning with the text-based feedback system.

H2: Participants who learn with the PA instructor will rate the instructor higher on the SATQ subscales than participants who learn with the text-based instructor system.

Based on our hypotheses, we believe the PA instructor's social cues will prime participants' conversation schema to a greater extent than the text-based instructor system. In turn, this conversation schema would have a motivating effect on participants, as described by Lin and colleagues [31]. We propose this process will result in both improved learning outcomes for participants, as well as higher subjective ratings of social agency for the instructor in the PA condition.

Table 2. Items and factors of the social agency theory questionnaire (SATQ)

Social Cues

*The **instructor** was easy to connect with.
I felt as if the **instructor was talking directly to me.
I felt as if the **instructor was interacting with me.
The **instructor was trying to be helpful.
‡The **instructor** communicated with me naturally.
‡My interaction with the **instructor** felt unnatural.
‡My interaction with the **instructor** felt like an interaction I would have with a tutor or teacher.
‡The **instructor** understood my learning needs.

Cooperation Principle

*The **instructor** focused me on the relevant information.
*The **instructor** communicated the main ideas clearly.
‡The **instructor** gave me too much information.
‡The **instructor** gave me sufficient information.
‡The **instructor** should have given me more information.
‡The **instructor** gave me inaccurate information.
‡The **instructor** gave me accurate information.
‡The **instructor** gave me the right information when I needed it.
‡I felt prepared for the next task after interacting with the **instructor**.
‡The **instructor** could have given me the same information in a simpler way.
‡The **instructor** gave me the material in a confusing manner.

Deep Cognitive Processing

*The **instructor** led me to think more deeply about the presentation.
*The **instructor** made the instruction interesting.
*The **instructor** encouraged me to reflect on what I was learning.
*The **instructor** helped me to concentrate on the presentation
*The **instructor** helped me learn the material
*The **instructor** was easy to learn from
‡The **instructor** helped me self-evaluate my performance.
‡The **instructor** helped me connect the new material to information I already knew.
‡The **instructor's** use of words and images helped me to better understand the task.
‡The **instructor** helped me connect main ideas.
‡The **instructor** helped me remember relevant information.
‡The **instructor** presented the material in a cohesive manner.

---

*Item adapted from API-R; Schroeder, Romine, & Craig [42]
**Item adapted from suggestions in Atkinson, Mayer, & Merrill [39]
‡Original items

The word **instructor** should be adapted for use in other domains as appropriate (i.e., agent, character, etc.). All items on this scale should be rated on a 1 (Strongly Disagree) to 5 (Strongly Agree) Likert scale.

4 Method

Participants will learn a radar detection task where they must perceive radar signals (i.e., emitters) in the environment, identify characteristics (e.g., emitter name, what type of platform the emitter is on, the type of threat to the mission, etc.) and report that information to the Mission Commander. Participants must determine that information from multiple information sources provided by the research testbed such as emitter parametric data (emitter frequency, scan times, scan types, etc.). They will also be required to classify the threat level of the emitters based on these variables and report this information in a timely manner to the Mission Commander. Participant training on the radar detection task will be presented in a text-based instructional PowerPoint and an experimenter-led demonstration of the task workflow. Participants will then complete a knowledge quiz to ensure they understand the content, followed by a 10-min pre-test scenario. Next, participants will be randomly assigned to a condition where they receive either PA-supported feedback (PA-Present) or text-only feedback (PA-Absent). Depending on their assigned condition, participants will be notified that an instructor (PA-Present) or an instructor system (PA-Absent) has connected to their system and will be providing feedback throughout their next three scenarios. During each of these three, 10-min scenarios, all participants will have four opportunities to receive feedback. The feedback they receive will be in video form, determined by an algorithm that detects the task component they are struggling with the most at the time the feedback is selected. The feedback video will demonstrate the necessary steps for completing the selected task component. Participants in the PA-Present group will receive narration accompanying their feedback video, whereas participants in the PA-Absent group will receive text. For each feedback topic, the feedback videos in the separate conditions will be identical regarding their instructional content; furthermore, the narration provided will be identical to the text presented in each video.

At the end of these scenarios, participants will be asked about their perceptions of the instructor or text-based instructor system via the Agent Persona Instrument-Revised (API-R) [42] and the Social Agency Theory Questionnaire (SATQ). We will also assess their subjective perceptions of flow (using the Core Flow Scale) [44] and engagement (using the Game Engagement Questionnaire) [45]. Afterward, participants will complete a post-test scenario without experimental feedback. Lastly, participants will complete a radar detection transfer scenario requiring the same skills as in previous scenarios, but those skills will be used in a different way with a different overall objective. Specifically, instead of reporting on all the emitters in the environment, participants will be told to search for specific emitters and report them immediately. Participants will be provided with incomplete emitter information (i.e., only frequency and scan time) and be asked to find the specific emitter based on this limited data, complete the emitter report, and report the emitter to the Mission Commander in a timely manner.

5 Results

Due to the COVID pandemic, data collection for this experiment has been delayed. However, data collection is currently underway, and we intend to provide results when this paper is presented. As data are collected, the proposed SATQ will be psychometrically evaluated and revised as appropriate.

To evaluate the SATQ, we will examine differences in each of the subscales between our PA-Present and PA-Absent groups. Per SAT, social cues are the prerequisite for eliciting the cooperation principle and deep cognitive processing. If the SATQ is an appropriate measure of SAT, we should expect the PA-Present group to have higher subjective ratings of social cues, the cooperation principle, and deep cognitive processing compared to the PA-Absent group. If these results emerge, we suggest they would offer preliminary evidence for construct validity of the SATQ.

We will also conduct exploratory analyses examining the correlation among SATQ subscales and subjective measures of flow and engagement. As previously mentioned, well-designed PAs can foster flow and engagement in the learning experience, which are assumed to confer benefits for learning. We posit that a PA that elicits the cooperation principle or deep cognitive processing should also facilitate flow and engagement.

Once sufficient data are collected, we intend to assess the internal consistency of our measure and its subscales. In addition, we plan to perform a confirmatory factor analysis to examine item factor loadings onto the latent variables of the SATQ to determine which items to retain, modify, or discard.

6 Applications and Extensions

This scale represents an evolution of the SAT suggestions outlined by Atkinson and colleagues [39], and a possible solution to their claim that social agency theory requires direct testing. As mentioned, studies directly testing SAT have been sparse without clear conclusions, and scholars have recently emphasized the dearth of research expanding SAT [46].

At a subscale level, the SATQ may be useful outside of the learning domain. According to the Computers are Social Actors (CASA) framework [47] humans tend to perceive their interactions with a computer through a human-to-human conversational lens. The SATQ may be beneficial in cases where subjective perceptions of non-instructional characters are important to a researcher or designer. For example, designers may be interested in user perceptions of an interactive non-instructional NPC in a simulation for evaluations of believability, which is important for an NPC to be effective [15, 16].

For analysis purposes, we believe it is worth considering other relevant variables that PAs influence in learning. Variables such as flow, engagement, or enjoyment should correlate to the social cues subscale of the SATQ, since these factors relate to interacting with well-designed PAs. The subscales of the SATQ may also be useful for regression analyses where researchers are interested in assessing process models about SAT for learning. Other researchers have noted the need for tools that can empower analyses beyond comparing group differences in means and standard deviations [46]. For example, it is possible that individual differences in perceptions of the social agency process will explain additional variance in learning outcomes.

For practitioners, we propose that the SATQ could be useful for anyone who designs a training system with PAs. If this scale can identify subjective aspects of SAT that are conducive to learning, it may be a useful tool for assessing learners' experiences with

PAs. Based on scale responses, practitioners could iteratively adjust the characteristics of their agent to ensure they foster social agency that leads to learning. For advanced training systems that rely on additional instructional features (such as AT systems), the exemplary PA would foster social agency without detracting from the benefits of adaptivity. Further, the use of PAs in AT systems that trigger the processes of social agency could be a means for students to learn as much as they would from a one-on-one human tutoring experience.

Acknowledgments. We gratefully acknowledge Dr. Jason Hochreiter, Mr. Rob Veira, and Ms. Rebecca Pharmer for their assistance with testbed development. This work was funded under the Naval Innovative Science and Engineering program established by the National Defense Authorization Act, Section 219. Due to difficulties with data collection brought on by the COVID-19 pandemic, data collection for this study is ongoing at time of writing. Data relevant to this project are to be presented virtually at the 2022 HCII conference. Interested readers are encouraged to contact the authors for recent developments and analyses with the SATQ. Presentation of this material does not constitute or imply its endorsement, recommendation, or favoring by the U.S. Navy or the Department of Defense (DoD). The views expressed herein are those of the authors and do not necessarily reflect the official position of the Department of Defense or its components. NAWCTSD Public Release 22-ORL016 Distribution Statement A – Approved for public release; distribution is unlimited.

References

1. Landsberg, C.R., Astwood Jr., R.S., Van Buskirk, W.L., Townsend, L.N., Steinhauser, N.B., Mercado, A.D.: Review of adaptive training system techniques. Mil. Psychol. **24**(2), 96–113 (2012)
2. Landsberg, C.R., Mercado, A., Van Buskirk, W.L., Lineberry, M., Steinhauser, N.: Evaluation of an adaptive training system for submarine periscope operations. In: Proceedings of the Human Factors and Ergonomics Society 56th Annual Meeting, pp. 2422–2426. SAGE Publications, Los Angeles (2012)
3. Marraffino, M.D., Johnson, C.I., Whitmer, D.E., Steinhauser, N.B., Clement, A.: Advise when ready for game plan: adaptive training for JTACs. In: Proceedings of the Interservice/Industry, Training, Simulation, and Education Conference (I/ITSEC), Orlando (2019)
4. Wickens, C.D., Hutchins, S., Carolan, T., Cumming, J.: Effectiveness of part-task training and increasing-difficulty training strategies: a meta-analysis approach. Hum. Factors **55**(2), 461–470 (2013). https://doi.org/10.1177/0018720812451994
5. Bloom, B.S.: The 2 sigma problem: the search for methods of group instruction as effective as one-to-one tutoring. Educ. Res. **13**, 4–16 (1984)
6. VanLehn, K.: The relative effectiveness of human tutoring, intelligent tutoring systems, and other tutoring systems. Educ. Psychol. **46**(4), 197–221 (2011)
7. Buff, W.L., Campbell, G.E.: What to do or what not to do?: identifying the content of effective feedback. In: Proceedings of the 46th Annual Meeting of the Human Factors and Ergonomics Society, pp. 2074–2078. HFES, Santa Monica (2002)
8. Landsberg, C.R., Bailey, S., Van Buskirk, W.L., Gonzalez-Holland, E., Johnson, C.I.: Designing effective feedback in adaptive training systems. In: Interservice/Industry, Training, Simulation, and Education Conference (I/ITSEC), Orlando (2016)

9. Veletsianos, G., Russell, G.S.: Pedagogical agents. In: Spector, J., Merrill, M., Elen, J., Bishop, M. (eds.) Handbook of Research on Educational Communications and Technology, pp. 759–769. Springer, New York (2014). https://doi.org/10.1007/978-1-4616-3185-5_61

10. Martha, A.S.D., Santoso, H.B.: The design and impact of the pedagogical agent: a systematic literature review. J. Educat. Online **16**(1) (2019)

11. Backlund, P., Engstrom, H., Hammar, C., Johannesson, M., Lebram, M.: SIDH–a game based firefighter training simulation. In: 2007 11th International Conference Information Visualization (IV 2007), pp. 899–907. IEEE (2007)

12. Halverson, R., Blakesley, C., Figueiredo-Brown, R.: Video game design as a model for professional learning. In: Learning to Play: Exploring the Future of Education with Video Games, pp. 9–28 (2011)

13. Muntean, C.H., Andrews, J., Muntean, G.M.: Final frontier: an educational game on solar system concepts acquisition for primary schools. In: 2017 IEEE 17th International Conference on Advanced Learning Technologies (ICALT), pp. 335–337. IEEE (2017)

14. Lankoski, P., Björk, S.: Gameplay design patterns for believable non-player characters. In: Akira, B. (ed.) Situated Play: Proceedings of the 2007 Digital Games Research Association Conference, pp. 416–423. The University of Tokyo, Tokyo (2007)

15. Riedl, M., Lane, H.C., Hill, R., Swartout, W.: Automated story direction and intelligent tutoring: towards a unifying architecture. U.S. Army Research, Development, and Engineering Command (2006). https://apps.dtic.mil/dtic/tr/fulltext/u2/a459187.pdf

16. Warpefelt, H., Johansson, M., Verhagen, H.: Analyzing the believability of game character behavior using the game agent matrix. In: DiGRA Conference, pp. 1–11 (2013)

17. Mitrovic, A., Suraweera, P.: Evaluating an animated pedagogical agent. In: Gauthier, G., Frasson, C., VanLehn, K. (eds.) ITS 2000. LNCS, vol. 1839, pp. 72–82. Springer, Berlin (2000). https://doi.org/10.1007/3-540-45108-0_11

18. Moreno, R., Mayer, R.E., Spires, H.A., Lester, J.C.: The case for social agency in computer based teaching: do students learn more deeply when they interact with animated pedagogical agents? Cogn. Instr. **19**(2), 177–213 (2001)

19. Zakharov, K., Mitrovic, A., Johnston, L.: Towards emotionally-intelligent pedagogical agents. In: Woolf, B.P., Aïmeur, E., Nkambou, R., Lajoie, S. (eds.) ITS 2008. LNCS, vol. 5091, pp. 19–28. Springer, Heidelberg (2008). https://doi.org/10.1007/978-3-540-69132-7_7

20. Adams, E.: Fundamentals of Game Design, 3rd edn. Pearson Education, Peachpit (2014)

21. Schroeder, N.L., Adesope, O.O., Gilbert, R.B.: How effective are pedagogical agents for learning? A meta-analytic review. J. Educ. Comput. Res. **49**(1), 1–39 (2013)

22. Atkinson, R.K.: Optimizing learning from examples using animated pedagogical agents. J. Educ. Psychol. **94**(2), 416–427 (2002)

23. Domagk, S.: Do pedagogical agents facilitate learner motivation and learning outcomes? J. Media Psychol. **22**(2), 82–95 (2010)

24. Wang, N., Johnson, W.L., Mayer, R.E., Rizzo, P., Shaw, E., Collins, H.: The politeness effect: pedagogical agents and learning outcomes. Int. J. Hum Comput Stud. **66**(2), 98–112 (2008)

25. Kim, Y., Baylor, A.L.: Research-based design of pedagogical agent roles: a review, progress, and recommendations. Int. J. Artif. Intell. Educ. **26**(1), 160–169 (2016)

26. Baylor, A.L., Kim, Y.: Simulating instructional roles through pedagogical agents. Int. J. Artif. Intell. Educ. **15**(1), 95 (2005)

27. Veletsianos, G.: Contextually relevant pedagogical agents: visual appearance, stereotypes, and first impressions and their impact on learning. Comput. Educ. **55**(2), 576–585 (2010)

28. Mayer, R.E., Sobko, K., Mautone, P.D.: Social cues in multimedia learning: role of speaker's voice. J. Educ. Psychol. **95**(2), 419–425 (2003). https://doi.org/10.1037/0022-0663.95.2.419

29. Mayer, R.E., DaPra, C.S.: An embodiment effect in computer-based learning with animated pedagogical agents. J. Exp. Psychol. Appl. **18**(3), 239–252 (2012). https://doi.org/10.1037/a0028616

30. Grice, H.P.: Logic and conversation. In: Cole, P., Morgan, J. (eds.) Syntax and Semantics, vol. 3, pp. 41–58. Academic Press (1975). https://doi.org/10.1163/9789004368811_003

31. Lin, L., Atkinson, R.K., Christopherson, R.M., Joseph, S.S., Harrison, C.J.: Animated agents and learning: does the type of verbal feedback they provide matter? Comput. Educ. **67**, 239–249 (2013). https://doi.org/10.1016/j.compedu.2013.04.017

32. Lester, J.C., Converse, S.A., Kahler, S.E., Barlow, S.T., Stone, B.A., Bhogal, R.S.: The persona effect: affective impact of animated pedagogical agents. In: Pemberton, S. (ed.) Proceedings of CHI 1997: Human Factors in Computing Systems, pp. 259–266. ACM Press (1997). https://doi.org/10.1145/258549.258797

33. Arner, T., McCarthy, K.S., McNamara, D.S.: iSTART stairstepper – using comprehension strategy to game the test. Computers **10**(4), 48 (2021). https://doi.org/10.3390/computers100 40048

34. Lester, J.C., Stone, B.A.: Increasing believability in animated pedagogical agents. In: Proceedings of the First International Conference on Autonomous Agents, pp. 16–21. ACM Press (1997)

35. van Mulken, S., Andre, E., Müller, J.: The persona effect: how substantial is it? In: Johnson, H., Nigay, L., Roast, C. (eds.) People and Computers XIII, pp. 53–66. Springer, London (1998). https://doi.org/10.1007/978-1-4471-3605-7_4

36. Walker, J.H., Sproull, L., Subramani, R.: Using a human face in an interface. In: Adelson, B., Dumais, S., Olson, J. (eds.) Proceedings of CHI 1994: Human Factors in Computing Systems, pp. 85–91. ACM Press (1994). https://doi.org/10.1145/191666.191708

37. Schroeder, N.L., Gotch, C.M.: Persisting issues in pedagogical agent research. J. Educ. Comput. Res. **53**(2), 183–204 (2015). https://doi.org/10.1177/0735633115597625

38. Louwerse, M.M., Graesser, A.C., Lu, S., Mitchell, H.H.: Social cues in animated conversational agents. Appl. Cogn. Psychol. **19**(6), 693–704 (2005)

39. Atkinson, R.K., Mayer, R.E., Merrill, M.M.: Fostering social agency in multimedia learning: examining the impact of an animated agent's voice. Contemp. Educ. Psychol. **30**(1), 117–139 (2005)

40. Moon, J., Ryu, J.: The effects of social and cognitive cues on learning comprehension, eye-gaze pattern, and cognitive load in video instruction. J. Comput. High. Educ. **33**(1), 39–63 (2020). https://doi.org/10.1007/s12528-020-09255-x

41. Baylor, A., Ryu, J.: The API (agent persona instrument) for assessing pedagogical agent persona. In: EdMedia+ Innovate Learning, pp. 448–451. Association for the Advancement of Computing in Education (AACE) (2003)

42. Schroeder, N.L., Romine, W.L., Craig, S.D.: Measuring pedagogical agent persona and the influence of agent persona on learning. Comput. Educ. **109**, 176–186 (2017)

43. Johnson, W.L., Rickel, J.W., Lester, J.C.: Animated pedagogical agents: face-to-face interaction in interactive learning environments. Int. J. Artif. Intell. Educ. **11**(1), 47–78 (2000)

44. Martin, A.J., Jackson, S.A.: Brief approaches to assessing task absorption and enhanced subjective experience: examining 'short' and 'core' flow in diverse performance domains. Motiv. Emot. **32**(3), 141–157 (2008)

45. Brockmeyer, J.H., Fox, C.M., Curtiss, K.A., McBroom, E., Burkhart, K.M., Pidruzny, J.N.: The development of the game engagement questionnaire: a measure of engagement in video game playing. J. Exp. Soc. Psychol. **45**(4), 624–634 (2009)

46. Schroeder, N.L., Yang, F., Banerjee, T., Romine, W.L., Craig, S.D.: The influence of learners' perceptions of virtual humans on learning transfer. Comput. Educ. **126**, 170–182 (2018)

47. Nass, C., Steuer, J., Tauber, E.R.: Computers are social actors. In: Adelson, B., Dumais, S., Olson, J. (eds.) CHI 1994: Proceedings of the SIGCHI Conference on Human Factors in Computing Systems, vol. 12, pp. 72–78. Association for Computing Machinery, New York (1994)

Author Index

Printed in the United States
by Baker & Taylor Publisher Services